This item must be returned or renewed by the last date shown below. The loan period may be shortened if it is reserved by another reader.   A fine will be due if it is not returned on time.

16 MAR 2009   **DATE OF RETURN**

UCL LIBRARY SERVICES
Gower Street London WC1E 6BT

KNOWLEDGE DISCOVERY IN BIOINFORMATICS

KNOWLEDGE DISCOVERY IN BIOINFORMATICS

Techniques, Methods, and Applications

Edited by

XIAOHUA HU
Drexel University, Philadelphia, Pennsylvania

YI PAN
Georgia State University, Atlanta, Georgia

WILEY-
INTERSCIENCE

A JOHN WILEY & SONS, INC., PUBLICATION

Library of Congress Cataloging-in-Publication Data:

Knowledge discovery in bioinformatics : techniques, methods, and applications
/ edited by Xiaohua Hu, Yi Pan.
    p. cm.
  ISBN 978-0-471-77796-0
1. Bioinformatics. 2. Computational biology. I. Hu, Xiaohua (Xiaohua Tony)
II. Pan, Yi, 1960–
  [DNLM: 1. Computational Biology–methods. 2. Medical Informatics–methods.
QU 26.5 K73 2007]
  QH506.K5564 2007
  570′ .285–dc22
                              2006032495

Printed in the United States of America
10 9 8 7 6 5 4 3 2 1

CONTENTS

CONTRIBUTORS

Hsinchun Chen, Department of Management Information Systems, University of Arizona, Tucson, AZ 85721

Young-Rae Cho, Department of Computer Science and Engineering, State University of New York at Buffalo, Buffalo, NY 14260-2000

Craig W. Codrington, Department of Physics, Purdue University, West Lafayette, IN 47907

Ron Davis, Stanford Genome Technology Center, Palo Alto, CA 94304

Kevin DeRonne, Department of Computer Science and Engineering, University of Minnesota–Twin Cities, Minneapolis, MN 55455

Robert W. Harrison, Department of Computer Science and Department of Biology, Georgia State University, Atlanta, GA 30302-4110

Hae-Jin Hu, Department of Computer Science, Georgia State University, Atlanta, GA 30302-4110

Xiaohua Hu, College of Information Science and Technology, Drexel University, Philadelphia, PA 19104

Woo-Chang Hwang, Department of Computer Science and Engineering, State University of New York at Buffalo, Buffalo, NY 14260-2000

Bo Jin, Department of Computer Science, Georgia State University, Atlanta, GA 30302-4110

Trupti Joshi, Department of Computer Science and Christopher Bond Life Sciences Center, University of Missouri–Columbia, Columbia, MO 65211-2060

Kara Juneau, Stanford Genome Technology Center, Palo Alto, CA 94304

George Karypis, Department of Computer Science and Engineering, University of Minnesota–Twin Cities, Minneapolis, MN 55455

Terran Lane, Department of Computer Science, University of New Mexico, Albuquerque, NM 87131

Jiexun Li, Department of Management Information Systems, University of Arizona, Tucson, AZ 85721

Chuan Lin, Department of Computer Science and Engineering, State University of New York at Buffalo, Buffalo, NY 14260-2000

Huiqing Liu, Department of Biochemistry and Molecular Biology, University of Georgia, Athens, GA 30602

Jianghui Liu, Bioinformatics and Life Science Informatics Laboratory, New Jersey Institute of Technology, Newark, NJ 07102-1982

B. Offmann, Laboratoire de Biochimie et Génétique Moléculaire, Université de La Réunion, BP 7151, 97715 Saint Denis Messag Cedex 09, La Réunion, France

Curtis J. Palm, Stanford Genome Technology Center, Palo Alto, CA 94304

Yi Pan, Department of Computer Science, Georgia State University, Atlanta, GA 30302-4110

Pengjun Pei, Department of Computer Science and Engineering, State University of New York at Buffalo, Buffalo, NY 14260-2000

Shibin Qiu, Department of Computer Science, University of New Mexico, Albuquerque, NM 87131

Wenny Rahayu, Department of Computer Science and Computer Engineering, La Trobe University, Bundoora, Victoria 3086, Australia

Katrina M. Ramonell, Department of Biological Sciences, University of Alabama, Tuscaloosa, AL 35487-0344

Huzefa Rangwala, Department of Computer Science and Engineering, University of Minnesota–Twin Cities, Minneapolis, MN 55455

Andrew Robinson, Department of Computer Science and Computer Engineering, La Trobe University, Bundoora, Victoria 3086, Australia

Min Song, Department of Information System, College of Computing Sciences, New Jersey Institute of Technology, Newark, NJ 07102-1982

Il-Yeol Song, College of Information Science and Technology, Drexel University, Philadelphia, PA 19104

Audrey Southwick, Stanford Genome Technology Center, Palo Alto, CA 94304

N. Srinivasan, Molecular Biophysics Unit, Indian Institute of Science, Bangalore 560012, India

Gary Stacey, Divisions of Plant Science and Biochemistry, Department of Molecular Microbiology and Immunology, Christopher S. Bond Life Sciences Center, University of Missouri–Columbia, Columbia, MO 65211-2060

Hua Su, Department of Management Information Systems, University of Arizona, Tucson, AZ 85721

L. S. Swapna, Molecular Biophysics Unit, Indian Institute of Science, Bangalore 560012, India

Phang C. Tai, Department of Biology, Georgia State University, Atlanta, GA 30302-4110

David Taniar, Clayton School of Information Technology, Monash University, Clayton, Victoria 3800, Australia

Jinrong Wan, Divisions of Plant Science and Biochemistry, Department of Molecular Microbiology and Immunology, Christopher S. Bond Life Sciences Center, University of Missouri–Columbia, Columbia, MO 65211-2060

Jason T. L. Wang, Department of Computer Science, New Jersey Institute of Technology, Newark, NJ 07102-1982

Dongrong Wen, Department of Computer Science, New Jersey Institute of Technology, Newark, NJ 07102-1982

Limsoon Wong, School of Computing, National University of Singapore, Singapore

Daniel D. Wu, College of Information Science and Technology, Drexel University, Philadelphia, PA 19104

Dong Xu, Department of Computer Science and Christopher Bond Life Sciences Center, University of Missouri–Columbia, Columbia, MO 65211-2060

Ying Xu, Department of Biochemistry and Molecular Biology, University of Georgia, Athens, GA 30602

Mary Qu Yang, National Human Genome Research Institute, National Institutes of Health, Bethesda, MD 20892

Jack Y. Yang, Harvard Medical School, Harvard University, Boston, MA 02115

Illhoi Yoo, College of Information and Technology, Drexel University, Philadelphia, PA 19104

Mohammed J. Zaki, Department of Computer Science, Rensselaer Polytechnic Institute, Troy, NY 12180

Aidong Zhang, Department of Computer Science and Engineering, State University of New York at Buffalo, Buffalo, NY 14260-2000

Xiaodan Zhang, College of Information and Technology, Drexel University, Philadelphia, PA 19104

Yan-Qing Zhang, Department of Computer Science, Georgia State University, Atlanta, GA 30302-4110

Lizhuang Zhao, Department of Computer Science, Rensselaer Polytechnic Institute, Troy, NY 12180

Xiaohua Zhou, College of Information and Technology, Drexel University, Philadelphia, PA 19104

PREFACE

Bioinformatics is the science of integrating, managing, mining, and interpreting information from biological data sets. Although tremendous progress has been made over the years, many of the fundamental problems in bioinformatics, such as protein structure prediction or gene finding, data retrieval, and integration, are still open. In recent years, high-throughput experimental methods in molecular biology have resulted in enormous amounts of data. Mining bioinformatics data is an emerging area of intersection between bioinformatics and data mining. The objective of this book is to facilitate collaboration between data mining researchers and bioinforrnaticians by presenting cutting-edge research topics and methodologies in the area of data mining for bioinformatics.

This book contains articles written by experts on a wide range of topics that are associated with novel methods, techniques, and applications of data mining in the analysis and management of bioinformatics data sets. It contains chapters on RNA and protein structure analysis, DNA computing, sequence mapping, genome comparison, gene expression data mining, metabolic network modeling, phyloinformatics, biomedical literature data mining, and biological data integration and searching. The important work of some representative researchers in bioinformatics is brought together for the first time in one volume. The topic is treated in depth and is related to, where applicable, other emerging technologies, such as data mining and visualization. The goal of the book is to introduce readers to the principal techniques of data mining in bioinformatics in the hope that they will build on them to make new discoveries of their own. The key elements of each chapter are summarized briefly below.

Progress in machine learning technology provided various advanced tools for prediction of protein secondary structure. Among the many machine learning approaches, support vector machine (SVM) methods are the most recently applied in the structure prediction. It shows successful performance compared with other machine learning schemes. However, compared to other machine learning approaches, there is no systematic review about this SVM approach applied to secondary-structure prediction. In Chapter 1, H.-J. Hu, R. W. Harrison, P. C. Tai, and Y. Pan present methods for predicting secondary structure based on support vector machines. Evaluation of the performance of SVM methods, challenges of these SVM approaches, and efforts to overcome the problems are also discussed.

The problem of mining hidden links from complementary and noninteractive biomedical literature was exemplified by Swanson's pioneering work on the Raynaud's disease/fish oils connection. Two complementary and noninteractive sets of articles (independently created fragments of knowledge), when considered together, can reveal useful information of scientific interest not apparent in either of the two sets alone. In Chapter 2, X. Hu, X. Zhang, and X. Zhou discuss a comprehensive comparison of seven methods for mining hidden links among medical concepts. A series of experiments using these methods are performed and analyzed. Their research works present a comprehensive analysis for mining hidden links and how different weighting schemes plus semantic information affect the knowledge discovery procedure.

In Chapter 3, B. Jin and Y.-Q. Zhang discuss voting scheme–based evolutionary kernel machines used in drug activity comparisons. The performance of support vector machines is affected primarily by kernel functions. With the growing interest in biological data prediction and chemical data prediction, more complicated kernels are designed to measure data similarities: ANOVA kernels, convolution kernels, string kernels, tree kernels, and graph kernels. These kernels are implemented based on kernel decomposition properties. Experimental results show that the new kernel machines are more stable than the traditional kernels.

Microarrays are a well-established technology for measuring gene expression levels at a large scale. A tiling array is a set of oligonucleotide microarrays for the entire genome. Tiling array technology has several advantages over other microarray technologies. In Chapter 4, T. Joshi, J. Wan, C. J. Palm, K. Juneau, R. Davis, A. Southwick, K. M. Ramonell, G. Stacey, and D. Xu discuss the whole-genome tiling array design and techniques to analyzing these data to obtain a wide variety of genomic scale information using bioinformatics techniques. They also discuss onto-logical analyses and antisense identification techniques using tiling array data.

Identification of marker genes is of great importance to provide more accurate, cost-effective prediction, and a better understanding of genes' biological functions. In Chapter 5, J. Li, H. Su, and H. Chen discuss gene selection from high-dimensional microarray data. They present a framework for gene selection methods, focusing on optimal search–based gene subset selection. A comparative study of gene selection methods on three cancer data sets is presented with algorithmic details. Evaluation of both statistical and expert analysis is also presented.

In current practice, physicians assess the risk profile of a cancer patient based primarily on various clinical characteristics. However, these factors do not fully reflect the molecular heterogeneity of the disease, and treatment stratification is difficult. In fact, in many cases, patients with a similar diagnosis respond very differently to the same treatment. In Chapter 6, H. Liu, L. Wong, and Y. Xu introduce a number of methods for predicting cancer survival based on gene expression data and provide some answers to a challenging question: Is there any relationship between gene expression profiles and patient survival?

In recent years, RNA interference (RNAi) has surged into the spotlight of pharmacy, genomics, and system biology. In Chapter 7, S. Qiu and T. Lane describe the biological mechanisms of RNAi, covering both secondary RNA and microRNA as interfering initiators, followed by in-depth discussions of a few computational topics, including RNAi specificity, microRNA gene and target prediction, and siRNA silencing efficacy estimation. The authors also highlight some open questions related to RNAi research.

Currently, our ability to produce sequence information far outpaces the rate at which we can produce structural and functional information. Consequently, researchers rely increasingly on computational techniques to extract useful information from known structures contained in large databases, although such approaches remain incomplete. Unraveling the relationship between pure sequence information and three-dimensional structure thus remains one of the great fundamental problems in molecular biology. In Chapter 8, H. Rangwala, K. DeRonne, and G. Karypis show several ways in which researchers try to characterize the structural, functional, and evolutionary nature of proteins.

Public genomic databases involve three main aspects of data use: data representation, data storage, and data access. In Chapter 9, A. Robinson, W. Rahayu, and D. Taniar present a comprehensive study of genomic databases covering these three aspects. Data representation in the biology domain consists of three principal structures: sequence-centric, annotation-centric, and XML formatting. Data storage is usually achieved by maintaining flat files or a management system such as a relational database management system. Genetic data access is provided by a varied range of user interfaces; the main groups are single-database access points, cross-referencing, multidatabase access points, and tool-based interfaces.

A vast number of unstructured data become a difficult challenge in text mining. To tackle this issue, M. Song and I.-Y. Song propose a novel technique that integrates information retrieval techniques with text mining. In Chapter 10, they present a new unsupervised query expansion technique that utilizes keyphrases and part-of-speech (POS) phrase categorization. The keyphrases are extracted from the documents retrieved and are weighted with an algorithm based on information gain and cooccurrence of phrases. The keyphrases selected are translated into disjunctive normal form based on the POS phrase categorization technique for better query refomulation. Additionally, the authors discuss whether ontologies such as WordNet and MeSH improve retrieval performance in conjunction with the keyphrases.

Understanding the molecular basis of complex formations between proteins as well as between modules and domains in multimodular protein systems is central to the development of strategies for human intervention in biological processes. In Chapter 11, L. S. Swapna, B. Offmann, and N. Srinivasan trace the drift of protein–protein interaction surfaces between families related at superfamily level or between subfamilies related at family level when their interacting surfaces are not equivalent. They have investigated such evolutionary drifts in protein–protein interfaces in the class I glutamine amidotransferase–like superfamily of proteins and its constituent families.

Comparing and aligning RNA secondary structures is fundamental to knowledge discovery in biomolecular informatics. In recent years, much progress has been made in RNA structure alignment and comparison. However, existing tools either require a large number of prealigned structures or suffer from high time complexities. This makes it difficult for the tools to process RNAs whose prealigned structures are unavailable or to process very large RNA structure databases. In Chapter 12, J. T. L. Wang, D. Wen, and J. Liu present an efficient method, called RSmatch, for comparing and aligning RNA secondary structures. RSmatch can find the optimal global or local alignment between two RNA secondary structures. Also presented is a visualization tool, called RSview, which is capable of displaying the output of RSmatch in a colorful and graphic manner.

Despite an extensive effort to use computational methods in deciphering transcriptional regulatory networks, research on translation regulatory networks has attracted little attention in the bioinformatics and computational biology community, due probably to the nature of data available and to a bias in the conventional wisdom, In Chapter 13, D. D. Wu and X. Hu present a global network analysis of protein translation networks in yeast, a first step in attempting to facilitate elucidation of the structures and properties of translation networks. They extract the translation proteome using the MIPS functional category and analyze it in the context of the full protein–protein interaction network and derive individual translation networks from a full interaction network using the proteome extracted. They show that in contrast to a full network, protein translation networks do not exhibit power-law degree distributions. These results have potential implications for understanding mechanisms of translational control from a systems perspective.

Membrane proteins account for roughly one-third of all proteins and play a crucial role in processes such as cell-to-cell signaling, transport of ions across membranes, and energy metabolism, and are a prime target for therapeutic drugs. In Chapter 14, M. Q. Yang, J. Y. Yang, and C. W. Codrington emphasize the prediction of α-helical transmembrane regions in proteins using variants of the self-organizing feature map algorithm. To identify features that are useful for this task, the authors have conducted a detailed analysis of the physiochemical properties of transmembrane and intrinsically unstructured proteins.

In Chapter 15, L. Zhao and M. J. Zaki present a novel, efficient, deterministic, triclustering method called triCluster that addresses the challenge issues in three-dimensional microarray data analysis.

Clustering in the PPI network context groups together proteins that share a higher number of interactions. The results of this process can illuminate the structure of the PPI network and suggest possible functions for members of the cluster that were previously uncharacterized. In Chapter 16, C. Lin, Y.-R. Cho, W.-C. Hwang, P. Pei, and A. Zhang begin with a brief introduction to the properties of protein–protein interaction networks, including a review of data generated by both experimental and computational approaches. A variety of methods that have been employed to cluster these networks are also presented. These approaches are broadly characterized as either distance- or graph-based clustering methods. Techniques for validating the results of these approaches are also discussed.

We would like to express our sincere thanks to all authors for their important contributions. We would also like to thank the referees for their efforts in reviewing the chapters and providing valuable comments and suggestions. We would like to extend our deepest gratitude to Paul Petralia (senior editor) and Whitney A. Lesch (editorial assistant) at Wiley for their guidance and help in finalizing the book. Xiaohua Hu would like to thank his parents, Zhikun Hu and Zhuanhui Wang; his wife, Shuetyue Tsang; and his son, Michael Hu, for their love and encouragement. Yi Pan would like to thank Sherry for friendship, help, and encouragement during preparation of the book.

XIAOHUA HU
YI PAN

1

CURRENT METHODS FOR PROTEIN SECONDARY-STRUCTURE PREDICTION BASED ON SUPPORT VECTOR MACHINES

HAE-JIN HU, ROBERT W. HARRISON, PHANG C. TAI, AND YI PAN

Department of Computer Science (H-J.H., R.W.H., Y.P.) and Department of Biology (R.W.H., P.C.T.), Georgia State University, Atlanta, Georgia

The desire to understand protein structure has produced many approaches over the last four decades since Blout et al. (1960) attempted to correlate the sequence information of amino acids with their structural elements (Casbon, 2002). Instead of costly and time-consuming experimental approaches, effective prediction methods have been developed continuously. With the help of growing databases and the evolutionary information available from multiple-sequence alignments, resources for secondary-structure prediction became abundant. Also, progress in machine learning technology provided various advanced tools for prediction. Among the many machine learning approaches, support vector machine (SVM) methods are the most recent to be used for structure prediction. SVMs perform successfully, but compared with other machine learning approaches, there is no systematic review in the SVM approach when applied to secondary-structure prediction. Therefore, this study focuses mainly on methods of predicting secondary structure based on support vector machines. The organization of this chapter is as follows. In Section 1.1, traditional secondary-structure prediction approaches are described. In Section 1.2, various SVM-based prediction methods are introduced. In Section 1.3, the performance of SVM methods is evaluated, and in Section 1.4, problems with the SVM approach and efforts to overcome them are discussed.

Knowledge Discovery in Bioinformatics: Techniques, Methods, and Applications, Edited by Xiaohua Hu and Yi Pan

1.1 TRADITIONAL METHODS

1.1.1 Statistical Approaches

The first attempts to predict secondary structure took place in the 1970s. These include the Chou–Fasman algorithm (Chou and Fasman, 1974) and the GOR method (Garnier et al., 1978), both based on empirical and statistical methods. In the Chou–Fasman algorithm, the conformational parameters P_α, P_β, and P_c for each amino acid are obtained by analyzing the x-ray-determined structures of 15 proteins including 2473 amino acid residues. These parameters represent a given amino acid's tendency to be found in α-helix, β-sheet, or coil, and they contain the physicochemical properties of amino acids. Based on these parameters, the authors established empirical rules for secondary-structure prediction. This relatively simple algorithm has been criticized (Kyngas and Valjakka, 1998) since it is based on a small database for statistical analysis; the original algorithm consists of 15 proteins with 2473 amino acid residues, and the revised version (Chou, 1989) has a database of 64 proteins with 11,445 amino acid residues.

The second statistical approach, that of Garnier, Osguthorpe, and Robson (*GOR I*; Garnier et al., 1978), is based on information theory. This method applies a sliding window of 17 residues and calculates the probabilities of eight neighboring residues on each side of the sliding window for predicting a residue in the middle. This algorithm has been revised during the past 20 years, and in 2002, version GOR V was developed (Kloczkowski et al., 2002). Similar to the Chou–Fasman algorithm, the first version of GOR I used a small database of 26 proteins with about 4500 residues. However, during the revision process, the database size was extended to 267 proteins for GOR IV and to 513 proteins for GOR V. A major improvement of GOR V over earlier versions was achieved by including evolutionary information through use of PSI-BLAST multiple-sequence alignments into the GOR method (Kloczkowski et al., 2002). According to the authors of GOR V, the average prediction accuracy of secondary structure is $Q_3 = 73.5\%$ on the CB513 data set (Cuff and Barton, 1999) using a jackknife test. In the test, each protein in the data set is singled out in turn for an independent test while the remaining proteins are used for training. This process is repeated until all proteins are selected for testing.

1.1.2 Machine Learning Approaches

Nearest-Neighbor Method   The nearest-neighbor approach predicts the secondary structure of the central residue of a segment based on the secondary structure of homologous proteins from a database with known three-dimensional structures (Nishikawa and Ooi, 1986). In other words, this method matches each segment of the query sequence against all sequences in the database. Once nearest neighbors (homologous segments) are found based on similarity criteria, the secondary structure of the central residue is determined using the frequencies (f_{helix}, f_{sheet}, f_{coil}) of secondary-structure states at the central position of its neighbors (Biou et al., 1988; Levin and Garnier, 1988). Even though the basic idea is simple, it requires adjusting lots of factors, such as similarity measures, window size of querying sequence, or the number of nearest neighbors. Therefore, there have been many studies with various ways of applying different parameters with various results (Yi and Lander, 1993; Salamov and Solovyev, 1995; Levin, 1997). For example, Yi and Lander (1993) matched their 19 residue segments against the database of 110 proteins with known tertiary structure. Based on the local structural environmental scoring metric of Bowie et al. (1991), 50 nearest neighbors are found. This metric contains environmental parameters such as secondary-structure state, polarity, and accessible surface area. The score of matching a residue R_i with a local structural environment E_i was defined as

$$\text{Score}(R_i, E_j) = \log_{10} \frac{P(R_i|E_j)}{P(R_i)} \tag{1.1}$$

where $P(R_i|E_j)$ is the probability of finding residue i in environment j, and $P(R_i)$ is the probability of finding residue i in any environment (Yi and Lander, 1993). The secondary structure predicted for a test residue was selected as the state of maximum frequency, $\max(f_{\text{helix}}, f_{\text{sheet}}, f_{\text{coil}})$, of secondary-structure states at the central position of its 50 neighbors. The authors tested various scoring schemes with different numbers of environment classes and obtained the optimal system. It consists of 15 environmental classes, including three secondary structures combined with five different accessibility or polarity classes based on the mutation matrix of Gonnet et al. (1992). To combine the results from six scoring schemes, the authors adopted the neural network for jury decision and attained an accuracy of $Q_3 = 68\%$ with a jackknife test.

Salamov and Solovyev (1995) revised Yi and Lander's scheme. Their improvements were first, changing the scoring scheme by considering the N- and C-terminal positions of α-helices and β-sheets and by taking beta turns as unique types of secondary structure. Second, the authors restricted the database to a smaller subset of proteins that are close to a test sequence in general properties, thus reducing the computation time. Third, by applying multiple-sequence alignments and a jury decision process, they achieved $Q_3 = 72.2\%$ accuracy on a 126-protein data set using a jackknife test.

In 1997, Levin modified his own method, called SIMPA (Levin and Garnier, 1988), by applying a larger database of known structures, the Blosum62 substitution matrix, and a regularization algorithm (Zimmermann, 1994). However, both prediction algorithms are the same. The algorithm compares every residue in a window with each residue of the same window size in a database of known structures. Based on the Blosum62 scoring matrix, the match score is calculated. If the match score is smaller than a cutoff value, the peptide is not considered. Otherwise, the conformation observed is allocated to the test residue with its similarity score. The prediction of a test sequence is made based on the highest score at each residue location by applying a regularization algorithm. The regularization algorithm restricts the minimum length of helix and strand to four and two residues, respectively. By including the evolutionary information, the updated version of SIMPA96 reached $Q_3 = 71.4\%$ accuracy in a jackknife test (Levin, 1997).

Hidden Markov Model The hidden Markov model (HMM) is a probabilistic finite-state machine applied to model stochastic sequences. In HMMs, domain information can be included in the topology of the HMM, while other information is learned by training the emission and transition probabilities on data (Won et al., 2005). Because of this merit, HMMs have been applied widely in computational biology, such as in gene finding, sequence alignment, and protein structure prediction (Krogh et al., 1994; Bystroff et al., 2000).

A HMM consists of a set of states, emission probabilities related to each state, and transitions that connect states. Each state has symbols characterizing an amino acid residue or a secondary-structure type. If the symbols are the set of 20 amino acids denoted by $\mathbf{A}$ and the set of HMM parameters are represented by π, HMM assigns a probability to a given sequence $\mathbf{s} = (s_1, s_2, \ldots, s_m)$ of length m:

$$\sum_{s \in A^m} P(s|\pi) = 1 \tag{1.2}$$

If the HMMs emit class labels as well as symbols from the 20 amino acids, a sequence $\mathbf{s}$ can be associated with a corresponding class labels $\mathbf{t} = (t_1, t_2, \ldots, t_m)$. If we let the set of states be represented by Q and denote a sequence of states $\mathbf{q} = (q_1, q_2, \ldots, q_m)$, the chance of a sequence $\mathbf{s}$ having class labels $\mathbf{t}$ can be written as the sum over all possible paths through the states:

$$P(s,t|\pi) = \sum_{q \in Q^m} P(s, t, q|\pi) \tag{1.3}$$

For a given sequence $\mathbf{s}$ and the corresponding class labels $\mathbf{t}$, the maximum likelihood (ML) set of parameters can be found based on the Baum–Welch algorithm (Rabiner, 1989):

$$\pi^{\mathrm{ML}} = \arg \max_{\pi} P(s, t|\pi) \tag{1.4}$$

Among the studies using HMM, Bystroff et al. (2000) introduced a new HMM model, HMMSTR, for general protein sequences based on basic local structural motifs called *I-sites*. These structural motifs are a set of sequence motifs of length 3 to 19 obtained from a nonredundant database of known structures. Each motif has information about the sequence and structure and was expressed as a chain of Markov states. By merging the linear chains of states (motifs) based on sequence and structure similarity, an HMM graph with branched topology was formed. The interesting feature of HMMSTR is that it models local motifs common to all protein families instead of modeling individual protein families. Based on this advanced HMM model, the authors attained $Q_3 = 74.3\%$ accuracy in secondary-structure prediction.

In 2005, Won et al. designed a new scheme to evolve a HMM with a genetic algorithm (GA). The authors applied a hybrid GA that adopts traditional GA operators to search the space of HMM topologies combined with the Baum–Welch algorithm to optimize the transition and emission probabilities. With this method, they achieved $Q_3 = 75\%$ accuracy with a fivefold cross-validation test.

Neural Network Methods Inspired by neurons in the brain, artificial neural networks are parallel information-processing structures frequently used in classification problems. For secondary-structure prediction, feedforward network architecture is used successfully. The feedforward network is organized with a layered structure, including input layer, output layer, or zero or more hidden layers, such as that shown in Figure 1.1. Each layer has one or more processing units called *nodes* or *neurons*. Each node in a layer is connected to nodes in a preceding layer by weights. In secondary-structure prediction with neural networks, the inputs are sliding windows with a 13- to 17-residue sequence. In the input layer each node takes a feature value representing the amino acid type. Based on the node's input value, the output value is calculated in each node. If we let the nodes of the preceding layer be i ($i = 1, 2, \ldots, M$), a node of the current layer be j, and the weight between nodes i and j be w_{ji}, the total weighted input to j, X_j, can be calculated using the formula (Eidhammer et al., 2004)

$$X_j = \sum_{i=1}^{M} w_{ji} a_i \tag{1.5}$$

where a_i is the activity level of node i in the preceding layer. The activity level of node j, a_j, can be computed based on some function of the total weighted input. Typically, the sigmoid function is applied as follows:

$$a_j = \frac{1}{1 + e^{-X_j}} \tag{1.6}$$

This value is sent to all lines connected out from the node j.

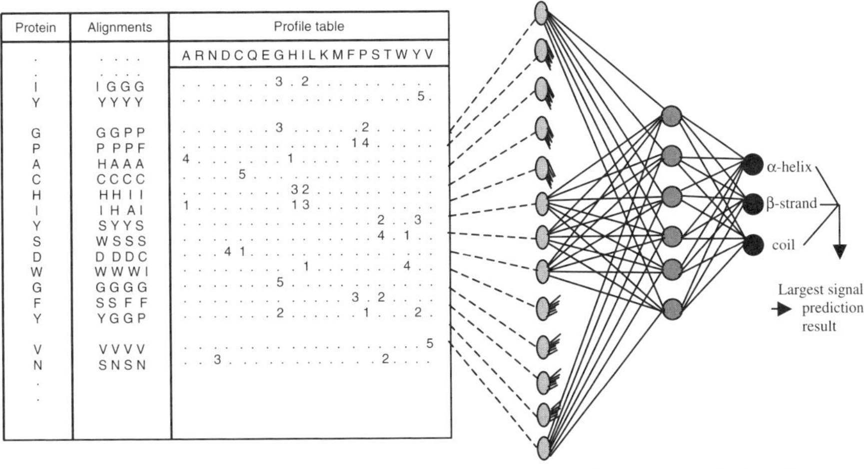

FIGURE 1.1 Neural network architecture. As an encoding profile for training the neural network, frequencies of each amino acid at each position are adopted. These frequencies are obtained from multiple-sequence alignments. The input signal is propagated through the network with one input layer, one hidden layer, and one output layer. (Based on Rost and Sander, 1993a.)

The neural network is trained using the supervised learning method. Here the training process is finding the appropriate value for the weight of each line in the network to make as many correct predictions as possible. In supervised learning, a training data set is formed as encoded feature vectors combined with correct class labels, such as helix, sheet, or coil. Before training the network, the weights are initialized to small random values. Once all training data have been fed into the network, the network generates output based on its current weights. By comparing the output with the correct class label, an error is computed for each node of the output layer. The classical back-propagation algorithm (Rumelhart et al., 1986) can be used to propagate the errors to the previous layers and to adjust the weights.

Rost and Sander's PHD (1993b) is one of the most popular methods for predicting secondary structure. PHD uses a two-layer feedforward neural network and incorporates the evolutionary information based on multiple sequence alignments. The initial version (Rost and Sander, 1993b) gives $Q_3 = 70.8\%$ accuracy for globular proteins with a seven-fold cross-validation test. Since then there have been many approaches to improving this result (Riis and Krogh, 1996; Chandonia and Karplus, 1999; Cuff and Barton, 1999; Petersen et al., 2000; Pollastri et al., 2002).

The PSIPRED method by Jones (1999) is another successful approach for predicting secondary structure. In PSIPRED, a greatly simplified two-stage neural network was used based on the position-specific scoring matrices generated by PSI-BLAST (Figure 1.2). The author set up a new cross-validation scheme by screening the training and testing sets based on a structural similarity criterion. Instead of removing a protein from a training set that has a high degree of sequence similarity to the testing set, the author discarded a protein with a fold similar to that of any of the testing set. Based on this test, the author attained an accuracy between $Q_3 = 76.5$ and 78.3% on 187 test proteins.

In 1999, Chandonia and Karplus (1999) used an enlarged database of 258 to 681 proteins to train the neural networks with more informative patterns. In addition, by applying second-level networks called *juries* they obtained $Q_3 = 74.9\%$ average accuracy on 681 proteins when tested with 15-fold cross-validation.

Petersen et al. (2000) combined 800 neural network predictions based on new methods called output expansion and balloting procedure. In the *output expansion process*, the secondary structures of a residue and its neighbors are predicted at the same time. The central idea of this approach is that these additional outputs give more information to the neural networks for optimizing the weights. To combine the results of multiple predictions, the authors adopted a statistical method called the *balloting scheme*. The balloting method is a variety of weighted-average scheme based on a mean and standard deviation. The authors reported that this balloting scheme enhances the performance better than straight averaging. An accuracy of $Q_3 = 80.2\%$ was claimed when the RS126 data set is used as an independent test set.

In 2002, Pollastri et al. introduced the ensembles of a bidirectional recurrent neural network (BRNN), which are the next version (SSpro2.0) of their previous BRNN architecture (Baldi et al., 1999). SSpro2.0 achieves a Q_3 value of 78.1% on an RS126 independent test set.

Position Specific Scoring Matrix

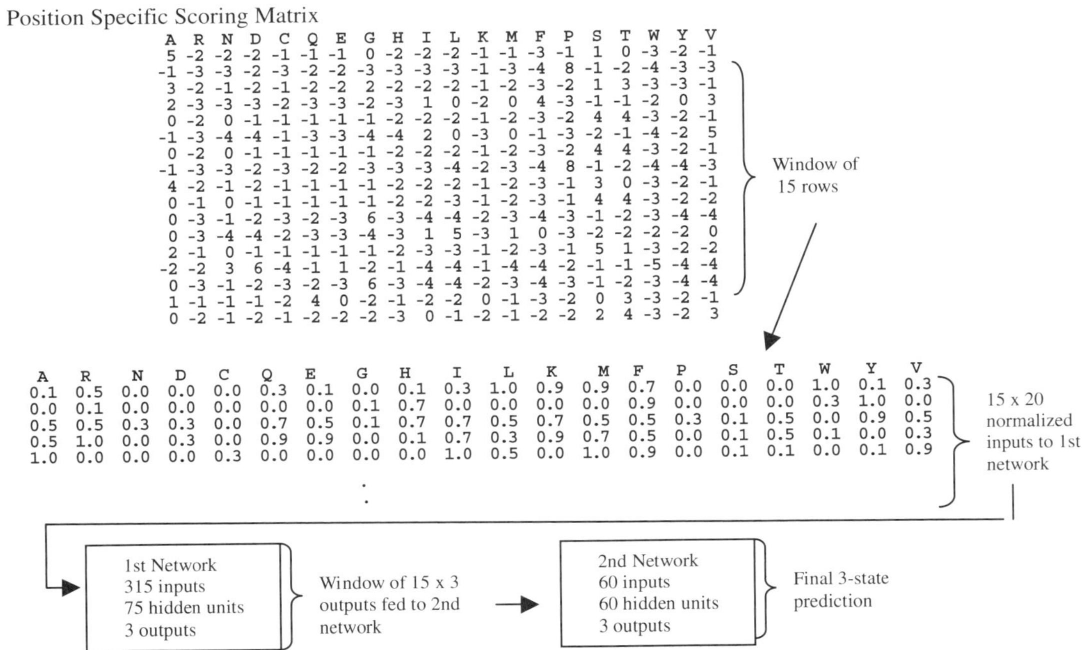

FIGURE 1.2   PSIPRED method, describing how the PSSMs are processed. (Based on Jones, 1999.)

Kernel-Based Methods   Recently, a number of kernel-based learning schemes, including the support vector machine (SVM) method, kernel Fisher discriminant (KFD), and kernel principal component analysis (KPCA), were introduced (Müller et al., 2001). The difference in these approaches is that they used different algorithms to handle the high dimensionality of kernel feature space. Among these schemes, the SVM method is the most widely used machine learning approach. Therefore, in the next section several SVM-based secondary-structure prediction methods are discussed in detail.

1.2   SUPPORT VECTOR MACHINE METHOD

1.2.1   Introduction to SVM

SVM is a modern learning system designed by Vapnik and his colleagues (Vapnik and Cortes, 1995). Based on statistical learning theory, which explains the learning process from a statistical point of view, the SVM algorithm creates a hyperplane that separates the data into two classes with the maximum margin. Originally, it was a linear classifier based on the optimal hyperplane algorithm developed by Vapnik in 1963. However, by applying the kernel method to a maximum-margin hyperplane, in 1992 Vapnik and his colleagues proposed a method to build a nonlinear classifier. In 1995, Cortes and Vapnik suggested a soft margin classifier, which is a modified maximum margin classifier that allows for misclassified data. If there is no hyperplane that can separate the data into two classes, the soft margin classifier selects a hyperplane that separates the data as cleanly as possible with a maximum margin.

SVM learning is related to recognizing the pattern from the training data (Burges, 1998; Cristianini and Shawe-Taylor, 2000). Namely, a function $f: R_N \rightarrow \{\pm 1\}$ is estimated based on the training data, which have an N-dimensional pattern $\mathbf{x}_i$ and class labels y_i. By imposing a restriction called *structural risk minimization* (SRM) on this function, it will correctly classify the new data $(\mathbf{x}, y)$ that have the same probability distribution $P(\mathbf{x}, y)$ as the training data. SRM is used to find the learning machine that yields a good trade-off between low empirical risk (mean error over the training data) and small capacity (a set of functions that can be implemented by the learning machine). In the linear soft margin SVM, which allows some misclassified points, the optimal hyperplane can be found by solving the following constrained quadratic optimization problem:

$$\min_{w,b,\varepsilon} \tfrac{1}{2} \| \mathbf{w} \|^2 + C \sum_{i=1}^{l} \varepsilon_i \tag{1.7}$$

$$\text{subject to} \quad y_i(\mathbf{w} \bullet \mathbf{x}_i + b) \geq 1 - \varepsilon_i, \qquad \varepsilon_i > 0, \quad i = 1, \ldots, l \tag{1.8}$$

where $\mathbf{x}_i$ is an input vector, $y_i = +1$ or -1 based on whether $\mathbf{x}_i$ is in a positive or a negative class, l is the number of training data, $\mathbf{w}$ is a weight vector perpendicular to the hyperplane, and b is a bias that moves the hyperplane parallel to itself. C is a cost factor (penalty for misclassified data) and ε is a slack variable for misclassified points. The resulting hyperplane decision function is

$$f(x) = \text{sign}\left[\sum_{i=1}^{SV} \alpha_i y_i \left(\mathbf{x} \bullet \mathbf{x}_i\right) + b\right] \qquad (1.9)$$

where, α_i is a Lagrange multiplier for each training data. The points $\alpha_i > 0$ lie on the boundary of the hyperplane and are called *support vectors*. In Eqs. (1.8) and (1.9) it is observed that both the optimization problem and the decision function rely on the dot products between each pattern.

In nonlinear SVM, the algorithm first maps the data into high-dimensional feature space via a kernel function and constructs the optimal separating hyperplane there using the linear algorithm. The common kernel functions are the following:

$$K(\mathbf{x}, \mathbf{x}') = \begin{cases} \left(\mathbf{x} \bullet \mathbf{x}' + 1\right)^p & (1.10) \\ e^{-\gamma\|\mathbf{x} - \mathbf{x}'\|^2} & (1.11) \\ \tanh(k\mathbf{x} \bullet \mathbf{x}' - \delta). & (1.12) \end{cases}$$

Equation (1.10) is a polynomial, Eq. (1.11) is a Gaussian radial basis function (RBF), and Eq. (1.12) is a two-layer sigmoidal neural network kernel. Based on one of the kernel functions above, the final nonlinear decision function has the form

$$f(x) = \text{sign}\left[\sum_{i=1}^{SV} \alpha_i y_i K\left(\mathbf{x} \bullet \mathbf{x}_i\right) + b\right] \qquad (1.13)$$

SVM[light] (Joachims, 1999) and LIBSVM (Chang and Lin, 2001) are widely used software implementations of SVM.

The SVM algorithm has the following outstanding features. First, it can avoid overfitting effectively with the use of structural risk minimization. Second, the formulation can be simplified to a convex quadratic programming (QP) problem; the training will converge to a global optimum. Note that the global optimum is the best solution for a given kernel and training data sets. Different qualities of results will be found with different kernels and data. Third, for a given data set, information can be condensed while training without losing useful information (Hua and Sun, 2001). Since this SVM has outperformed most other learning systems in most pattern recognition problems (Hua and Sun, 2001), it has gradually been applied to pattern classification problems in biology.

One recent study adopting this SVM learning machine for secondary-structure prediction adopted frequency profiles with evolutionary information as an encoding scheme for SVM (Hua and Sun, 2001). Another approach applied two layers of SVM with a weighted cost function for balanced training (Casbon, 2002). Also, there were methods that incorporated PSI-BLAST PSSM profiles as an input vector and that applied new tertiary classifiers (Kim and Park, 2003; Hu et al., 2004). Based on the results of these studies, the success of SVM methods depends on the proper choice of encoding profile, kernel function, and tertiary classifier. In the following section we investigate how these factors are optimized in each study.

1.2.2 Encoding Profile

In the studies above, the SVM was trained with different sequence and structural information using a sliding window scheme, a method that has been used in many other secondary-structure prediction schemes, including a later version of GOR methods, neural network methods, and nearest-neighbor algorithms. In the sliding window method, a window becomes one training pattern for predicting the structure of the residue at the center of the window. In each training pattern, information about local interactions among neighboring residues is embedded. The feature value of each amino acid residue in a window represents the weight (costs) of each residue in a pattern.

Orthogonal Encoding In orthogonal encoding, each residue has a unique binary vector, such as $(1,0,0,\ldots)$, $(0,1,0,\ldots)$, $(0,0,1,\ldots)$, and so on. Each binary vector is 20-dimensional. In this method, the weights of all residues in a window are assigned to 1. As a simple example, if we take a window size as 5, the dimension of one input pattern becomes:

$$\text{one vector dimension} = (20 \text{ binary bits})(5 \text{ residues}) = 100$$

$$\longrightarrow \text{window size}$$

Therefore, the amino acid segment STAAD can be written as the following vector based on the indices of Eq. (1.14).

$$\underbrace{(0,0,\ldots,1,0,0,0,0)}_{S\ (16)}, \underbrace{(0,0,\ldots,1,0,0,0)}_{T\ (37)}, \underbrace{(1,0,0,\ldots)}_{A\ (41)}, \underbrace{(1,0,0,\ldots)}_{A\ (61)}, \underbrace{(0,0,0,1,0,\ldots)}_{D\ (84)}$$

Hydrophobicity Encoding Hu et al. (2004) examined hydrophobicity encoding as one of their encoding profiles. Among the many different hydrophobicity measures, they adopted the Radzicka and Wolfenden scale (Radzicka and Wolfenden, 1988). These index values are as follows:

$$\text{amino acids } [\cdot] = \{A, R, N, D, C, Q, E, G, H, I, L, K, M, F, P, S, T, W, Y, V\} \quad (1.14)$$

$$\begin{aligned}
\text{hydrophobicity index } [\cdot] = \{&1.81, -14.92, -6.64, -8.72, 1.28, -5.54, -6.81, \\
&0.94, -4.66, 4.92, 4.92, -5.55, 2.35, 2.98, 4.04, \\
&-3.40, -2.57, 2.33, -0.14, 4.04\} \quad (1.15)
\end{aligned}$$

The hydrophobicity matrix is formulated based on the values above, and by using the following function:

$$\text{hydrophobicity matrix}[i][j] = \frac{\text{abs}(\text{hydrophobicity index}[i] - \text{hydrophobicity index}[j])}{20.0}$$

$$(1.16)$$

The denominator, 20, is used to convert the data range into [0,1] since SVM feature values are within this range. According to the function above, hydrophobicity matrix[2][3] means the absolute value of the difference of the hydrophobicity indices of two amino acids: for example, R (-14.92) and N (-6.64). With the range adjustment, it becomes 0.414. Based on this method, a 20×20 hydrophobicity matrix can be formulated.

BLOSUM Encoding   In BLOSUM coding (Hu et al., 2004) each amino acid is represented by values from the BLOSUM62 amino acid replacement cost matrix. The BLOSUM62 matrix represents the "log-odds" scores for the likelihood that a given amino acid pair will interchange with another. It is expected that this would partially account for the structural conservation of residue upon replacement.

Hybrid Encoding   Since each of these coding schemes captures different aspects of the properties of amino acids, Hu et al. (2004) tested combinations of two encodings, such as the following: orthogonal matrix + hydrophobicity matrix, BLOSUM62 matrix + hydrophobicity matrix, and orthogonal matrix + BLOSUM62 matrix.

Frequency Matrix Encoding   In this coding (Hua and Sun, 2001), the frequency of occurrence of 20 amino acid residues at each position in the multiple sequence alignment is calculated for each residue. This encoding was initially applied in Rost and Sander's neural network approach (Rost and Sander, 1993a) (Figure 1.1).

PSSM Encoding   PSSM coding applies the position-specific scoring matrix (PSSM) generated by PSI-BLAST (Kim and Park, 2003; Hu et al., 2006). In this coding the individual profiles were used to reflect detailed conservation of amino acids in a family of homologous proteins. This scheme was originally adopted by Jones (1999) to perform the prediction of protein secondary structure with his neural network (Figure 1.2). According to the author, PSI-BLAST is a very effective sequence query method, due to three aspects. First, the alignments generated by PSI-BLAST are based on pairwise local alignments. The previous study (Frishman and Argos, 1997; Salamov and Solovyev, 1997) reported that by using reliable local alignments, the prediction accuracy could be improved. Next, based on the iterated profiles, the sensitivity of PSI-BLAST was enhanced. Finally, the author tried many automatic multiple-sequence alignments. Among them, the PSI-BLAST alignments performed best.

1.2.3 Kernel Functions

The choice of kernel function is critical to the success of SVM. Most studies (Hua and Sun, 2001; Casbon, 2002; Kim and Park, 2003; Hu et al., 2004) adopted the Gaussian radial basis function (RBF) kernel after testing the common kernel functions from Section 1.2.1. Recently, an approach to designing new kernel functions based on a substitution matrix for protein secondary-structure prediction was developed (Vanschoenwinkel and Manderick, 2004). In another approach, Altun et al. (2006a,b) designed hybrid kernels that combined a substitution matrix–based kernel, an edit kernel, or a self-organizing map (SOM)–based kernel with the RBF kernel. Even though both approaches are not very successful in outperforming the best secondary-structure prediction methods, these are decent examples to give some idea of designing new kernel functions. These approaches are introduced in detail in this section.

Substitution Matrix–Based Kernels Substitution matrix–based kernels were developed by Vanschoenwinkel and Manderick (2004). The authors introduced a pseudo inner product (PI) between amino acid sequences based on the Blosum62 substitution matrix values. PI is defined as follows:

Definition 1.1 Let M be a 20×20 symmetric substitution matrix with entries $M(a_i, a_j) = m_{ij}$, where a_i and a_j are components of the 20-tuple $\mathbf{A} = (A, C, D, E, F, G, H, I, K, L, M, N, P, Q, R, S, T, V, W, Y) = (a_1, \ldots, a_{20})$. Then the inner product of two amino acid sequences $\mathbf{s}$, $\mathbf{s}' \in \sum^n$ with $\mathbf{s} = (a_{i1}, \ldots, a_{in})$ and $\mathbf{s}' = (a_{j1}, \ldots, a_{jn})$, with $a_{ik}, a_{jk} \in \mathbf{A}$, $i,j \in \{1, \ldots, 20\}$ and $k = 1, \ldots, n$, is defined as

$$\langle \mathbf{s}|\mathbf{s}' \rangle = \sum_{k=1}^{n} M(a_{ik}, a_{jk}) \tag{1.17}$$

By applying the PI above, the authors defined a substitution matrix–based distance function between amino acid sequences as follows:

Definition 1.2 Let $\mathbf{s}, \mathbf{s}' \in \sum^n$ be two amino acid sequences with $\mathbf{s} = (a_{i1}, \ldots, a_{in})$ and $\mathbf{s}' = (a_{j1}, \ldots, a_{jn})$, and let $\langle \mathbf{s}|\mathbf{s}' \rangle$ be the inner product as defined by Eq. (1.17); then the substitution distance d_{sub} between $\mathbf{s}$ and $\mathbf{s}'$ is defined as

$$d_{\text{sub}}(\mathbf{s}, \mathbf{s}') = \sqrt{\langle \mathbf{s}|\mathbf{s} \rangle - 2\langle \mathbf{s}|\mathbf{s}' \rangle + \langle \mathbf{s}'|\mathbf{s}' \rangle} \tag{1.18}$$

Based on both the PI and the substitution distance d_{sub} above, three different kernel functions were developed:

$$K_{\text{PIK}}(\mathbf{s}, \mathbf{s}') = (\langle \mathbf{s}|\mathbf{s}' \rangle + c)^d \tag{1.19}$$

$$K_{\text{SRBK}}(\mathbf{s}, \mathbf{s}') = \exp[-\gamma d_{\text{sub}}(\mathbf{s}, \mathbf{s}')^2] \tag{1.20}$$

$$K_{\text{NSDK}}(\mathbf{s}, \mathbf{s}') = -[d_{\text{sub}}(\mathbf{s}, \mathbf{s}')]^\beta \qquad \text{with} \quad 0 < \beta \leq 2 \tag{1.21}$$

The pseudo inner product kernel (K_{PIK}) has the form of a polynomial kernel. However, instead of the inner product in Eq. (1.10), PI was used. The substitution radial basis kernel (K_{SRBK}) is formulated based on the radial basis kernel of Eq. (1.11) and the substitution distance of Eq. (1.18). The negative substitution distance kernel (K_{NSDK}) is created based on the negative distance kernel (K_{ND}) and the substitution distance of Eq. (1.18). K_{ND} has the form

$$K_{\text{ND}}(\mathbf{s}, \mathbf{s}') = -\| \mathbf{s} - \mathbf{s}' \|^\beta \qquad \text{with} \quad 0 < \beta \leq 2 \tag{1.22}$$

The authors reported that the kernel functions based on the substitution matrix outperformed the counterparts that do not apply the matrix when tested with a sixfold cross-validation test based on a CB513 data set (Cuff and Barton, 1999). Also, those kernel functions performed slightly better (about 1.5% higher accuracy) than did the radial basis kernel function.

Hybrid Kernels Altun et al. (2006a) designed three different hybrid kernels by combining a substitution matrix (SM)–based kernel, an edit kernel, or a self-organizing map (SOM)–based kernel with an RBF kernel. An example and the algorithm of the hybrid kernel of an SM-based kernel and an RBF kernel (SVM_{SM+RBF}) are given in Figure 1.3, which explain how a sequence segment is given to the hybrid kernel for finding distances. The data encoding given to SVM_{SM+RBF} is shown in Figure 1.4. The data input for each sequence is the PSSM encoding of the sequence and the sequence combined.

The second hybrid kernel is a combination of an edit kernel and an RBF kernel, $SVM_{EDIT+RBF}$ (Altun et al., 2006a). The edit kernel was devised by Li and Jiang (2004) to predict translation initiation sites in eukaryotic mRNAs with SVM. It is based on the string edit distance, which contains biological and probabilistic information. The edit distance is the minimum number of edit operations (e.g., insertion, deletion, substitution) that transform one sequence to another. These edit operations can be considered as a series of evolutionary events. In nature, evolutionary events

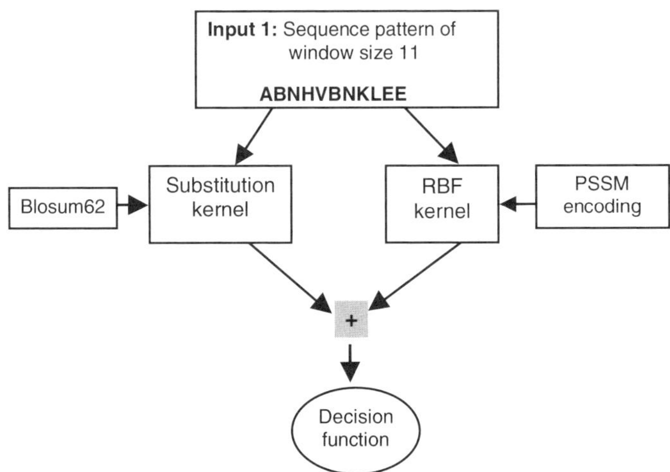

FIGURE 1.3 Example of the SVM_{SM+RBF} algorithm.

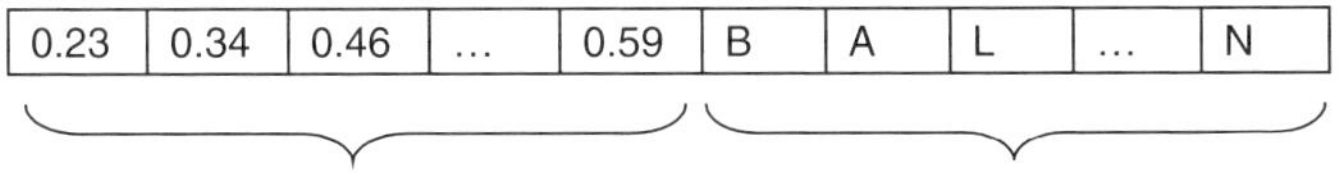

FIGURE 1.4 Data encoding for SVM_{SM+RBF} and $SVM_{EDIT+RBF}$.

happen with different probabilities. The authors (Li and Jiang, 2004) defined the edit kernel as follows:

$$K(x, x') = e^{-\gamma \bullet \, \text{edit}\,(x, x')} \tag{1.23}$$

$$\text{edit}\,(x, x') = -\tfrac{1}{2}\left[\sum_i \log P(x_i|x_i') + \sum_i \log P(x_i'|x_i)\right] \tag{1.24}$$

where the edit distance is the average of the negative log probability of mutating x into x' and that of mutating x' into x. They modified the 1-PAM matrix to get an asymmetric substitution cost matrix (SCM) for the edit kernel above. Altun et al. combined this edit kernel with the RBF kernel. An example of $\text{SVM}_{\text{EDIT+RBF}}$ is given in Figure 1.5, which explains how a sequence segment is given to the hybrid kernel for finding distances.

The third hybrid kernel is the combination of a self-organizing map (SOM)–based kernel and an RBF kernel, $\text{SVM}_{\text{SOM+RBF}}$ (Altun et al., 2006b). Self-organizing maps are a data visualization technique invented by T. Kohonen, which reduce the dimensions of data through the use of self-organizing neural networks (Kohonen, 1997) and provide a way of representing multidimensional data in lower-dimensional spaces (usually, two dimensions). In SOM, usually a two-dimensional grid of neurons or nodes is used where the grid forms the output space and input data patterns form the input space. Each node has a specific topological position (an x,y coordinate in the grid) and contains a vector of weights of the same dimension as that of the input vectors. SOM initializes its nodes or clusters by random sampling of the data. Then the network of artificial neurons is trained by providing information about the input. When applied this algorithm, vectors that are similar to each other in the initial space remain close on the two-dimensional grid. The input sequence patterns (PSSM

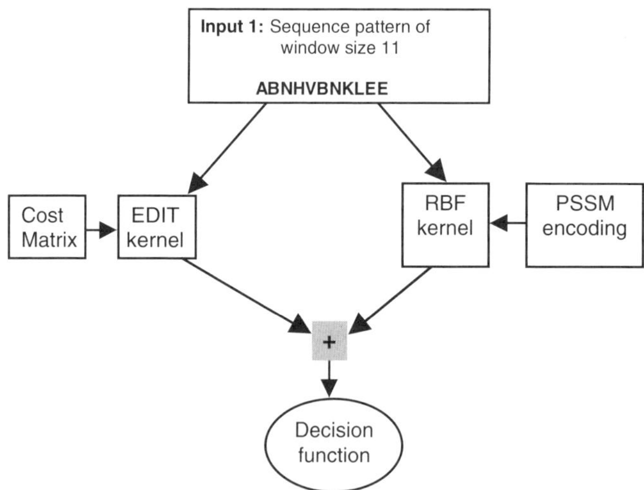

FIGURE 1.5   Example of an $\text{SVM}_{\text{EDIT+RBF}}$ algorithm.

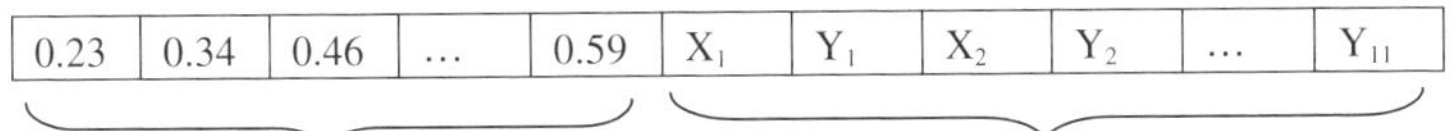

FIGURE 1.6   Data encoding for SVM$_{\mathrm{SOM+RBF}}$.

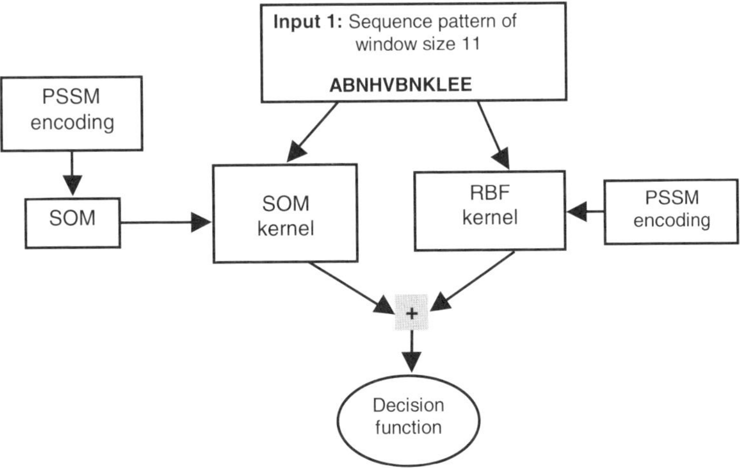

FIGURE 1.7   Example of an SVM$_{\mathrm{SOM+RBF}}$ algorithm.

encoding per residue) have been given to SOM and an output value of x and y data points on the grid have been assigned to each residue. Later, these x and y coordinates are used in a new encoding scheme. In this encoding scheme, x and y coordinates and the PSSM encoding of 11 sequence samples are combined. The new encoding scheme is shown in Figure 1.6.

Each encoded input is sent to the hybrid kernel of SOM and RBF. The PSSM encoding part is sent to the RBF kernel, and the x- and y-coordinate part is sent to the SOM kernel (Figure 1.7). The SOM kernel calculates the distance based on the following distance function:

$$d_{\mathrm{som}}\left(x, x'\right) = \sum_{i=1}^{11} \left(x_i - x_i'\right)^2 + \left(y_i - y_i'\right)^2 \tag{1.25}$$

The algorithmic details of the SOM–RBF hybrid kernel are shown in Figure 1.7. The authors reported that SVM$_{\mathrm{SOM+RBF}}$ produced results similar to those of the SVM$_{\mathrm{SM+RBF}}$ and SVM$_{\mathrm{RBF}}$ methods.

1.2.4   Tertiary Classifier Design

SVM is a binary classifier. To apply this binary classifier to three class problems (helix, sheet, coil) of secondary-structure prediction, many studies have attempted to

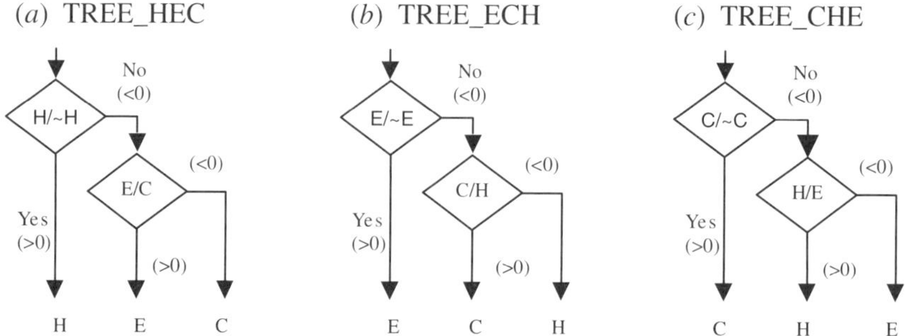

FIGURE 1.8   Tree-based tertiary classifiers. (Based on Hua and Sun, 2001.)

design effective ways of combining the output from binary classifiers. Some methods depend on the result from the six SVM binary classifiers, including three one-versus-rest classifiers ("one" positive class; "rest" negative class): H/~H, E/~E, and C/~C; and three one-versus-one classifiers: H/E, E/C, and C/H. For example, the classifier H/E is constructed on training samples having helices and sheets and classifies the testing sample as a helix or sheet. Some methods depend either on three one-versus-the rest binary classifiers or on three one-versus-one binary classifiers.

Instead of combining the result of binary classifiers, Nguyen and Rajapakse applied direct multiclass SVM for secondary-structure prediction (Nguyen and Rajapakse, 2003). In this section, several tertiary classifiers based on binary classifiers and multiclass SVMs are introduced.

Tree-Based Tertiary Classifier   This method (Hua and Sun, 2001) is based on three one-versus-rest binary classifiers (H/~H, E/~E, and C/~C) and three one-versus-one classifiers (H/E, E/C, and C/H). With these classifiers, three cascade tertiary classifiers, TREE_HEC (H/~H, E/C), TREE_ECH (E/~E, C/H), and TREE_CHE (C/~C, H/E), were created. These tree-based classifiers are shown in Figure 1.8.

Simple Voting Tertiary Classifier (SVM_VOTE)   In this method (Hua and Sun, 2001), all six binary classifiers are combined by using a simple voting scheme in which the testing sample is predicted to be state i (i is among H, E, or C) if the largest number of the six binary classifiers classify it as state i. If a testing sample has two classifications in each state, it is considered to be a coil. Examples are given in Table 1.1. In the first example case, there are three votes for class C (C/~C: $+2.0$,

TABLE 1.1   SVM_VOTE Scheme

Example Cases	H/~H	E/~E	C/~C	H/E	E/C	C/H	Final Class
1	-1.3	-2.7	$+2.0$	-1.4	-2.3	$+1.9$	C
2	-0.5	$+1.1$	-2.6	-0.6	-1.1	$+1.8$	C
3	$+1.9$	-0.2	$+1.4$	$+2.2$	$+1.0$	-0.4	H
4	-2.5	$+1.7$	$+2.3$	-0.3	$+2.5$	-0.1	E

TABLE 1.2 SVM_MAX_D Scheme

Example Cases	H/$\sim$H	E/$\sim$E	C/$\sim$C	Final Class
1	-1.3	-2.7	$+2.0$	C
2	-0.5	$+1.1$	$+0.6$	E
3	$+1.9$	$+0.2$	-1.4	H
4	-2.5	$+1.7$	$+2.3$	C
5	$+1.1$	-0.1	$+1.9$	C

E/C: -2.3, and C/H: $+1.9$). Therefore, the final class was assigned as C. In the second example case, there are two votes for class E (E/$\sim$E: $+1.1$ and H/E: -0.6) and class C (E/C: -1.1 and C/H: $+1.8$), respectively. In this case, the final class is assigned as C.

SVM_MAX_D In this classifier (Hua and Sun, 2001), three one-versus-rest classifiers (H/$\sim$H, E/$\sim$E, and C/$\sim$C) are combined for handling the multiclass case. The class of a testing sample (H, E, or C) was assigned to the one that presents the largest positive distance from the optimal separating hyperplane. For example, if the distance values of each one-versus- the rest classifiers (H/$\sim$H, E/$\sim$E, and C/$\sim$C) are -1.7, 1.2, and 2.5, respectively, the negative distance of the H/$\sim$H binary classifier does not give any information for the decision. Only two positive values (1.2, 2.5) are compared. Finally, the class for the test sample is assigned to a coil because 2.5 is larger between the two values. More examples are given in Table 1.2.

Tertiary Classifier Combined with a Neural Network (SVM_NN) The outputs of the six binary classifiers are fed into the neural network (NN) (Hua and Sun, 2001). The NN has six units in the input layer, 20 units in the hidden layer, and three units in the output layer. The size of the hidden layer was optimized with various tests. All the parameters, including weights and bias, are decided during the training procedure.

Combined Classifier of All the Results of the Tertiary Classifiers (SVM_JURY)
SVM_JURY combines the results of tertiary classifiers using a jury technique (Hua and Sun, 2001). The jury technique is just a majority vote of tertiary classifiers. This technique was used initially in the PHD method to reduce the bad effects of incomplete optimization of a single network by combining the results of a set of networks. Based on the tertiary classifiers that participated in the decision, different versions of SVM_JURY can be created. For example, Hua and Sun (2001) combined all tertiary classifiers, including SVM_MAX_D, SVM_TREEs, SVM_VOTE, and SVM_NN, for a jury decision.

Directed Acyclic Graph–Based Tertiary Classifier The directed acrylic graph (DAG) tertiary classifier (Kim and Park, 2003) is based on three one-versus-one classifiers (H/E, E/C, and C/H) (Figure 1.9). Many test results show that one-versus-one classifiers are more accurate than one-versus-rest classifiers, due to the fact that the one-versus-rest scheme often needs to handle two data sets with very different

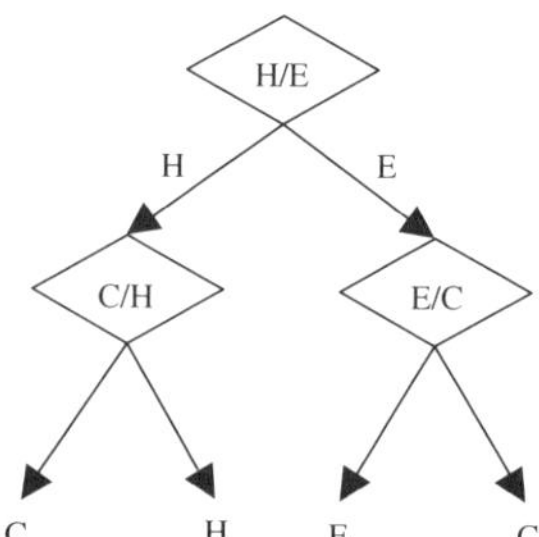

FIGURE 1.9   DAG scheme. (Based on Kim and Park, 2003.)

sizes (i.e., unbalanced training data) (Chang and Lin, 2001; Heiler, 2002). In this scheme, if the testing point is predicted to be H (not E) from the H/E classifier, the C/H classifier is used. If the point is predicted from the H/E classifier, not to be a helix ($\sim$H), the E/C classifier is used to determine if it is a sheet or coil. This DAG scheme is illustrated in Figure 1.9.

SVM_Representative   This method is similar to the SVM_MAX_D classifier in that the maximum distance is used for the decision. But unlike the SVM_MAX_D classifier, this scheme (Hu et al., 2004) combines the three one-versus-one binary classifiers (H/E, E/C, and C/H), which give more information than the one-versus-one binary classifier provides. In a one-versus-one classifier, both positive and negative values are meaningful to assign a final class, but in a one-versus-rest classifier, the negative value does not provide specific information for the decision. In this scheme, regardless of whether the distance values are positive or negative, the classifier with the absolute maximum distance is chosen as the representative classifier for the final decision of the class. For example, if the distance values of the one-versus-one classifiers (H/E, E/C, and C/H) are -1.7, 0.4, and -2.5, respectively, the binary classifier with the highest absolute value, here the C/H classifier, can be chosen for deciding the final class. Once this representative classifier is selected, the final class is assigned based on the value of this classifier. In this example, since the value of the C/H classifier shows negative, the final class is assigned as a helix.

Hu et al. (2004) designed this tertiary classifier to combine the result of one-versus-one binary classifiers that are trained with an orthogonal and Blosum62 matrix hybrid encoding profile. The authors claimed that the improved Q_3 accuracy was based on this new tertiary classifier, which captured the information effectively from the binary classifiers. Their scheme was overestimated while combining the results of a seven-fold cross-validation test by failing to exclude preknowledge from the test data. Therefore, for a fair comparison the result was not included in the performance comparison with other typical machine learning approaches.

SVM_Maxpoint   The first step of the SVM_Maxpoint scheme (Hu et al., 2006) is to divide all the value pairs into eight different cases based on the possible sign of each classifier. Since there are three classifiers and each one can have a positive or a

TABLE 1.3 Description of SVM_Maxpoint

Possible Cases	Cases in Which Values from Three One-Versus-One Binary Classifiers Fall			Classes Assigned Based on the Signs to the Left		
	H/E	E/C	C/H	H/E	E/C	C/H
1	—	—	—	E	C	H
2	—	—	+	E	C	C
3	—	+	—	E	E	H
4	+	—	—	H	C	H
5	—	+	+	E	E	C
6	+	—	+	H	C	C
7	+	+	—	H	E	H
8	+	+	+	H	E	C

negative sign, the three value pairs can fall into one of eight cases in Table 1.3. If we write a result class based on each sign in the Table 1.3, the second part of the table can be obtained. Since the sign is already used for assigning classes as in Table 1.3, the only thing to be considered now are the absolute decision function values, which indicate the distance from the hyperplane. As can be observed from Table 1.3, except for cases 1 and 8, all cases consist of two of the same class and one different class. Based on this observation, SVM_Maxpoint is formulated as follows:

- *Cases 1 and 8*: Assign the class that has the largest absolute decision function value.
- *Cases 2 through 7*: Add the absolute values of the decision function of two of the same class and compare this added value with the absolute decision function value of one different class. Assign the class that has the larger value.

Example: In case 2, E $= -1.2$, C $= -0.3$, and C $= 1.0$. Then the E point $= 1.2$ and the C point $= 0.3 + 1.0 = 1.3$. Since the C point is greater than the E point, the final class is C.

Multiclass SVM   In multiclass SVM, instead of combining the results of binary classifiers [Figure 1.10 (*a*)], all classes are considered in one step [Figure 1.10 (*b*)]. Nguyen and Rajapakse (2003) applied the multiclass SVM method for secondary-structure prediction. The authors examined two multiclass SVM algorithms proposed by Vapnik and Weston (Vapnik, 1998; Weston and Watkins, 1999) and by Crammer and Singer (2000). Even though the authors reported that Vapnik and Weston's multiclass SVM showed better performance than other schemes, including combined binary SVMs, the accuracy improvement is trivial (less than 0.5%). Considering the mathematical complexity of the multiclass SVM, it is not clear that this scheme is more suitable for protein secondary-structure prediction than are combined binary SVMs.

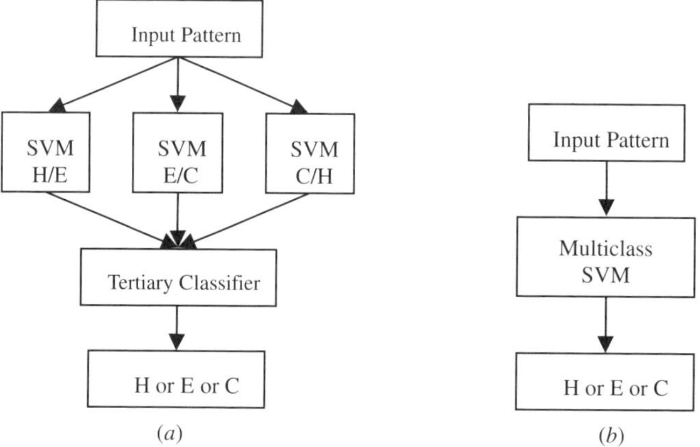

FIGURE 1.10 Combined scheme of binary classifiers and multiclass SVM: (*a*) combined scheme of one-versus-one binary classifiers; (*b*) multiclass SVM.

1.2.5 Accuracy Measure of SVM

There are several standard evaluation methods of secondary-structure prediction; Q_3, Matthew's correlation coefficient, and segment overlap measure (SOV) are widely used assessing methods.

Q_3 Q_3, one of the most commonly used performance measures in protein secondary-structure prediction, refers to the three-state overall percentage of residues predicted correctly. This measure is defined as

$$
Q_3 = \frac{\sum\limits_{i \in \{H,E,C\}} \text{no. of residues predicted correctly}}{\sum\limits_{i \in \{H,E,C\}} \text{no. of residues in class } i} \times 100 \qquad (1.26)
$$

Based on Eq. (1.26) the per-residue accuracy for each type of secondary structure (Q_H, Q_E, Q_C) can be obtained as

$$
Q_I = \frac{\text{no. of residues correctly predicted in state } I}{\text{no. of residues in state } I} \times 100 \qquad I \in \{H, E, C\}
$$

$$(1.27)$$

Matthew's Correlation Coefficient, C_i Matthew's correlation coefficient, another measure used in protein secondary-structure prediction, shows how closely the prediction is correlated with the results (Casbon, 2002). Matthew's correlation coefficient is given by

$$
C_i = \frac{\text{TP} \times \text{TN} - \text{FP} \times \text{FN}}{\sqrt{(\text{TP} + \text{FN})(\text{TP} + \text{FP})(\text{TN} + \text{FP})(\text{TN} + \text{FN})}} \qquad i \in \{H, E, C\} \quad (1.28)
$$

where, TP, FP, FN, TN are the number of true positives, false positives, false negatives, and true negatives for class i, respectively, and for clarity the dependency on i has been dropped on the right-hand side. This coefficient value falls on the range between -1 and 1, with 1 showing complete agreement, -1 complete disagreement, and 0 showing that the prediction was uncorrelated with the results.

Segment Overlap Measure   Segment overlap measure was developed by Rost et al. (1994) and modified by Zemla et al. (1999) to evaluate the quality of a prediction in a more realistic manner. While all the previous measures are general statistics that can be used in any classification problem, SOV is the specially designed measure for secondary-structure prediction. In secondary-structure prediction, it is important that the continuous structural elements are predicted to exist (Casbon, 2002). Even though Q_3 shows high accuracy value, if a continuous element is not predicted as existing continuously, this cannot be a good prediction. By including this knowledge, SOV became a measure suitable for evaluation of the secondary-structure segment rather than individual residues. SOV is calculated as (Zemla et al., 1999)

$$\text{SOV} = \frac{1}{N} \sum_{i \in \{H,E,C\}} \sum_{s(i)} \left[\frac{\min \text{ov}(s_1,s_2) + \delta(s_1,s_2)}{\max \text{ov}(s_1,s_2)} \times \text{len}(s_1) \right] \times 100 \qquad (1.29)$$

where N is the normalization value and a sum of $N(i)$ over all three states $= \sum_{i \in \{H,E,C\}} N(i)$

$$N(i) = \sum_{s(i)} \text{len}(s_1) + \sum_{s'(i)} \text{len}(s_1)$$

$S(i)$ is the set of all overlapping pairs of segments (s_1,s_2) in state $i = \{(s_1, s_2) : s_1 \cap s_2 \neq \emptyset, \text{in state } i\}$

$S'(i)$ is the set of of segments (s_1, s_2) in state i for which there is no overlapping, $= \{s_1 : \forall s_2, s_1 \cap s_2 = \emptyset, \text{in state } i\}$

$\text{len}(s_1)$ is the number of residues in segment $s_1 = e(s_1) - b(s_1) + 1$

$b(s_1)$ *is the position at which segment* s_1 *begins and* $e(s_1)$ *is the position at which segment* s_1 *ends*

$\text{minov}(s_1,s_2)$ is the length of the actual overlap $= \min(e(s_1), e(s_2)) - \max(b(s_1), b(s_2)) + 1$

$\text{maxov}(s_1, s_2)$ is the total extent of the segment expressed as $\max(e(s_1), e(s_2)) - \min(b(s_1), b(s_2)) + 1$

$\delta(s_1, s_2) = \min\{(\text{maxov}(s_1, s_2) - \text{minov}(s_1, s_2)), \text{minov}(s_1, s_2), \text{int}(\text{len}(s_1)/2), \text{int}(\text{len}(s_2)/2)\}$

The quality of the matching of each segment pair is taken as a ratio of the overlap of the two segments $\text{minov}(s_1, s_2)$ and the total extent of the pair $\text{maxov}(s_1, s_2)$. The definition of δ and the normalization factor N differs between SOV94 (Rost et al., 1994) and SOV99 (Zemla et al., 1999).

1.3 PERFORMANCE COMPARISON OF SVM METHODS

Table 1.4 is a performance comparison among typical machine learning approaches with different SVM methods. As can be noted, SVM methods and typical machine learning approaches show comparable performance.

TABLE 1.4 Accuracy Comparison Based on the RS126 Data Set

Method[a]	Q_3 (%)	Q_H (%)	Q_E (%)	Q_C (%)	SOV94 (%)	SOV99 (%)
NNSSP (Salamov and Solovyev, 1995)	72.2	—	—	—	—	—
PHD(93) (Rost and Sander, 1993b)[b]	70.8	72.0	66.0	72.0	73.5	—
PHD(94) (Rost and Sander, 1994)[b]	71.6	—	—	—	—	—
NNetwork_CK (Chandonia and Karplus, 1999)	76.6	—	—	—	—	—
NNetwork_P (Petersen et al., 2000)	80.2	—	—	—	—	—
SSpro2.0 (Pollastri et al., 2002)	78.1	82.4	66.2	81.3	—	—
SVMfreq (Hua and Sun, 2001)[b]	71.2	73.0	58.0	75.0	74.6	—
SVMpsi (Kim and Park, 2003)[b]	76.1	77.2	63.9	81.5	79.6	72.0
SVMpssm_RP (Hu et al., 2006)[b]	75.0	68.7	59.9	86.1	—	66.5
SVMpssm_MP (Hu et al., 2006)[b]	76.4	72.1	62.5	85.1	—	68.3

[a]NNSSP is the prediction result of the nearest-neighbor method by Salamov and Solovyev (1995); PHD(93) is the neural network prediction result obtained by Rost and Sander (1993b); PHD(94) is the neural network prediction result obtained by Rost and Sander (1994); NNetwork_CK is the neural network prediction result of Chandonia and Karplus (1999); NNetwork_P is the neural network prediction result of Petersen et al. (2000); SSpro2.0 is the neural network prediction result of Pollastri et al. (2002); SVMfreq is the prediction result of SVM by Hua and Sun (2001); SVMpsi is the prediction result of SVM obtained by Kim and Park (2003); SVMpssm_RP is the prediction result of SVM with PSSM encoding and SVM_Representative tertiary classifier by Hu et al. (2006); SVMpssm_MP is the prediction result of SVM with PSSM encoding and SVM_Maxpoint tertiary classifier by Hu et al. (2006).
[b]The combined results of a sevenfold cross-validation based on RS126 are shown. In other cases, RS126 is used as an independent test set.

1.4 DISCUSSION AND CONCLUSIONS

SVM schemes are successful and promising machine learning methods for secondary-structure prediction. With strong generalization ability, they outperformed most machine learning methods. However, similar to neural network methods, the SVMs are black box models. They could not generate logical models that explain the predictions they make. In the area of bioinformatics, the ability to explain the underlying principles of the decision is especially crucial, since based on this principle, "wet experiments" are guided. In addition, the rules extracted can be applied to make decisions in the future. Therefore, poor comprehensibility has been considered a big challenge to the success of SVM. To overcome this drawback, recently, much research has attempted to extract rules from an SVM decision (Mitsdorffer et al., 2002; Barakat and Diederich, 2004; He et al., 2006a,b). With the help of these efforts, SVM is gradually obtaining comprehensibility. Even though it will take more time and effort to obtain rules that are biologically meaningful, considering the achievements so far, the future of the SVM approach for secondary-structure prediction is optimistic.

REFERENCES

Altun, G., H. Hu, D. Brinza, R. Harrison, A. Zelikovsky, and Y. Pan (2006a). Hybrid SVM kernels for protein secondary structure prediction, *IEEE International Conference on Granular Computing* (GrC'06), Atlanta, GA.

Altun, G., H. Hu, D. Brinza, R. Harrison, A. Zelikovsky, and Y. Pan (2006b). SOM kernel based SVM for protein secondary structure prediction, Technical Report, Computer Science Department, Georgia State University, Atlanta, GA.

Baldi, P., S. Brunak, P. Frasconi, G. Pollastri, and G. Soda (1999). Exploiting the past and the future in protein secondary structure prediction, *Bioinformatics*, 15, pp. 937–946.

Barakat, N., and J. Diederich (2004). Learning-based rule-extraction from support vector machine, *Third Conference on Neuro-computing and Evolving Intelligence* (NCEI'04).

Biou, V., J. F. Gibrat, J. M. Levin, B. Robson, and J. Garnier (1988). Secondary structure prediction: combination of three different methods, *Protein Eng.*, 2, pp. 185–191.

Blout, E., C. de Loz, S. Bloom, and G. D. Fasman, (1960). Dependence of the conformation of synthetic polypeptides on amino acid composition, *J. Am. Chem. Soc.*, 82, pp. 3787–3789.

Bowie, J. U., R. Luthy, and D. Eisenberg (1991). A method to identify protein sequences that fold into a known three-dimensional structure, *Science*, 253, pp. 164–170.

Burges, C. J. C. (1998). A tutorial on support vector machines for pattern recognition, *Data Min. Knowledge Discovery*, 2, pp. 121–167.

Bystroff, C., V. Thorsson, and D. Baker (2000). HMMSTR: a hidden Markov model for local sequence-structure correlations in proteins, *J. Mol. Biol.*, 301, pp. 173–190.

Casbon, J. (2002). *Protein Secondary Structure Prediction with Support Vector Machines*, M. Sc. thesis, University of Sussex, Brighton, UK.

Chandonia, J. M., and M. Karplus (1999). New methods for accurate prediction of protein secondary structure, *Proteins: Struct. Funct. Genet.*, 35, pp. 293–306.

Chang, C. C., and C. J. Lin (2001). LIBSVM: a library for support vector machines, software available at http://www.csie.ntu.edu.tw/~cjlin/libsvm.

Chou, P. Y. (1989). *Prediction of Protein Structure and the Principles of Protein Conformation*, Plenum Press, New York.

Chou, P. Y., and G. D. Fasman (1974). Prediction of protein conformation, *Biochemistry*, 13, pp. 211–215.

Crammer, K., and Y. Singer (2000). On the learnability and design of output codes for multiclass problems, *Comput. Learn. Theory*, pp. 35–46.

Cristianini, N., and J. Shawe-Taylor (2000). *An Introduction to Support Vector Machines*, Cambridge University Press, New York.

Cuff, J. A., and G. J. Barton (1999). Evaluation and improvement of multiple sequence methods for protein secondary structure prediction, *Proteins: Struct. Funct. Genet.*, 34, pp. 508–519.

Eidhammer, I., I. Jonassen, and W. R. Taylor (2004). *Protein Bioinformatics: An Algorithmic Approach to Sequence and Structure Analysis*, Wiley, Hoboken, NJ.

Frishman, D., and P. Argos (1997). 75% accuracy in protein secondary structure prediction, *Proteins*, 27, pp. 329–335.

Garnier, J., D. J. Osguthorpe, and B. Robson (1978). Analysis and implications of simple methods for predicting the secondary structure of globular proteins, *J. Mol. Biol.*, 120, pp. 97–120.

Gonnet, G. H., M. A. Cohen, and S. A. Benner (1992). Exhaustive matching of the entire protein sequence database, *Science*, 256, pp. 1143–1445.

He, J., H. Hu, R. Harrison, P. C. Tai, and Y. Pan (2006a). Rule generation for protein secondary structure prediction with support vector machines and decision tree, *IEEE Trans. NanoBiosci.*, 5(1), pp. 46–53.

He, J., H. Hu, R. Harrison, P. C. Tai, and Y. Pan (2006b). Transmembrane segments prediction and understanding using support vector machine and decision tree, *Expert Syst. Appl.* (Special Issue on Intelligent Bioinformatics Systems), 30(1), pp. 64–72.

Heiler, M. (2002). *Optimization Criteria and Learning Algorithms of Large Margin Classifiers*, University of Mannheim, Mannheim, Germany.

Hu, H., Y. Pan, R. Harrison, and P. C. Tai (2004). Improved protein secondary structure prediction using support vector machine with a new encoding scheme and an advanced tertiary classifier, *IEEE Trans. NanoBiosci.*, 3(4), pp. 265–271.

Hu, H., P. C. Tai, J. He, R. W. Harrison, and Y. Pan (2006). A novel tertiary classifier for protein secondary structure prediction based on support vector machine and a PSSM profile, Technical Report, Computer Science Department, Georgia State University, Atlanta, GA.

Hua, S., and Z. Sun (2001). A novel method of protein secondary structure prediction with high segment overlap measure: support vector machine approach, *J. Mol. Biol.*, 308, pp. 397–407.

Joachims, T. (1999). *Making Large Scale SVM Learning Practical*, MIT Press, Cambridge, MA.

Jones, D. T. (1999). Protein secondary structure prediction based on position-specific scoring matrices, *J. Mol. Biol.*, 292(2), pp. 195–202.

Kim, H., and H. Park (2003). Protein secondary structure prediction based on an improved support vector machines approach, *Protein Eng.*,16, pp. 553–560.

Kloczkowski, A., K. L. Ting, R. L. Jernigan, and J. Garnier (2002). Combining the GOR V algorithm with evolutionary information for protein secondary structure prediction from amino acid sequence, *Proteins: Struct. Funct. Genet.*, 49, pp. 154–166.

Kohonen, T. (1997). *Self-Organizating Maps*, Springer-Verlag, New York.

Krogh, A., M. Brown, I. S. Mian, K. Sjolander, and D. Haussler (1994). Hidden Markov models in computational biology: applications to protein modeling, *Biology*, 235, pp. 1501–1531.

Kyngas, J., and J. Valjakka (1998). Unreliability of the Chou–Fasman parameters in predicting protein secondary structure, *Protein Eng.*, 11(5), pp. 345–348.

Levin, J. M. (1997). Exploring the limits of nearest neighbor secondary structure prediction, *Protein Eng.*, 10(7), pp. 771–776.

Levin, J. M., and J. Garnier (1988). Improvements in a secondary structure prediction method based on a search for local sequence homologies and its use as a model building tool, *Biochim. Biophys. Acta*, 955(3), pp. 283–295.

Li, H., and T. Jiang (2004). A class of edit kernels for SVMs to predict translation initiation sites in eukaryotic mRNAs, *J. Comput. Biol.*, 12, pp. 702–718.

Nishikawa, K., and T. Ooi (1986). Amino acid sequence homology applied to the prediction of protein secondary structure, and joint prediction with existing methods, *Biochim. Biophys. Acta*, 871, pp. 45–54.

Nguyen, M. N., and J. C. Rajapakse (2003). Multi-class support vector machines for protein secondary structure prediction, *Genome Inf.*, 14, pp. 218–227.

Mitsdorffer, R., J. Diederich, and C. Tan (2002). Rule-extraction from Technology IPOs in the U.S. stock market, ICONIP'02, Singapore.

Müller, K. R., S. Mika, G. Rätsch, K. Tsuda, and B. Schölkopf (2001). An introduction to kernel-based learning algorithms, *IEEE Trans. Neural Networks*, 12(2), pp. 181–201.

Petersen, T. N., C. Lundegaard, M. Nielsen, H. Bohr, J. Bohr, S. Brunak, P. G. Gippert, and O. Lund (2000). Prediction of protein secondary structure at 80% accuracy, *Proteins: Struct. Funct. Genet.*, 41, pp. 17–20.

Pollastri, G., D. Przybylski, B. Rost, and P. Baldi (2002). Improving the prediction of protein secondary structure in three and eight classes using recurrent neural networks and profiles, *Proteins: Struct. Funct. Genet.*, 47, pp. 228–235.

Rabiner, L. R. (1989). A tutorial on hidden Markov models and selected applications in speech recognition, *Proc. IEEE*, 77, pp. 257–286.

Radzicka, A., and R. Wolfenden (1988). Comparing the polarities of the amino acids: side-chain distribution coefficients between the vapor phase, cyclohexane, 1-octanol, and neutral aqueous solution, *Biochemistry*, 27, pp. 1664–1670.

Riis, S. K., and A. Krogh (1996). Improving prediction of protein secondary structure using structured neural networks and multiple sequence alignments, *J. Comput. Biol.*, 3, pp. 163–183.

Rost, B., and C. Sander (1993a). Improved prediction of protein secondary structure by use of sequence profiles and neural networks, *Proc. Natl. Acad. Sci. USA*, 90, pp. 7558–7562.

Rost, B., and C. Sander (1993b). Prediction of protein secondary structure at better than 70% accuracy, *J. Mol. Biol.*, 232, pp. 584–599.

Rost, B., and C. Sander (1994). Combining evolutionary information and neural networks to predict protein secondary structure, *Proteins*, 19, pp. 55–72.

Rost, B., C. Sander, and R. Schneider (1994). Redefining the goals of protein secondary structure prediction, *J. Mol. Biol.*, 235, pp. 13–26.

Rumelhart, D. E., G. E. Hinton, and R. J. Williams (1986). Learning representations by back-propagating errors, *Nature*, 323, pp. 533–536.

Salamov, A. A., and V. V. Solovyev (1995). Prediction of protein secondary structure by combining nearest-neighbor algorithms and multiple sequence alignments, *J. Mol. Biol.*, 247, pp. 11–15.

Salamov, A. A., and V. V. Solovyev (1997). Protein secondary structure prediction using local alignments, *J. Mol. Biol.*, 268, pp. 31–36.

Vanschoenwinkel, B., and B. Manderick (2004). Substitution matrix based kernel functions for protein secondary structure prediction, *Proc. ICMLA*.

Vapnik, V., and C. Cortes (1995). Support vector networks, *Mach. Learn.*, 20, pp. 273–293.

Vapnik, V. (1998). *Statistical Learning Theory*, Wiley, New York.

Weston, J., and C. Watkins (1999). Multi-class support vector machines, in M. Verleysen, Ed., *Proc. ESANN'99*, De Facto Press, Brussels, Belgium.

Won, K. J., T. Hamelryck, P. B. Adam, and A. Krogh (2005). Evolving hidden Markov models for protein secondary structure prediction, *Proc. IEEE Congress on Evolutionary Computation*, pp. 33–40.

Yi, T. M., and E. S. Lander (1993). Protein secondary structure prediction using nearest-neighbor methods, *J. Mol. Biol.*, 232, pp. 1117–1129.

Zemla, A., C. Venclovas, K. Fidelis, and B. Rost (1999). A modified definition of SOV, a segment-based measure for protein secondary prediction assessment, *Proteins: Struct. Funct. Genet.*, 34, pp. 220–223.

Zimmermann, K. (1994). When awaiting "bio" champollion: dynamic programming regularization of the protein secondary structure predictions, *Protein Eng.*, 7, pp. 1197–1202.

2

COMPARISON OF SEVEN METHODS FOR MINING HIDDEN LINKS

XIAOHUA HU, XIAODAN ZHANG, AND XIAOHUA ZHOU

College of Information and Technology, Drexel University, Philadelphia, Pennsylvania

In this chapter we compare seven methods for mining hidden links among medical concepts. These seven methods include one existing method and six new methods. To discover hidden links among concepts, it is very important to judge the relevance of two terms according to certain information measures and to rank hidden links among concepts. So the seven methods are different combinations of three information measures and four ranking methods. The three information measures include an association rule, mutual information, and chi-square proposed as a novel information measure on mining hidden links; four methods of ranking hidden links are discussed, including three existing ranking methods (information measure ranking, shared connections ranking, semantic knowledge–based ranking); and a novel method called B-term centered ranking (BCR) is described. Semantic information as a support for mining hidden links is also discussed. A series of experiments of seven methods are performed and analyzed. Our research works present a comprehensive analysis for mining hidden links and how different weighting schemes plus semantic information affect the knowledge discovery procedure.

2.1 ANALYSIS OF THE LITERATURE ON RAYNAUD'S DISEASE

The problem of mining hidden links from complementary and noninteractive biomedical literature was exemplified by Swanson's pioneering work on Raynaud's

Knowledge Discovery in Bioinformatics: Techniques, Methods, and Applications, Edited by Xiaohua Hu and Yi Pan
Copyright © 2007 John Wiley & Sons, Inc.

disease in 1986 [14]. Two complementary and noninteractive sets of articles (independently created fragments of knowledge), when considered together, can reveal useful information of scientific interest not apparent in either of the two sets alone [14,15]. Swanson formalizes the procedure as the ABC model of finding undiscovered public knowledge (UPK) from the biomedical literature as follows: Consider two separate literature sets, CL and AL, where the documents in CL discuss concept C and documents in AL discuss concept A. Both sets of literature discuss their relationship with intermediate concepts B (also called *bridge concepts*). However, their possible connection by means of concepts B is not discussed in either of the literature sets, as shown in Figure 2.1. For example, Swanson tried to uncover novel suggestions for which B concepts cause Raynaud's disease (C) or which B concepts are symptoms of the disease, and which A concepts might treat the disease, as shown in Figure 2.1. Through analyzing the document set that in which Raynaud's disease is discussed, he found that the disease (C) is a peripheral circulatory disorder aggravated by high platelet aggregation (B), high blood viscosity (B), and vasoconstriction (B). Then he searched these three concepts (B) against Medline to collect a document set relevant to them. In analyzing the document set, he found that the articles showed that the ingestion of fish oils (A) can reduce these phenomena (B); however, no single article from both document sets mentions Raynaud's disease (C) and fish oils (A) together. Putting the two sets of literature together, Swanson hypothesized that fish oils (A) may be beneficial to people suffering from Raynaud's disease (C). This novel hypothesis was confirmed clinically by DiGiacomo in 1989 [3]. Later, Swanson used the same approach to uncover 11 connections between migraine and magnesium [13]. One of the drawbacks of Swanson's method is that it requires a great deal of manual intervention and very strong domain knowledge, especially in the process of qualifying the intermediate concepts: Swanson's B concepts. Much work has been done to overcome the limitations of Swanson's approach; most explore various

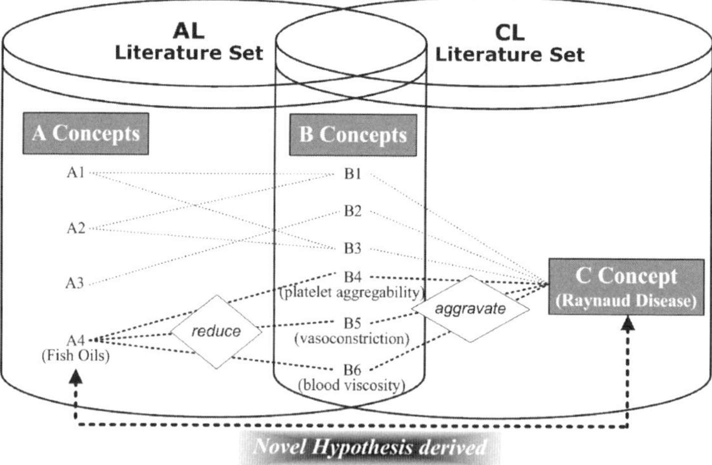

FIGURE 2.1 Swanson's ABC model: the connection between fish oils and Raynaud's disease.

information measures to generate implicit connections between concepts [3,15] and apply ontology to reduce discovery space [8–10, 14]. Some work focused on term relationship implication measures without applying domain knowledge [3,15]. However, they usually had trouble ranking new concepts because certain measures favor only certain types of concepts. Other work focused on reducing the search space [9,10], but that normally led to an inability to explore relationships among concepts. Moreover, no approach was comprehensive in evaluating different information measures combined with semantic knowledge from domain ontology.

In this chapter, seven fully automated methods for mining hidden links from biomedical literature are compared. Our work is similar to work [15] that applies a mutual information measure to discover inferred novel connections among concepts. Whereas [15] focuses primarily on mining an inferred connection by calculating the number and strength of connections shared by the two concepts through a mutual information measure and a ranking algorithm, we focus on evaluating a combination of information measures and ranking methods. Also, we evaluate an information measure chi-square that had never been used in an ABC model. Moreover, a new ranking method for ranking implicit connections among concepts is developed. In our approach, instead of querying certain medical concepts through PubMed, PubMed data as a whole within a certain time range are analyzed. Certain statistical methods such as chi-square are used to help determine whether the two terms are statistically interdependent and how strong the relationship is. Some terms not only cooccur frequently with a certain term, but also cooccur with many others. For example, while "humans" cooccurs with almost every "Raynaud's disease," it also cooccurs with many other MeSH concepts. After analyzing the statistical significance, "human" is ranked much lower than to be ranked only by the query term returned data set. As the PubMed database stores a huge number of documents each year, analyzing a data set within a certain year provides a global picture of medical concept relationships. We believe that our work helps explain how combinations of methods affect the knowledge discovery process.

2.2   RELATED WORK

Several algorithms have been developed to overcome the limitations of Swanson's approach. Hristovski et al. [3] used MeSH descriptors rather than the title words of documents. They use association rule algorithms to find cooccurrences of the words. Their methods find all B concepts as bridges related to the starting C concept. Then all A concepts related to B concepts are found through Medline searching. But because in Medline each concept can be associated with many other concepts, the possible number of B → C and A → B combinations can be extremely large. To deal with this combinatorial problem, the algorithm incorporates filtering and ordering capabilities [8,9]. Pratt and Yetisgen-Yildiz [8] used unified medical language system (UMLS) concepts instead of the MeSH terms assigned to Medline documents. Similar to Swanson's method, their search space is limited only by the titles of documents for the starting concept. They can reduce the number of terms (B and A concepts) by limiting the search space before generating association rules and tried to group the concepts (B or A concepts) to get a much coarser level of synonyms. Their methods still require strong domain knowledge, especially in selecting semantic types for A and B concepts, and some vague parameters on defining concepts categorized as being "too general." Srinivasan [9] and Weeber et al. [14] both employ UMLS concepts to reduce the size of the search space. Srinivasan viewed Swanson's method as working in two dimensions. The first dimension is about identifying relevant concepts for a given concept. The second dimension is about exploring the specific relationships between concepts. However, Srinivasan and Libbus [10] deal only with the first dimension. Weeber uses a metamap program to map to UMLS concepts the raw text in titles and abstracts. He reduces the search space further by what he calls *semantic filtering*, where he collects all terms that come under a particular UMLS concept and ranks them. Srinivasan and Libbus [10] first build a topic profile for A concepts after querying PubMed based on their occurrence frequency using the TF*IDF ranking scheme. They also employ input semantic types to reduce the number of B and C concepts. Finally, they rank terms under selected semantic types. One advantage of this approach is that it filters out terms of irrelevant semantic type. However, it does not explore the specific relationships between concepts. In contrast, Wren [15] takes an open-ended discovery approach that uses no domain knowledge. A mutual information measure is used to calculate a term's relationship to inferred connections. New associations are inferred based on commonalities; that is, the more associations that A and B share, the closer they are. Since every pair of terms can have both direct and inferred mutual information, the closer their rankings, the better the ranking measure is. However, when ranking by the number of shared relationships, the entries scoring highest tend to be more general and vague in nature [15]. Gordon and Lindsay [6] used lexical statistics (TF*IDF) and manual clustering to identify groups of terms within the B-term list, and then the groups are used in manual query expansion to locate additional hidden links. Although Swanson's experiments were replicated successfully, their methods require a high level of human involvement. Gordon and Dumais [7] use latent semantic indexing (LSI) as the underlying technique for hidden link discovery. However, when replicating Swanson's experiment, the term "fish oils" is not highly ranked. LSI is also a very computation intensive technique, so it is not feasible for a large data set.

All of the research cited has furthered Swanson's method significantly. However, none of the approaches are comprehensive in evaluating various information measures and considering specific semantic relationships. The association problem should be tackled not only by an information measure but also by the semantic information that exists among the concepts. In this chapter, seven methods of mining implicit connections between concepts are evaluated.

2.3 METHODS

In this section we discuss three information measures and four ranking methods used in mining implicit connections between concepts. A detailed algorithm is presented for each method.

2.3.1 Information Measures

Chi-Square   Chi-square is often used to measure dependency between variables. This is the first attempt to discover implicit connections between concepts. For a term cooccurrence matrix, let O be the frequency observed and E be the frequency expected; then the formula for chi-square is $\chi^2 = \sum (O - E)^2 / E$. Chi square is more likely to establish significance when (1) the relationship is strong, (2) the sample size is large, and/or (3) the number of values of two associated variables is large. A chi-square probability of 0.05 or less is commonly interpreted as justification for rejecting the null hypothesis that the two terms are independent of each other. The null hypothesis is very useful when filtering out those terms that are statistically independent of a given term. In this chapter, for chi-square the α value is set to 1%, and accordingly, the critical value is 6.63.

Mutual Information   Mutual information is widely used to calculate dependencies between variables and is shown as $\mathrm{MI}(A, B) = \log(P_{AB}/P_A P_B)$, where P_A and P_B indicate the probability that terms A and B occur, respectively, and P_{AB} is the probability that the terms A and B cooccur. To avoid negative weighting, following the ideas in [15], in our design the log function is removed and the equation becomes $\mathrm{MI}(A, B) = P_{AB}/P_A \cdot P_B$. It is possible that rare associations might receive a very high MI score [15]. In real applications especially, these rare associations may sometimes indicate truly golden luggage.

Association Rule   An association rule is frequently used in data mining applications [1,3]. An association rule is of the form $A \rightarrow B$, where B and A are disjoint conjunctions of term pairs. The confidence of the rule is the conditional probability of B given A, $\Pr(B \mid A)$, and the support of the rule is the prior probability of A and B, $\Pr(A$ and $B)$. The mining procedure can be described as follows: Given a document collection, a minimal confidence threshold, and a minimal support threshold, find all association rules whose confidence and support are above the corresponding thresholds. Here concept A is taken as input, then all $A \rightarrow B$ rules are found in one data set. Then from a separate data set, $B \rightarrow C$ rules are found. Finally, a transitive law is applied to get the hidden link: $A \rightarrow C$. It must be noted that an associate rule algorithm cannot be used to get $A \rightarrow C$ directly because A and C do not occur in the same data set. A and C are connected through bridge concept B. The following symbols are used to describe an association rule: Support (B) indicates the probability that B occurs. Accordingly, Support$(B \cap C)$ indicates the possibility that B and C occur together. "Conf" indicates confidence; Conf $= \mathrm{Sup}(B \cap C)/\mathrm{sup}(B)$ is the confidence that B implies $C(B \rightarrow C)$. $F(B \rightarrow C)$ is a measure of the confidence that $B \rightarrow C$. The larger the value, the greater the confidence that $B \rightarrow C$.

2.3.2 Ranking Methods

Information Measure Ranking   This approach ranks a term according to the term's relationships. In this case, a B term is ranked according to its closeness to an A term by a certain information measure. The same ranking method is used to connect a C term to a B term. A heuristic approach is used to pick the top k closest B terms to A terms. Then from each B term selected, the top M C terms are selected.

Semantic Knowledge–Based Ranking   Semantic knowledge–based ranking is used as a support for information measure ranking. A common approach is to choose only terms of certain semantic types defined in UMLS. In this chapter we use this approach to reduce unnecessary terms by checking the relationship between the semantic types of two terms.

Shared Connections (Links) Ranking   The ranking of C terms is judged by normalizing the scored connections between C and A terms. The more connections there are and the larger the scores of connections, the better the ranking of C terms will be.

B-Term-Centered Ranking   In addition to the methods mentioned above, we provide a novel ranking method called B-term-centered ranking. C terms are ranked according to their cooccurrence with B terms. Some C terms cooccur with most B terms very frequently, some C terms cooccur with most B terms but less frequently, and only some C terms cooccur with some B terms very frequently. A mixture of weight ranking formulas is used to assign weights to a portion of the types of terms that are required. This is discussed in detail in the following section.

2.3.3 Seven Methods

As shown in Table 2.1, seven methods have been designed on the basis of three information measures and four ranking methods. In this section a detailed algorithm is presented for each method. Here, ABC refers to Swanson's ABC model, discussed in Section 2.2.

Chi-Square ABC   In this algorithm, chi-square values between a given term and all cooccurring terms are checked. If a chi-square value is smaller than a certain critical value, the cooccurring term is filtered out. All the filtered terms are considered to be statistically independent of term A. This improves the performance of the data mining process dramatically by reducing the number of implicit connections among concepts through chi-square filtering.

Input: (1) Concept A, (2) k (the top B concept), (3) M (the top C concept for each B concept), and (4) the time range of the PubMed articles.

1. Download PubMed articles for a certain time range to build a term document matrix.
2. A chi-square dependency graph matrix is built based on the term document matrix. If the chi-square value between two terms is smaller than a critical value such as 6.63, it means that it accepts the null hypothesis that the two terms are

TABLE 2.1 Seven Information and Ranking Methods

Method Name	Short Description
Chi-square ABC	Chi-square is used to measure the closeness between terms. The degree of closeness is thus used to rank the terms.
Mutual information ABC	Mutual information is used to measure the closeness of two terms. The degree of closeness is thus used to rank the terms.
Chi-square association rule ABC	Chi-square is used to filter out terms that are not statistically dependent on given terms. An association rule algorithm is used to find A $\rightarrow$ B and B $\rightarrow$ C rule sets.
Semantic chi-square ABC	Chi-square is used as in chi-square ABC. In addition, a semantic type check is used to filter out semantically unrelated terms and to show only terms of certain semantic types.
Chi-square link ABC	The ranking of C terms is judged by the shared links between A and C terms. Chi-square is used as a measure of closeness.
Mutual information link ABC	The ranking of C terms is judged by the shared links between A and C terms. Mutual information is used as a measure of closeness.
B-term-centered ABC	C terms are ranked according to their cooccurrence with B terms.

statistically independent of each other ($p < 0.01$). Thus, the cell value of the dependency matrix is set at zero.

3. All the terms that have a chi-square value higher than the critical value are sorted. The higher the chi-square between the two concepts, the more interdependent the concepts are. This term list is that of the candidate B terms. The top k terms from this list are selected as a final B concept list.

4. As there is a dependency matrix, the corresponding C concept list is generated from each B concept selected. Each C concept list contains M highly scored terms. It is noted that all the terms cooccurring with the A term, including the A term itself, cannot be in the C concept list.

Mutual Information ABC

Input: (1) Concept A, (2) k (the top B concept), (3) M (the top C concept for each B concept), and (4) the time range of the PubMed articles.

1. Download PubMed articles for a certain time range to build a term document matrix.

2. A mutual information matrix is built based on term cooccurrence information.

3. All the terms are sorted according to their mutual information scores. The larger the mutual information score between two concepts, the closer they are. This term list is that of the candidate B terms as well as being a stop word list for finding C terms. The top k terms are selected from this candidate B term list to make the final B concept list.

4. A corresponding C concept list is generated from each of the B concepts selected. Note that the C concepts list does not contain all the concepts that cooccur with A concepts, because A and C concepts need to be literally disjointed.

Chi-Square Association Rule ABC   An association rule is frequently used in data mining applications to find the connection between two items. However, using this method it is difficult to explain the statistical significance of the relationship between two variables. For example, the term *humans* cooccurs with the term *Raynaud's disease*. With regard to the association rule formula, the term *humans* yields the highest score in our experiment. However, when judging the statistical significance of the relationship between the two terms, the term *humans* is ranked much lower. As a result, terms statistically unrelated by chi-square are filtered out to help improve the generation of novel implicit association rules. In other words, the association rule method itself can generate many nonmeaningful rules. Other methods, such as chi-square filtering, can be used to help pick out meaningful rule sets from the a huge number of rules. As noted in ref. 15, when considering the scale-free distribution of objects in the literature, it is apparent that some objects mentioned frequently could often be comentioned with other terms without actual biological association being implied. Therefore, chi-square is an important statistical means to measure the strength of association and can help improve the performance of ABC and eliminate a lot of unnecessary work.

Input: (1) Concept A, (2) k (the top B concept), (3) M (the top C concept for each B concept), (4) the time range of the PubMed articles, and (5) the direction of the rule generating A $\rightarrow$ B $\rightarrow$ C or C $\rightarrow$ B $\rightarrow$ A.

1. Download PubMed articles for a certain time range to build a term document matrix.

2. A chi-square dependency graph matrix is built based on the term document matrix. If the chi-square value between two terms is smaller than the critical value 6.63, it means that it accepts the null hypothesis that the two terms are statistically independent of each other ($p < 0.01$). Thus, the cell value of the dependency matrix is set at zero.

3. Generate B $\rightarrow$ A rules or A $\rightarrow$ B rules. The dependency graph matrix helps to generate rule sets within which each term is statistically significant dependent of another. According to the association rule, let Confidence $[P(\text{B} \rightarrow \text{A})] = P(\text{AB})/P(\text{B})$ and support(B)$=P(\text{B})$. The F measure 2 (Confidence)(Support)/ (Confidence + Support) is used to calculate the rule strength.

4. All the B terms are sorted according to the rule strength. The top k B terms are chosen from the list.

5. Accordingly, all the C → B or B → C rules are generated if the chi-square between B and C is over the critical value. All the B terms cannot be in the candidate C terms list, because all these terms cooccur with concept A. Similarly, only the top M C terms are picked from each C term list.

Semantic Chi-Square ABC This procedure is the same as the chi-Square ABC except that we check whether terms A and B are related semantically based on the biomedical ontology UMLS, as well as on B and C. For PubMed concepts, 135 semantic types are defined in the UMLS. Each term can have one or more concept IDs, and each concept ID has one particular semantic type. One semantic type may have a certain relationship with another. Given two terms A and B, their corresponding medical concepts in the UMLS database are retrieved, and from these concepts, semantic types are gathered. Next, the relationship of the semantic types between terms A and B is checked. If there is no relationship, the B term will be filtered out. Figure 2.2 shows the semantic relationships among the concepts *Raynaud's disease*, *blood viscosity*, and *fish oils*.

Moreover, how to rank C terms according to semantic type information is examined. Srinivasan [9] tried his open discovery process on the term *curcumin*, a spice used widely in Asia. One point of his work was to pick up terms of certain semantic types. However, picking up such terms still requires strong domain knowledge. For a fully automated semantic approach, a further step is made to rank C terms from a particular B term by both chi-square value and semantic type information. The procedure is described as follows: When generating C terms from each B term, C terms are ranked based on chi-square value and then divided into different semantic groups based on the semantic type information. Because 135 semantic types are defined by UMLS, there are at most 135 semantic groups for each B term. Thus, all the C terms fall into different subsets.

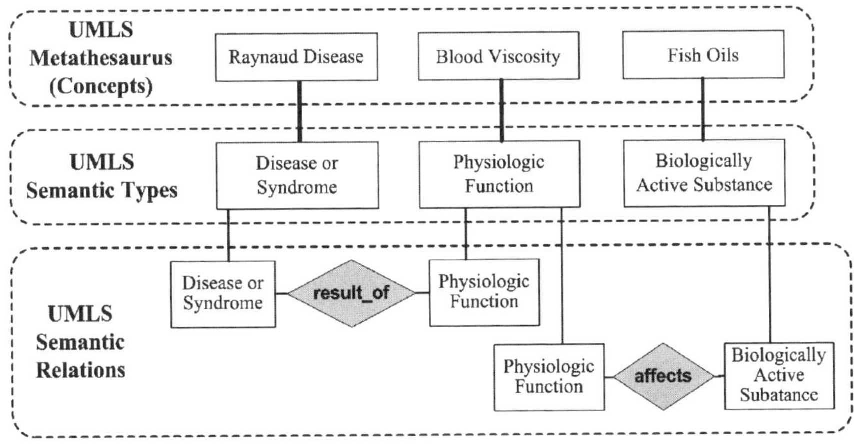

FIGURE 2.2 Example of the UMLS.

Chi-Square Link ABC Here the method indicated in ref. 15 is evaluated. Instead of mutual information [15], chi-square is used to measure the strength of associations, which has critical value for reference to check the significance level of the terms' dependency.

Input: (1) Concept A and (2) the time range of the PubMed articles.

1. Download PubMed articles for a certain time range to build a term document matrix.
2. A chi-square dependency matrix is built based on the term document matrix. If the chi-square value between two terms is smaller than the critical value 6.63, it means that it accepts the null hypothesis that the two terms are statistically independent of each other $(p < 0.01)$. Thus, the cell value of the dependency matrix is set at zero.
3. All the terms that have a chi-square value above 6.63 with concept A are sorted. The higher the chi-square between the two concepts, the more interdependent the concepts are. A candidate B term list is thus generated. Here all the terms that do not cooccur in data sets are treated as candidate C terms.
4. The chi-square value of each C term is checked with each B term. If the chi-square is over the critical value, the weight is added up as well as the chi-square values between A and B. Two times the number of B $\rightarrow$ C connections is used to normalize the weight. The equation is

$$\text{Score}(A, C) = \sum_{n=1}^{t} \frac{X^2(A, B) + X^2(B, C)}{2t}$$

where t stands for the number of associations shared by A and C.

Mutual Information Link ABC Except for using mutual information to calculate the association strength and the fact that there is no critical value limit, the procedure is the same as for chi-square ABC. The equation for weighting the inferred association between A and C is

$$\text{Score}(A, C) = \sum_{n=1}^{t} \left(\frac{P_{AB}/P_A P_B + P_{BC}/P_B P_C}{2t} \right)$$

where t stands for the number of associations that A and C share. For details, refer to ref. 15.

B-Term-Centered ABC One weakness of ref. 15 is in the ranking of C terms, which favors more general terms to more specific terms. If a C term has a strong relationship with only one or two B terms, it will rank lower than a C term that has a relationship with many B terms. For example, the term *fish oils* was discovered through only a small number of B concepts. To overcome this limitation, a more flexible ranking method called B-term-centered ranking has been developed, which can favor either general terms or special terms, as selected by the user.

Input: (1) Concept A, (2) k (the top B concept), and (3) the time range of the PubMed articles.

1. Build a B term list in which all B terms cooccur with the starting term A.
2. All the terms that do not cooccur with the A term can be candidate C terms. All these terms are put in a list called a candidate C term list.
3. Build a C term B term cooccurrence matrix. The ranking formula is

$$W(t) = w_1 \times \text{mean} + w_2 \times \text{mean}/\sqrt{\text{sd}} + w_3 \times \text{sd}/\text{mean}(w_1 + w_2 + w_3 = 1)$$

where sd stands for "standard derivation." A larger *mean* score implies that the term tends to cooccur many more times with more terms in the list. $\text{mean}/\sqrt{\text{sd}}$ favors those terms that cooccur fewer times than mean measure, but still cooccur with many terms in the list. However, sd/mean favors those terms that cooccur with only a few terms in the list. We assign different weights to w_1, w_2, and w_3 according to the discovery strategy.

2.4   EXPERIMENT RESULTS AND ANALYSIS

To evaluate the performance of the methods proposed for mining implicit connections among concepts, a series of experiments is performed on PubMed data. The experiments are divided into three groups according to methodology: (1) experiments with chi-square ABC, chi-square association rule ABC, and mutual information ABC are put together because it is easy to compare the performance of different information measures on ranking implicit connections; (2) methods with semantic support, such as semantic chi-square ABC are put together, which helps evaluate how semantic information affects the results; and (3) chi-square, mutual information ABC, and B-term-centered ABC are put together because they use all the terms that do not cooccur with the A term as candidate C terms, which is different from that of all the methods described above. As the pool of C terms is the same, it would be better to compare the three methods as a group. For better understanding of the performance of the various methods, we evaluate all seven methods using the same data set: Swanson's Raynaud's disease/fish oils discovery.

2.4.1   Data Set

For replication of Swanson's Raynaud's disease/fish oils discovery, 1,737,439 citations of PubMed articles from 1980 to 1985 are downloaded. MeSH terms are extracted from each article and a large term document matrix, $(16,045)(1,737,439)$

is built. As the sparse matrix technique is applied, although it is a large matrix, most algorithms performed on it can be finished within several minutes. Based on the term document matrix, the approaches described below are evaluated.

2.4.2 Chi-Square, Chi-Square Association Rule, and Mutual Information Link ABC Methods Compared

The result of our chi-square filtering is promising. A huge number of B terms have been filtered through chi-square filtering. All these terms have no statistically significant dependent relationship to the term *Raynaud's disease*. It is clear from Tables 2.2 and 2.3 that the chi-square association rule favors general terms over special terms. In contrast, the mutual information method ranks rare terms better than the other two, which is expected based on their formulas. For example, among these 37 B terms, the term *fish oils* is ranked within the top 100 fifteen times and within the top 50 six times. Chi-square stands in the middle between the two methods. However, when combining with the B term's ranking, chi-square ABC produces a better result than do the other two. Even the chi-square association rule does not rank term *fish oils* very well; however, a lot of computation time is saved by avoiding generating unnecessary association rules. Taking *Raynaud's disease* as an example, only 425 useful B rules needed to be generated, whereas the number of C rules needed to discover certain C terms such as *fish oils* ranged from 106 to 4558, thus dramatically reducing computation time and complexity.

2.4.3 Chi-Square ABC Method: Semantic Check for Mining Implicit Connections

In this section, we evaluate how semantic information affects the ranking of B and C terms. In Table 2.4, entries in italic type are those B terms that do not have a direct semantic relationship with the A term *Raynaud's disease*. There are 386 terms left after chi-square filtering followed by semantic filtering from all 1250 B terms. A semantic check helps filter out 38 B terms. As shown in the table, the rank of term *blood viscosity* is changed from 22 in chi-square ABC to 21 in chi-square semantic ABC. This is because skin temperature, which has no direct semantic relationship with the A term, is filtered out. Although 38 B terms and ranking ahead by one are only a small improvement, however, these facts give us at least two important items of information: (1) chi-square filtering can identify most semantically related terms, and (2) semantic checks combine with certain information measures to filter out semantically unrelated terms.

There are very few terms that have the same semantic type as the term *fish oils*. No doubt, it will stand out if you happen to pick this semantic type. As human experts were involved in ref. 10 in evaluating implicit connections, for comparison we take curcumin as an example to check how our work matched human experts' choice of C terms. For the experiment, all MeSH terms of articles from 1986 to 2004 from PubMed were downloaded. Here the top 10 B terms and the top 400 C terms were picked, ranked according to the chi-square ABC weighting scheme. Also as a result of

TABLE 2.2 Ranking B Terms According to their Closeness to Term A

	B Terms from Which *Fish Oils* Is Discovered	Chi-Square ABC	Chi-Square Association Rule ABC	Mutual Information ABC
1	6-Ketoprostaglandin F1 α	269	209	319
2	Adolescent	397	11	861
3	Alprostadil	66	88	108
4	Angina pectoris	237	108	462
5	Animals	61	53	1242
6	Arteriosclerosis	83	51	258
7	Arthritis	119	82	252
8	β-Thromboglobulin	142	188	139
9	Blood platelets	320	86	582
10	Blood vessels	419	199	522
11	**Blood viscosity**	**22**	**34**	**49**
12	Body weight	396	349	1239
13	Cardiovascular diseases	305	146	477
14	Child	235	41	1137
15	Clinical trials	80	22	413
16	Double-blind method	49	23	241
17	Epoprostenol	47	42	136
18	Erythrocyte deformability	110	226	64
19	Forearm	144	150	201
20	Humans	44	1	811
21	Liver	380	113	1154
22	Lupus erythematosus, systemic	21	17	80
23	Male	118	4	823
24	Muscle, smooth, vascular	374	145	551
25	Platelet aggregation	174	78	426
26	Platelet function tests	187	274	159
27	Prostaglandins	155	73	399
28	Prostaglandins E	127	60	352
29	Rats	138	155	1247
30	Rats, inbred strains	199	344	1250
31	Thrombosis	283	117	492
32	Thromboxane A2	231	214	279
33	Thromboxane B2	420	233	471
34	Thromboxanes	401	235	454

comparing with ref. 10, only the top-ranked terms from three semantic types are listed (see Table 2.5).

The C terms *pleurisy*, *retinal neoplasm*, and *Hodgkin's disease* are also identified by human experts as being the top three within each semantic group. It is obvious that it is easier to get a much better ranking from terms that are ranked within a certain semantic group. The result confirms that most of these terms are not top-ranked in our experiments.

TABLE 2.3   Ranking comparison of *Fish Oils* from Three Methods[a]

All B Terms from Which *Fish Oils* Is Discovered		Chi-Square ABC	Chi-Square Association Rule ABC	Mutual Information ABC
1	6-Ketoprostaglandin F1α	128/1070	263/1070	73/1070
2	Adolescent	6236/11,325	5794/11,325	8965/11,325
3	Alprostadil	159/1539	326/1539	116/1539
4	Angina pectoris	148/1372	181/1372	103/1372
5	Animals	5891/13,291	4662/13,291	3774/13,291
6	Arteriosclerosis	107/2760	2532760	92/2760
7	Arthritis	391/2418	431/2418	301/2418
8	β-Thromboglobulin	93/509	158/509	54/509
9	Blood platelets	97/3835	305/3835	74/3835
10	Blood vessels	154/2501	350/2501	110/2501
11	**Blood viscosity**	**111/1084**	**203/1084**	**71/1084**
12	Body weight	1080/6752	12476752	643/6752
13	Cardiovascular diseases	429/2801	475/2801	233/2801
14	Child	6077/10,923	5816/10,923	8983/10,923
15	Clinical trials	1268/6538	1351/6538	862/6538
16	Double-blind method	819/4613	959/4613	648/4613
17	Epoprostenol	70/1847	193/1847	60/1847
18	Erythrocyte deformability	60/363	182/363	34/363
19	Forearm	279/1128	260/1128	159/1128
20	Humans	10,725/14,166	6779/14,166	7910/14,166
21	Liver	1979/8309	2008/8309	1385/8309
22	Lupus erythematosus, systemic	593/2615	4642615	324/2615
23	Male	6893/13,760	5186/13,760	3016/13,760
24	Muscle, smooth, vascular	726/2470	581/2470	508/2470
25	Platelet aggregation	32/2896	124/2896	41/2896
26	Platelet function tests	71/405	162/405	49/405
27	Prostaglandins	151/3402	418/3402	105/3402
28	Prostaglandins E	237/3427	542/3427	152/3427
29	Rats	3685/10,844	3050/10,844	2165/10,844
30	Rats, inbred strains	1663/8626	2197/8626	1036/8626
31	Thrombosis	229/2357	364/2357	158/2357
32	Thromboxane A2	66/882	151/882	54/882
33	Thromboxane B2	33/1336	107/1336	26/1336
34	Thromboxanes	23/1228	87/1228	27/1228

[a]The cell value for these three values takes the following format: number 1/number 2. Number 1 indicates the rank of *fish oils* and number 2 shows the total number of C terms.

2.4.4   Chi-Square and Mutual Information Link ABC Methods

Three weighting schemes for the ranking of C terms are evaluated. The experimental results (Tables 2.6 and 2.7) show that both chi-square link ABC and mutual information ABC can not give *fish oils* a good ranking.

TABLE 2.4　Chi-Square ABC

B Terms[a]	Cooccurrences[b]	Chi-Square[c]	Rank
Scleroderma, systemic	205	53,531.21	1/386/424/1250
Fingers	173	23,993.94	2/386/424/1250
Vibration	95	13,134	3/386/424/1250
Telangiectasis	42	11,402.45	4/386/424/1250
Skin temperature	*66*	*7195.4*	*5/386/424/1250*
Mixed connective tissue disease	32	5925.3	5/386/424/1250
Sympathectomy	41	4827.788	6/386/424/1250
Thromboangiitis obliterans	22	4479.649	7/386/424/1250
Plethysmography	45	3794.035	8/386/424/1250
Antibodies, antinuclear	56	3581.977	9/386/424/1250
Connective tissue diseases	35	3540.37	10/386/424/1250
Cold	86	3482.94	11/386/424/1250
Hand	68	3065.012	12/386/424/1250
Scleroderma, localized	20	2864.835	13/386/424/1250
Centromere	21	2491.958	14/386/424/1250
Nails	25	2462.647	15/386/424/1250
Vascular diseases	50	2090.442	16/386/424/1250
Thoracic outlet syndrome	14	1895.985	17/386/424/1250
Occupational diseases	98	1838.723	18/386/424/1250
Calcinosis	51	1637.209	19/386/424/1250
Lupus erythematosus, systemic	66	1604.934	20/386/424/1250
Blood viscosity	35	1535.653	21/386/424/1250

[a]These are the 21 terms closest to the A term *Raynaud's disease*.

[b]The number of times that the B term cooccurs with the A term. These terms are ranked discerningly based on the chi-square value.

[c]This column has the format number1/number2/number3/number4 where number1, number2, number3, and number4 show the rank of the B term, the number of terms left after a semantic check, the number of terms left after chi-square filtering, and the total number of B terms for the term *Raynaud's disease*, respectively.

TABLE 2.5　C Terms Sorted by Semantic Type

Diseases or Syndrome	Organ, Body Part	Neoplastic Process
Coronary Disease	Retinal degeneration	Choriocarcinoma
Inflammatory bowel diseases	**Retinal neoplasm**	**Hodgkin's disease**
Plant diseases	Pulmonary alveoli	
Rheumatic diseases		
Neurodegenerative diseases		
Crohns disease		
Stomach diseases		
Pleurisy		

TABLE 2.6   Chi-Square Link ABC (Minimum Linkage)

C Terms	Weight	Ranking
Amputation, traumatic	482.9731	1/14794
Hand deformities, congenital	387.6049	2/14794
Syndactyly	355.3439	3/14794
Immersion foot	324.9525	4/14794
Lupus nephritis	319.5185	5/14794
Fish oils	**42.4119**	**4913/14,794**

TABLE 2.7   Mutual Information Link ABC (Minimum Link ABC)

C Terms	Weight	Ranking
Phenalenes	2.48×10^{-9}	1/14,794
Chromosomes, human, pair 6	2.24×10^{-9}	2/14,794
Tissue fixation	1.95×10^{-9}	3/14,794
Galanthus	1.53×10^{-9}	4/14,794
North Sea	1.53×10^{-9}	5/14,794
Fish oils	$\mathbf{3.41 \times 10^{-12}}$	**2797/14,794**

The Table 2.8 shows how our B-term-centered ranking scheme improved the weighting scheme described above. *Fish oils* is ranked 942 out of 14,794. In our experiment, for $W(t)$ we set $w_1 = 0.25$, $w_2 = 0.25$, and $w_3 = 0.5$ because we desire to discover those terms that especially favored a small number of B terms, which can help find rare associations that are easily ignored by people. In Table 2.8, *fish oils* has a much better ranking than it does using the chi-square and mutual link ABC methods. As it is an option for the user to decide the weight of various types of C terms, the method

TABLE 2.8   B-Term-Centered Ranking Scheme

C Term	Number of B Terms Cooccuring	Weight	Ranking
Pancreatic elastase	101	20,057.6	1/14,794
Mitochondria, muscle	90	8,199.675	2/14,794
Polylysine	65	7,296.759	3/14,794
Electron probe microanalysis	100	4,564.169	4/14,794
Clozapine	39	3,650.804	5/14,794
Cytarabine	93	3,190.168	6/14,794
Androstane-3,17-diol	39	2,739.69	7/14,794
Hemorrhoids	56	2,426.454	8/14,794
Stomatitis	82	2,284.51	9/14,794
Iofetamine	30	1,827.826	10/14,794
Triethyltin compounds	29	1,827.826	11/14,794
Acarbose	28	1,827.826	12/14,794
Fish oils	**60**	**30.97666**	**942/14,794**

provides a more flexible framework for ranking C terms. When only a small number of B terms of certain semantic types are selected, our method can help examine more precisely the connections between B and C terms, because there are fewer common terms (noise) to affect the ranking.

2.5 DISCUSSION AND CONCLUSIONS

A comparison of seven methods can help establish a global picture of relationships among concepts. A variety of statistical measures, such as chi-square and mutual information, can be used to judge the strength of a relationship, which leads to a better understanding of terms' relationships and thus to discovering more meaningful implicit relationships among concepts. Checking the semantic relationships between B and C terms after chi-square filtering, it was found that most are related to each other semantically, which means that chi-square filtering helps pick up more meaningful target terms. However, although chi-square filtering can help find more meaningful implicit relations, some rare or common associations might be ignored, depending on how well chi-square values can describe the dependency relationships between two terms. Although this statistical method provides a view of the strength of terms' relationships, semantic relationships between terms may help filter out unrelated terms, especially when statistical threshold filtering is not used. However, since the number of semantic types is very limited from UMLS, it is easy to find semantic relationships among many semantic types. Improving knowledge discovery requires more detailed semantic information.

It is really difficult to find a universal ranking method for C terms because different strategies can favor only certain types of C terms. Reevaluating Swanson's experiment can provide some information as to how our automatic approach matches the discovery of certain knowledge, such as the connection between Raynaud's disease and fish oils. However, as knowledge discovery is a process whose end we do not know, it is really difficult to judge to which target we should pay most attention. Our experiments show that the best way to make meaningful C terms stand out is to rank C terms both according to an information measure such as chi-square and their semantic types, which can make certain bad ranking C terms stand out (even though some C terms have less significant relationships with certain B terms, they may have stronger semantic relationships), since it is easy to pay more attention to certain types of medical concepts. Our approach presents an open discovery framework for ABC discovery and how different weighting schemes plus semantic information affect knowledge discovery.

Acknowledgments

This work was supported in part by National Science Foundation career grant IIS 0448023 and CCF 0514679, Pennsylvania Department of Health Tobacco Settlement Formula grants 240205 and 240196, and Pennsylvania Department of Health grant 239667.

REFERENCES

1. R. Agrawal et al., Fast discovery of association rules, in *Advances in Knowledge Discovery and Data Mining*, U. Fayyad et al., Eds., AAAI/MIT Press, Cambridge, MA, 1995.

2. R. A. DiGiacome, J. M. Kremer, and D. M. Shah, Fish oil dietary supplementation is patients with Raynaud's phenomenon: a double-blind, controlled, prospective study, *Am. J. Med.*, 8, 158–164, 1989.

3. D. Hristovski, J. Stare, B. Peterlin, and S. Dzeroski, Supporting discovery in medicine by association rule mining in Medline and UMLS, *Medinfo*, 10(pt. 2), 1344–1348, 2001.

4. X. Hu, Mining novel connections from large online digital library using biomedical ontologies, *Lib. Manage. J.* (Special Issue on Libraries in the Knowledge Era: Exploiting the Knowledge Wealth for Semantic Web Technology), May 2005.

5. M. D. Gordon, Literature-based discovery by lexical statistics, *J. Am. Soc. Inf. Sci.*, 50, 574–587, 1999.

6. M. D. Gordon and R. L. Lindsay, Towards discovery support systems: a replication, re-examination, and extension of Swanson's work on literature-based discovery of a connection between Raynaud's and fish oil, *J. Am. Soc. Inf. Sci.*, 47, 116–128, 1996.

7. M. D. Gordon and S. Dumais, Using latent semantic indexing for literature based discovery, *J. Am. Soc. Inf. Sci.*, 48, 674–685, 1998.

8. W. Pratt and M. Yetisgen-Yildiz, LitLinker: capturing connections across the biomedical literature, K-CAP'03, Sanibel Island, FL, Oct. 23–25, 2003, pp. 105–112.

9. P. Srinivasan, Text mining: generating hypotheses from Medline, *J. Am. Soc. Inf. Sci.*, 55(4), 396–413, 2004.

10. P. Srinivasan and B. Libbus, Mining Medline for implicit links between dietary substances and diseases.

11. D. R. Swanson, Migraine and magnesium: eleven neglected connections, *Perspect. Biol. Med.*, 31(4), 526–557, 1988.

12. D. R. Swanson, Fish-oil, Raynaud's syndrome, and undiscovered public knowledge, *Perspect. Biol. Med.*, 30(1), 7–18, 1986.

13. D. R. Swanson, Undiscovered public knowledge, *Libr. Q.*, 56(2), 103–118, 1986.

14. M. Weeber, R. Vos, H. Klein, L. T. W. de Jong-Van den Berg, A. Aronson, and G. Molema, Generating hypotheses by discovering implicit associations in the literature: a case report for new potential therapeutic uses for thalidomide, *J. Am. Med. Inf. Assoc.*, 10(3), 252–259, 2003.

15. J. D. Wren, Extending the mutual information measure to rank inferred literature relationships, *BMC Bioinf.*, 5, 145, 2004.

3

VOTING SCHEME–BASED EVOLUTIONARY KERNEL MACHINES FOR DRUG ACTIVITY COMPARISONS

Bo Jin and Yan-Qing Zhang

Department of Computer Science, Georgia State University, Atlanta, Georgia

The performance of support vector machines (SVMs) (Boser et al., 1992; Cortes and Vapnik, 1995; Vapnik, 1998) is affected primarily by kernel functions. With the growing interest in biological data prediction and chemical data prediction, more complicated kernels are designed to measure data similarities: ANOVA kernels (Vapnik, 1998), convolution kernels (Haussler, 1999), string kernels (Cristianini and Shawe-Taylor, 1999; Lodhi et al., 2001), tree kernels (Collins and Duffy, 2002; Kashima and Koyanagi, 2002), and graph kernels (Kashima and Inokuchi, 2002; Gärtner et al., 2003), for example. A detailed review is given by Gärtner (2003). These kernels are implemented based on kernel decomposition properties (Vapnik, 1998; Haussler, 1999).

Jin et al. (2005) used granular computing concepts to redescribe kernel decomposition properties and proposed a type of evolutionary granular kernel tree (EGKT) for drug activity comparisons. In EGKTs, features within an input vector are grouped into feature granules according to the possible substituent locations of compounds. The similarity between two feature granules is measured using a granular kernel, and all granular kernels are fused together by trees. The parameters of granular kernels and the connection weights of granular kernel trees (GKTs) are optimized by genetic algorithms (GAs).

Knowledge Discovery in Bioinformatics: Techniques, Methods, and Applications, Edited by Xiaohua Hu and Yi Pan

Sometimes, due to the lack of prior knowledge, it is difficult to predefine kernel tree structures. Considering such types of challenging problems, Jin and Zhang (2006) presented the granular kernel tree structure evolving system (GKTSES) to evolve the structures of GKTs. In a GKTSES, a population of granular kernel trees is first generated randomly. Crossover and mutation are then used to generate new populations of kernel trees. Finally, the kernel tree with the best structure is selected for data classification. In a GKTSES, k-fold cross-validation is used for fitness evaluation.

However, threefold cross-validation shows that GKTSESs are not stable. In this chapter, voting scheme–based evolutionary kernel machines called *evolutionary voting kernel machines* (EVKMs) are presented for drug activity comparisons. In EVKMs, the evolving procedure of GKTs is the same as in GKTSESs while the decision is made by several weighted SVMs during optimization. Experimental results show that EVKMs are more stable than GKTSESs in cyclooxygenase-2 inhibitor activity comparisons.

A genetic programming–based kernel tree was presented by Howley and Madden (2004) in which input vectors can operate as units under sum, minus, magnitude, or product operation. The approach does not, however, guarantee that the function obtained is a kernel function.

The remainder of the chapter is organized as follows. Definitions and properties of kernels and granular kernels are given in Section 3.1, GKTSESs are introduced in Section 3.2, EVKMs are presented in Section 3.3, simulations of cyclooxygenase-2 inhibitor activity comparisons are shown in Section 3.4, and Section 3.5 concludes and suggests future work.

3.1 GRANULAR KERNEL AND KERNEL TREE DESIGN

3.1.1 Definitions

Definition 3.1 (Cristianini and Shawe-Taylor, 1999)   A kernel is a function K that for all $\vec{x}, \vec{z} \in X$ satisfies

$$K(\vec{x}, \vec{z}) = \langle \phi(\vec{x}), \phi(\vec{z}) \rangle \tag{3.1}$$

where ϕ is a mapping from input space $X = R^n$ to an inner product feature space $F = R^N$:

$$\phi: \ \vec{x} \mapsto \phi(\vec{x}) \in F \tag{3.2}$$

Definition 3.2   A feature granule space G of input space $X = R^n$ is a subspace of X where $G = R^m$ and $1 \leq m \leq n$.

From input space we may generate many feature granule spaces, and some of them may overlap on some feature dimensions.

Definition 3.3   A feature granule $\vec{g} \in G$ is a vector that is defined in the feature granule space G.

Definition 3.4   A granular kernel gK is a kernel that for all $\vec{g}, \vec{g}' \in G$ satisfies

$$gK(\vec{g}, \vec{g}') = \langle \varphi(\vec{g}), \varphi(\vec{g}') \rangle \tag{3.3}$$

where φ is a mapping from feature granule space $G = R^m$ to an inner product feature space R^E.

$$\varphi : \vec{g} \mapsto \varphi(\vec{g}) \in R^E \tag{3.4}$$

3.1.2 Granular Kernel Properties

Property 3.1   Granular kernels inherit the properties of traditional kernels such as closure under sum, product, and multiplication, with a positive constant over the granular feature spaces.

Let G be a feature granule space and $\vec{g}, \vec{g}' \in G$. Let gK_1 and gK_2 be two granular kernels operating over the same space $G \times G$. The following functions are also granular kernels.

$$
\begin{align}
cgK_1(\vec{g}, \vec{g}'), \qquad c \in R^+ \tag{3.5} \\
gK_1(\vec{g}, \vec{g}') + c, \qquad c \in R^+ \tag{3.6} \\
gK_1(\vec{g}, \vec{g}') + gK_2(\vec{g}, \vec{g}') \tag{3.7} \\
gK_1(\vec{g}, \vec{g}')gK_2(\vec{g}, \vec{g}') \tag{3.8} \\
f(\vec{g})f(\vec{g}'), \qquad f : X \to R \tag{3.9} \\
\frac{gK_1(\vec{g}, \vec{g}')}{gK_1(\vec{g}, \vec{g})gK_1(\vec{g}', \vec{g}')} \tag{3.10}
\end{align}
$$

These properties can be elicited directly from the traditional kernel properties.

Property 3.2 (Berg et al., 1984; Haussler, 1999)   A kernel can be constructed with two granular kernels defined over different granular feature spaces under sum operation.

To prove it, let $gK_1(\vec{g}_1, \vec{g}_1')$ and $gK_2(\vec{g}_2, \vec{g}_2')$ be two granular kernels, where $\vec{g}_1, \vec{g}_1' \in G_1$, $\vec{g}_2, \vec{g}_2' \in G_2$, and $G_1 \neq G_2$. We may define new kernels like this:

$$gK((\vec{g}_1, \vec{g}_2), (\vec{g}_1', \vec{g}_2')) = gK_1(\vec{g}_1', \vec{g}_1')$$
$$gK'((\vec{g}_1, \vec{g}_2), (\vec{g}_1', \vec{g}_2')) = gK_2(\vec{g}_2, \vec{g}_2')$$

gK and gK' can operate over the same feature space $(G_1 \times G_2) \times (G_1 \times G_2)$. We get $gK_1(\vec{g}_1, \vec{g}_1') + gK_2(\vec{g}_2, \vec{g}_2') = gK((\vec{g}_1, \vec{g}_2), (\vec{g}_1', \vec{g}_2')) + gK'((\vec{g}_1, \vec{g}_2), (\vec{g}_1', \vec{g}_2'))$ According to the sum closure property of kernels (Cristianini and Shawe-Taylor, 1999), $gK_1(\vec{g}_1, \vec{g}_1') + gK_2(\vec{g}_2, \vec{g}_2')$ is a kernel over $(G_1 \times G_2) \times (G_1 \times G_2)$.

Property 3.3 (Berg et al., 1984; Haussler, 1999)   A kernel can be constructed with two granular kernels defined over different granular feature spaces under product operation.

To prove it, let $gK_1(\vec{g}_1, \vec{g}_1')$ and $gK_2(\vec{g}_2, \vec{g}_2')$ be two granular kernels, where $\vec{g}_1, \vec{g}_1' \in G_1, \vec{g}_2, \vec{g}_2' \in G_2$, and $G_1 \neq G_2$. We may define new kernels like this:

$$gK((\vec{g}_1, \vec{g}_2), (\vec{g}_1', \vec{g}_2')) = gK_1(\vec{g}_1, \vec{g}_1')$$
$$gK'((\vec{g}_1, \vec{g}_2), (\vec{g}_1', \vec{g}_2')) = gK_2(\vec{g}_2, \vec{g}_2')$$

So gK and gK' can operate over the same feature space $(G_1 G_2)(G_1 G_2)$. We get

$$gK_1(\vec{g}_1, \vec{g}_1')gK_2(\vec{g}_2, \vec{g}_2') = gK((\vec{g}_1, \vec{g}_2), (\vec{g}_1', \vec{g}_2'))gK'((\vec{g}_1, \vec{g}_2), (\vec{g}_1', \vec{g}_2'))$$

According to the product closure property of kernels (Cristianini, 1999), $gK_1(\vec{g}_1, \vec{g}_1')gK_2(\vec{g}_2, \vec{g}_2')$ is a kernel over $(G_1 \times G_2) \times (G_1 \times G_2)$.

3.2   GKTSESs

In GKTSESs, a population of individuals is generated in the first generation. Each individual encodes a granular kernel tree. For example, three-layer GKTs (GKTs-1 and GKTs-2) are shown in Figure 3.1. In GKTs-1, each node in the first layer is a granular kernel. Granular kernels are combined by sum and product connection operations in layers 2 and 3. Each granular kernel tree is encoded into a chromosome. For example, GKTs-1 and GKTs-2 are encoded in chromosomes c_1 and c_2 (see Figure 3.2), respectively. To generate an individual, features are randomly shuffled and then feature granules are generated randomly. Granular kernels are preselected from the candidate kernel set. Some traditional kernels, such as RBF and polynomial kernels, can be chosen as granular kernels, since these kernels have been used successfully in many real problems. In GKTSESs, RBF kernels are chosen as granular kernels and each feature as a feature granule. Granular kernel parameters and kernel connection operations are generated randomly for each individual.

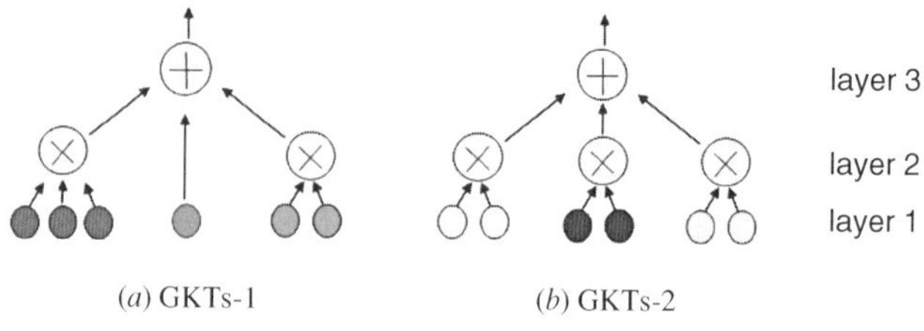

FIGURE 3.1   Three-layer GKTs.

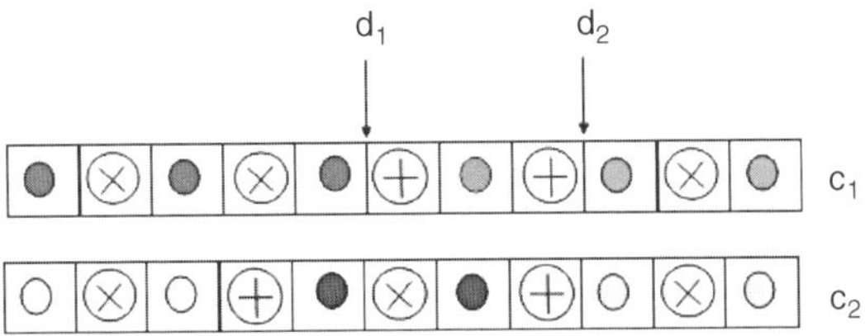

FIGURE 3.2 Chromosomes.

Chromosome Let P_i denote the population in generation G_i, where $i = 1, \ldots, m$ and m is the total number of generations. Each population P_i has p chromosomes $c_{ij}, j = 1, \ldots, p$. Each chromosome c_{ij} has $2q + 1$ genes $g_t(c_{ij})$, where $t = 1, \ldots, 2q + 1$. In each chromosome, genes $g_{2x-1}(c_{ij}), x = 1, \ldots, q + 1$, represent granular kernels, and genes $g_{2x}(c_{ij})$, $x = 1, \ldots, q$, represent sum or product operations. Here each granular kernel gene includes random connection weight and kernel parameters. GKTs(c_{ij}) is used to represent GKTs configured with genes $g_t(c_{ij})$, $t = 1, \ldots, 2q + 1$.

Fitness In GKTSESs, k-fold cross-validation is used to evaluate SVM performance in the training phase. In k-fold cross-validation, the training data set $\tilde{S}$ is separated into k mutually exclusive subsets $\tilde{S}_v$. For $v = 1, \ldots, k$, data set Λ_v is used to train SVMs with $GKTs(c_{ij})$ and $\tilde{S}_v$ is used to evaluate the SVM model:

$$\Lambda_v = \tilde{S} - \tilde{S}_v, \qquad v = 1, \ldots, k \tag{3.11}$$

After k training tests on different subsets, we get k prediction accuracies. The fitness f_{ij} of chromosome c_{ij} is calculated by

$$f_{ij} = \frac{1}{k} \sum_{v=1}^{k} \text{Acc}_v \tag{3.12}$$

where Acc_v is the prediction accuracy of GKTs(c_{ij}) on $\tilde{S}_v$.

Selection In the algorithm, the roulette wheel method (Michalewicz, 1996) is used to select individuals for the new population. Before selection, the best chromosome in generation G_{i-1} will replace the worst chromosome in generation G_i if the best chromosome in G_i is worse than the best chromosome in G_{i-1}. The sum of fitness values F_i in population G_i is calculated first:

$$F_i = \sum_{j=1}^{p} f_{ij} \tag{3.13}$$

Cumulative fitness $\tilde{q}_{ij}$ is then calculated for each chromosome:

$$\tilde{q}_{ij} = \sum_{t=1}^{j} \frac{f_{it}}{F_i} \tag{3.14}$$

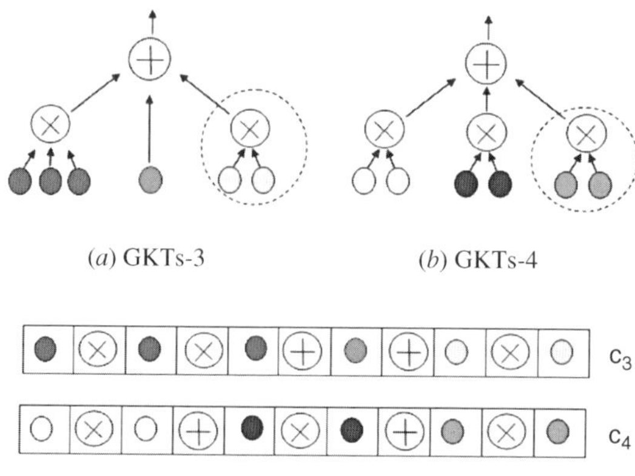

(*a*) GKTs-3   (*b*) GKTs-4

(*c*) Chromosomes of GKTs-3 and GKTs-4

FIGURE 3.3   Crossover at point d_2.

The chromosomes are then selected as follows. A random number r is generated within the range [0,1]. If r is smaller than $\tilde{q}_{i1}$, chromosome c_{i1} is selected; otherwise, chromosome c_{ij} is selected according to the inequality

$$\tilde{q}_{i\,j-1} < r \le \tilde{q}_{i\,j} \tag{3.15}$$

After running the foregoing selecting procedure p times, a new population is generated.

Crossover   Two GKTs are first selected from the current generation as parents and then the crossover point is selected randomly to separate GKTs. Subtrees of two GKTs are exchanged at the crossover point to generate two new GKTs. For example, two new GKTs generated from GKTs-1 and GKTs-2 through crossover operation at point d_2 (see Figure 3.2) are shown in Figure 3.3. In this example, two new GKTs (GKTs-3 and GKTs-4) have the same structures as their parents. In Figure 3.4, a new GKTs-5 and GKTs-6 are generated from GKTs-1 and GKTs-2 through crossover operation at point d_1. The structures of the two new GKTs in Figure 3.4 are different from those of the GKTs in Figure 3.1. The related chromosomes of new GKTs are shown in Figures 3.3*c* and 3.4*c*.

In each population, one chromosome has the same probability of being selected to cross over with another chromosome.

Mutation   In mutation, some genes of one chromosome are selected with the same probability. The values of the genes selected are replaced by random values. In GKTSESs, only connection operation genes are selected to do mutation. Figure 3.5 shows an example of mutation. In Figure 3.5*a*, the new chromosome c_7 is generated by changing the eighth gene of chromosome c_1 from sum operation to

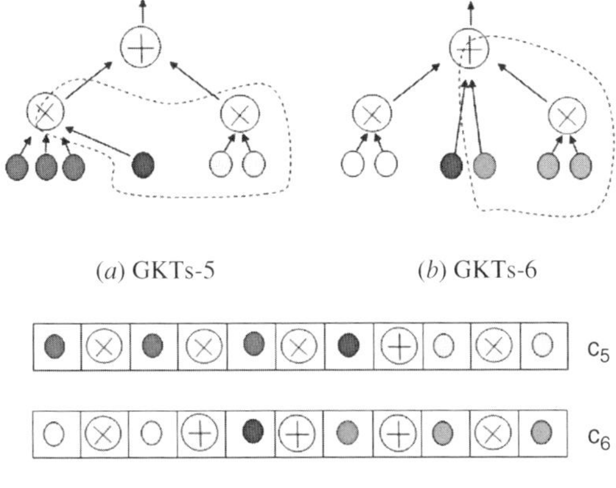

(c) Chromosomes of GKTs-5 and GKTs-6

FIGURE 3.4 Crossover at point d_1.

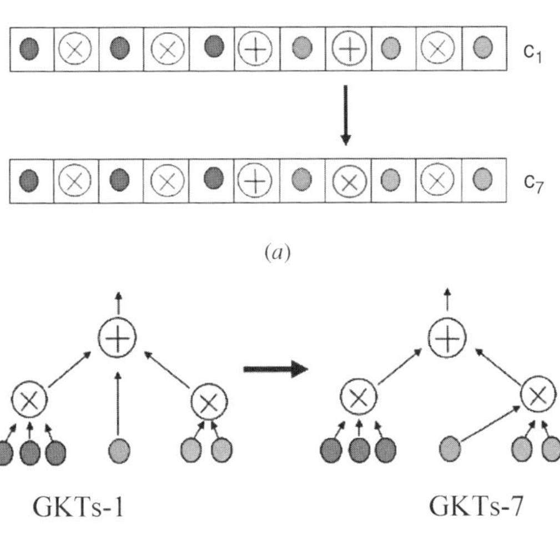

FIGURE 3.5 Example of mutation.

product operation, which is equivalent to transforming GKTs-1 to GKTs-7 (see Figure 3.5b).

3.3 EVOLUTIONARY VOTING KERNEL MACHINES

In EVKMs, selection, crossover, and mutation operations are the same as in GKTSESs. The difference is in fitness evaluation. In each GKTs(c_{ij}) evaluation, the training data set $\tilde{S} = \{(\vec{x}_1, y_1), \ldots, (\vec{x}_l, \vec{y}_l)\}$ is first separated into k mutually exclusive subsets $\tilde{S}_v$. For $v = 1, \ldots, k$, $\tilde{S}_v$ is used to train SVMs with GKTs, and k SVM decision functions $d_v(\vec{x})$ are then generated:

$$d_v(\vec{x}) = \sum_{i_v} \alpha_{i_v} y_{i_v} \, \text{GKTs}(\vec{x}, \vec{x}_{i_v}) \; + b_v \tag{3.16}$$

where $0 < \alpha_{i_v} \le C$, C is the regularization parameter, $\vec{x}_{i_v}$ are support vectors, and b_v is the threshold for the vth SVMs. The number of misclassified data in each training is calculated, and the weighted voting decision function $d(\vec{x})$ is defined as

$$d(\vec{x}) = \sum_v c_v d_v(\vec{x}) \tag{3.17}$$

In Eq. (3.17), c_v is a cost factor which is either the accuracy of the positive class of the vth SVMs if $d_v(\vec{x})$ is positive, or the accuracy of the negative class of if $d_v(\vec{x})$ is negative. Data set $\tilde{S}$ is then predicted by $d(\vec{x})$, and the accuracies (t_v^+ and t_v^-) of two classes are calculated, respectively. The prediction accuracy is chosen as fitness.

Finally, the optimized GKTs$'$ is generated and the decision function $d'(\vec{x})$ is calculated and used for the unseen testing set prediction:

$$\begin{aligned}
d'(\vec{x}) &= \sum_v c_v' d_v'(\vec{x}) \\
&= \sum_v c_v' \left[\sum_{i_v} \alpha_{i_v} y_{i_v} \, \text{GKTs}'(\vec{x}, \vec{x}_{i_v}) \; + b_v \right]
\end{aligned} \tag{3.18}$$

where the cost factor c_v' is either t_v^+ or t_v^-. The algorithm learning procedure follows.

```
Initialization
For each generation Gᵢ
   For each cᵢⱼ in Gᵢ
      Repeat v from 1 to k
         Train SVMs on Λᵥwith GKTs(cᵢⱼ) and generate dᵥ(x⃗)
```

$$d_v(\vec{x}) = \sum_{i_v} \alpha_{i_v} y_{i_v} \, \text{GKTs}(\vec{x}, \vec{x}_{i_v}) + b_v$$

```
         Evaluate SVMs on S̃ᵥ and calculate cᵥ
      Repeat end
      Generate d(x⃗) =
```

$$d(\vec{x}) = \sum_v c_v d_v(\vec{x})$$

```
      Evaluate SVMs on S̃ using d(x⃗) and calculate fitness
   For End
   Selection operation
   Crossover operation
   Mutation operation
For End
```

3.4 SIMULATIONS

3.4.1 Data Set and Experimental Setup

The data set of cyclooxygenase-2 inhibitors (Kauffman and Jurs, 2001) includes 314 compounds. The point of $\log(IC_{50})$ units used to discriminate active compounds from inactive compounds is set to 2.5. The number of active compounds is 153, and the number of inactive compounds is 161. One hundred nine features are selected to describe each compound, and each feature's absolute value is scaled to the range [0,1]. The data set is shuffled randomly and split evenly into three mutually exclusive parts. Each time, one part is chosen as the unseen testing set and the other two as the training set. We use threefold cross-validation to evaluate GKTSES performance on the training set. In EVKMs, the training set is split into three mutually exclusive parts. An RBF kernel is chosen as each granular kernel function. In each system, the ranges of all RBFs' γ are set to [0.00001,1], and the range of regularization parameter C is set to [1,256]. The probability of crossover is 0.8 and the mutation ratio is 0.2. The population size is set to 300, and the number of generations is set to 50. The software package of SVMs used in the experiments is LibSVM (Chang and Lin, 2001). In the first generation, the probability of sum operation is 0.5.

3.4.2 Experimental Results and Comparisons

Table 3.1 shows the performances of four types of systems. CV-1, CV-2, and CV-3 are three evaluations. In the table, GAs-RBF-SVMs are SVMs with traditional RBF kernels and the systems' regularization parameter C and the RBFs' γ are optimized with GAs. GAs-RBF-SVMs-1 uses the same fitness evaluation as that used by GKTSESs and GAs-RBF-SVMs-2 used the same voting scheme as that used by EVKMs. Table 3.2 shows four systems' average prediction accuracies and the standard deviations of testing accuracies.

From Table 3.1 we can see that in three evaluations, the testing accuracies of GKTSESs are always higher than those of GAs-RBF-SVMs-1 by about 2.9 to 3.9%, and EVKMs are always higher than those of GAs-RBF-SVMs-1 by about 1.9 to 2.9%.

TABLE 3.1 Prediction Accuracies in Three Evaluations (Percent)

		GAs-RBF-SVMs-1	GKTSESs	GAs-RBF-SVMs-2	EVKMs
CV-1	Fitness	83.7	87.1	91.4	90.9
	Training accuracy	90.9	92.3	—	—
	Testing accuracy	64.8	68.6	71.3	74.2
CV-2	Fitness	80.4	82.3	90	85.2
	Training accuracy	97.1	88.0	—	—
	Testing accuracy	78.1	81	76.8	78.7
CV-3	Fitness	81.4	82.4	90.1	85.6
	Training accuracy	95.2	88.6	—	—
	Testing accuracy	74	77.9	74.2	77

TABLE 3.2 Average Prediction Accuracies and Standard Deviation of Testing Accuracy (Percent)

	GAs-RBF-SVMs-1	GKTSESs	GAs-RBF-SVMs-2	EVKMs
Fitness	81.8	83.9	90.5	87.2
Training accuracy	94.4	89.6	—	—
Testing accuracy	72.3	75.8	74.1	76.6
Standard deviation of testing accuracy	6.8	6.5	2.8	2.3

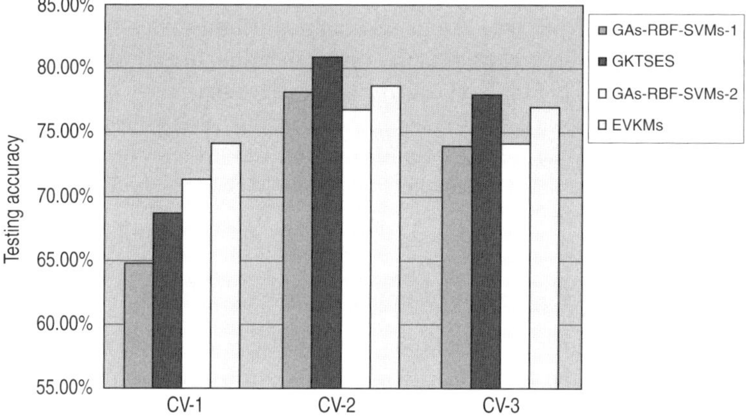

FIGURE 3.6 Testing accuracies in three evaluations.

In CV-1, the testing accuracies of EVKMs and GAs-RBF-SVMs-2 are higher than those of GKTSESs and GAs-RBF-SVMs-1. In CV-2 and CV-3, the testing accuracies of GKTSESs are higher than those of EVKMs, and the testing accuracies of GAs-RBF-SVMs-1 are higher than those of GAs-RBF-SVMs-2. Figure 3.6 shows the testing accuracies of three evaluations in one picture.

From Table 3.2 we can find that the average testing accuracy of EVKMs is a little bit higher than that of GKTSESs, and the average testing accuracy of GAs-RBF-SVMs-2 is higher than that of GAs-RBF-SVMs-2 by 1.8%. While comparing the standard deviations of testing accuracies, we can find that the standard deviations of EVKMs and GAs-RBF-SVMs-2 are much lower than those of GKTSESs and GAs-RBF-SVMs-1. It means that EVKMs and GAs-RBF-SVMs-2 are more stable than GKTSESs and GAs-RBF-SVMs-1.

3.5 CONCLUSIONS AND FUTURE WORK

Voting scheme–based evolutionary kernel machines (EVKMs) were proposed is this chapter. Experimental results show that EVKMs are more stable than GKTSESs in cyclooxygenase-2 inhibitor activity comparisons in terms of the standard deviation of the testing accuracy. We will continue our study on kernel evolving systems in the future.

Acknowledgments

This work was supported in part by the National Institutes of Health grant P20 GM065762. Bo Jin is supported by the Molecular Basis for Disease (MBD) Doctoral Fellowship Program. Special thanks to Peter C. Jurs and Rajarshi Guha for providing the data set of cyclooxygenase-2 inhibitors.

REFERENCES

Berg, C., J. P. R. Christensen, and P. Ressel, (1984). *Harmonic Analysis on Semigroups: Theory of Positive Definite and Related Functions*, Springer-Verlag, New York.

Boser, B., I. Guyon, and V. N. Vapnik, (1992). A training algorithm for optimal margin classifiers, *Proc. 5th Annual Workshop on Computational Learning Theory*, pp. 144–152.

Chang, C.-C., and C.-J. Lin (2001). LIBSVM: a library for support vector machines, software available at http://www.csie.ntu.edu.tw/~cjlin/libsvm.

Collins, M., and N. Duffy, (2002). Convolution kernels for natural language, in *Advances in Neural Information Processing Systems*, Vol. 14, T. G. Dietterich, S. Becker, and Z. Ghahramani, Eds., MIT Press, Cambridge, MA.

Cortes, C., and V. N. Vapnik, (1995). Support-vector networks, *Mach. Learn.*, 20, pp. 273–293.

Cristianini, N., and J. Shawe-Taylor, (1999). *An Introduction to Support Vector Machines and Other Kernel-Based Learning Methods*, Cambridge University Press, New York.

Gärtner, T. (2003). A survey of kernels for structured data, *ACM SIGKDD Explor. Newsl.* 5, pp. 49–58.

Gärtner, T., P. A. Flach, and S. Wrobel, (2003). On graph kernels: hardness results and efficient alternatives, *Proc. 16th Annual Conference on Computational Learning Theory and the 7th Kernel Workshop*.

Haussler, D. (1999). Convolution kernels on discrete structures, *Technical Report UCSC-CRL-99-10*, Department of Computer Science, University of California at Santa Cruz, Santa Cruz, CA.

Howley, T., and M. G. Madden, (2004). The genetic evolution of kernels for support vector machine classifiers, *Proc. 15th Irish Conference on Artificial Intelligence and Cognitive Science*.

Jin, B., and Y.-Q. Zhang, (2006). Evolutionary construction of granular kernel trees for cyclooxygenase-2 inhibitor activity comparison, *LNCS Trans. Comput. Syst. Biol.*, V, LNBI 4070.

Jin, B., Y.-Q. Zhang, and B. H. Wang, (2005). Evolutionary granular kernel trees and applications in drug activity comparisons, *Proc. IEEE Symposium on Computational Intelligence in Bioinformatics and Computational Biology*, San Diego, CA, pp. 121–126.

Kashima, H., and A. Inokuchi, (2002). Kernels for graph classification, *ICDM Workshop on Active Mining*.

Kashima, H., and T. Koyanagi, (2002). Kernels for semi-structured data, *Proc. 19th International Conference on Machine Learning*, pp. 291–298.

Kauffman, G. W., and P. C. Jurs, (2001). QSAR and k-nearest neighbor classification analysis of selective cyclooxygenase-2 inhibitors using topologically-based numerical descriptors, *J. Chem. Inf. Comput. Sci.*, 41(6), pp. 1553–1560.

Lodhi, H., J. Shawe-Taylor, N. Christianini, and C. Watkins, (2001). Text classification using string kernels, in *Advances in Neural Information Processing Systems*, Vol. 13, T. Leen, T. Dietterich, and V. Tresp, Eds, MIT Press, Cambridge, MA.

Michalewicz, Z. (1996). *Genetic Algorithms + Data Structures = Evolution Programs*, Springer-Verlag, Berlin.

Vapnik, V. (1998). *Statistical Learning Theory*, Wiley, New York.

4

BIOINFORMATICS ANALYSES OF *ARABIDOPSIS thaliana* TILING ARRAY EXPRESSION DATA

TRUPTI JOSHI AND JINRONG WAN

Department of Computer Science (T.J.) and Department of Molecular Biology (J.W.), University of Missouri-Columbia, Columbia, Missouri

CURTIS J. PALM, KARA JUNEAU, RON DAVIS, AND AUDREY SOUTHWICK

Stanford Genome Technology Center, Palo Alto, California

KATRINA M. RAMONELL

Department of Biological Sciences, University of Alabama, Tuscaloosa, Alabama

GARY STACEY AND DONG XU

Department of Molecular Biology (G.S.) and Department of Computer Science (D.X.), University of Missouri-Columbia, Columbia, Missouri

Microarrays are a well-established technology for measuring gene expression levels at a large scale. Recently, high-density oligonucleotide-based whole-genome microarrays have emerged as an important platform for genomic analysis beyond simple gene expression profiling. Tiling array is a set of oligonucleotide microarrays for the entire genome. Nonoverlapping or partially overlapping probes may be tiled to cover the entire genome end to end, or the probes may be spaced at regular intervals [1]. The tiling array technology has several advantages over other microarray technologies. For example, tiling arrays do not depend on genome annotation, thus covering the entire genome in an unbiased manner, facilitating the discovery of novel genes and

Knowledge Discovery in Bioinformatics: Techniques, Methods, and Applications, Edited by Xiaohua Hu and Yi Pan
Copyright © 2007 John Wiley & Sons, Inc.

antisense transcripts. Potential uses for such whole-genome arrays include empirical annotation of the transcriptome, novel gene discovery, DNA protein-binding studies, chromatin-immunoprecipitation-chip studies, analysis of alternative splicing, characterization of the methylome (the methylation state of the genome), polymorphism discovery and genotyping, comparative genome hybridization, and genome resequencing [2]. Tiling arrays are also useful for localization of the genomic binding sites of transcription factors and other chromatin-associated proteins.

For the purposes of this chapter, we discuss the whole-genome tiling array design and techniques for analyzing these data to obtain a wide variety of genomic scale information using bioinformatics techniques. We also discuss the ontological analyses and antisense identification techniques using tiling array data.

4.1 TILING ARRAY DESIGN AND DATA DESCRIPTION

A key issue in designing tiling arrays is to arrive at a single-copy tile path, as significant sequence cross-hybridization can result from the presence of nonunique probes on the array. This becomes increasingly problematic in complex eukaryotic genomes that contain many thousands of interspersed repeats. The tiling arrays design involves filtering out the repeats, followed by systematic scanning across the genome to find an optimal partitioning of nonrepetitive subsequences over a prescribed range of tile sizes, on a DNA sequence comprising repetitive and nonrepetitive regions.

4.1.1 Data

The first eukaryotic genome to be entirely represented on arrays was the *Arabidopsis* genome. The whole-genome tiling array for *Arabidopsis thaliana* developed at Stanford Genome Technology Center uses 12 chips to tile both strands of the *Arabidopsis* genome, end to end, with 25-bp nonoverlapping probes. This design is repeated twice with 8-bp offset each time to make a set of whole-genome tiling chips with a resolution of 8 bp (offset 1,9,17). The architecture of tiling array chips is shown in Table 4.1.

Tiling arrays representing the entire genome were used to profile labeled cRNA targets derived from different tissues and/or developmental stages of *Arabidopsis* plants [3]. The five experimental conditions are as follows: (1) chitin 8-mer, (2) floral

TABLE 4.1   Architecture of Tiling Array Chips

Positive Strand Chips			Negative Strand Chips			Chromosomes Covered
1–1	9–1	17–1	1-rc-1	9-rc-1	17-rc-1	1
1–2	9–2	17–2	1-rc-2	9-rc-2	17-rc-2	1 and 2
1–3	9–3	17–3	1-rc-3	9-rc-3	17-rc-3	2 and 3
1–4	9–4	17–4	1-rc-4	9-rc-4	17-rc-4	3 and 4
1–5	9–5	17–5	1-rc-5	9-rc-5	17-rc-5	4 and 5
1–6	9–6	17–6	1-rc-6	9-rc-6	17-rc-6	5

(FL), (3) root (RT), (4) culture (SC), and (5) control (7-day-leaf samples). The raw data (intensity values) and. 1lq files are available for all 12 chips with offset 1.

4.1.2 Tiling Array Expression Patterns

The expression levels of probes on the positive and negative strands are plotted against the genomic location. For clarity, the signs of the expression levels of the probes on the negative strand have been inversed and plotted against the genomic axis. Figure 4.1 shows the expression pattern at the detailed level for the first 50,000 bases of the genomic axis and the first few genes on chromosome 1. We calculated the correlation between two replicates for the 7-day-leaf tiling array data. Figure 4.2 shows a strong correlation of 0.95 between the replicates, indicating excellent quality and reproducibility of the data. Figure 4.3 shows the difference in the expression pattern between the coding and noncoding regions of the genomic axis. It clearly shows high expression in the exon region of the gene AT2G01180, in contrast to the surrounding noncoding or intron regions. This example shows that the tiling array technology can be utilized to identify novel coding genes and to refine the gene boundaries in terms of exons and introns based on the expression patterns observed [4].

4.1.3 Tiling Array Data Analysis

The initial step in the analysis of tiling array data is normalization. The intensity values for the probes are normalized against the median per chip. Later, the expression values for the probes were calculated using the following natural logarithmic conversion:

$$\text{probe expression} = \log \frac{\text{intensity}}{\text{median}}$$

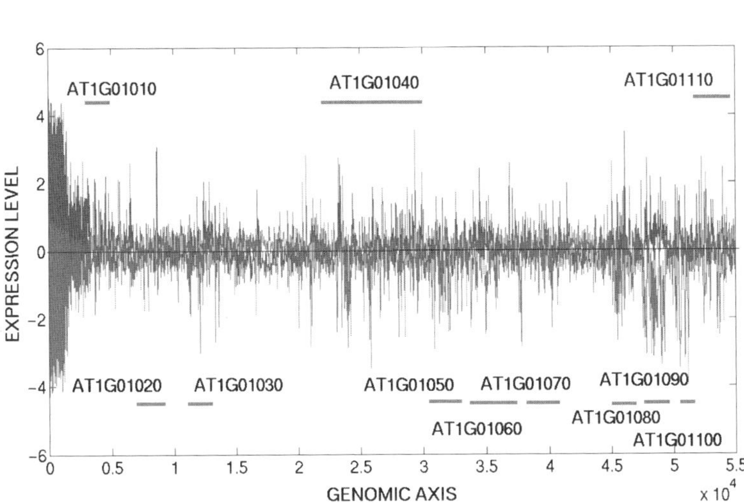

FIGURE 4.1 Expression patterns of some genes on chromosome 1.

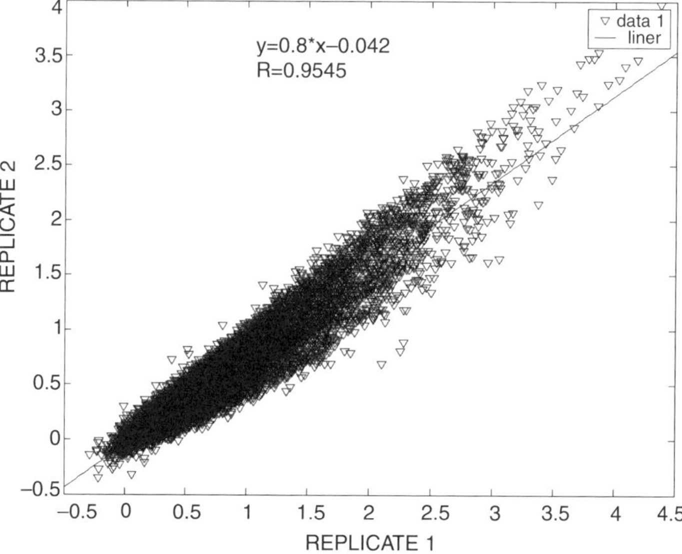

FIGURE 4.2   Correlation between two replicates for a 7-day-leaf sample for chip 1–2 for a tiling array.

The expression intensity for the gene was calculated by taking an average of the expression values of all the probes tiling the exon regions of the gene. To include the probe in the intensity calculation for the gene, its start position should lay within the gene's exon boundaries. Figure 4.4 shows the schematic representation of the tiling array architecture. It also shows an example of intensity calculation for gene A

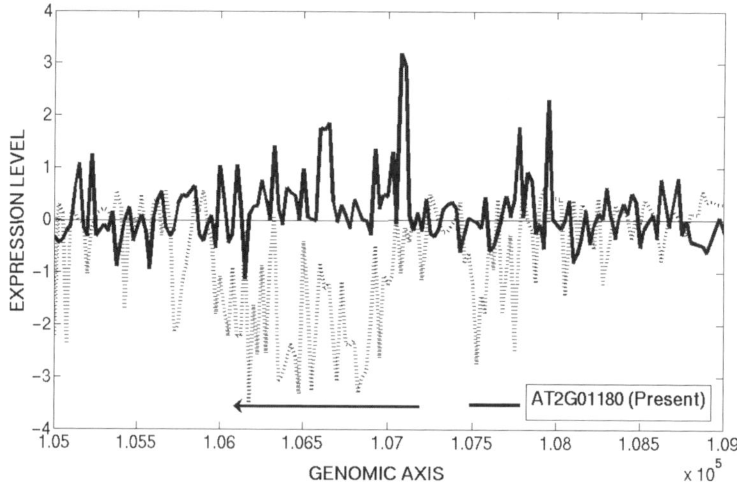

FIGURE 4.3   Expression pattern of a coding region with high expression in the exon region for an AT2G01180 gene.

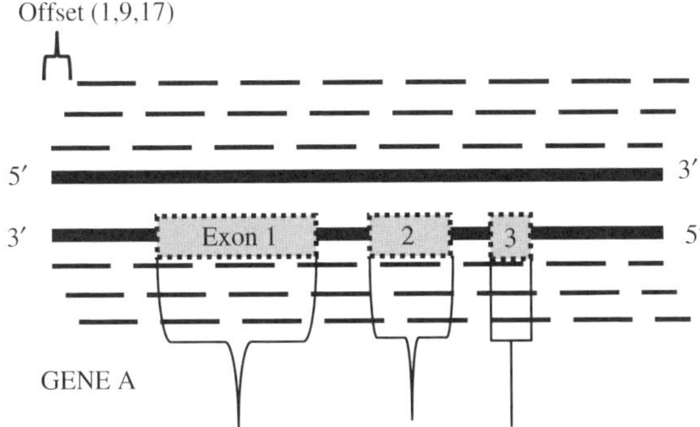

FIGURE 4.4 Architecture of tiling array data and calculation of expression values for the coding genes.

present on the negative strand. Further, we identified the up- and down-regulated genes in under various experimental conditions. We are interested in studying the effects on chitin 8-mer treatment against other treatment types. Toward this end, we calculated the postnormalization fold change (twofold) for the chitin 8-mer experiments against all other experiment types.

In all, 3256 genes were identified as up-regulated and 9986 as down-regulated in chitin 8-mer in comparison to the control sample experiments. Table 4.2 shows the number of 8-mer genes that are up- and down-regulated against all the other experimental conditions. There are 44 genes that are commonly up-regulated and 832 that are commonly down-regulated in chitin 8-mer against all the other conditions.

4.2 ONTOLOGY ANALYSES

A particular gene can be characterized with respect to its molecular function at the biochemical level (e.g., cyclase or kinase, whose annotation is often more related to sequence similarity and protein structure) or the biological process to which it contributes (e.g., pyrimidine metabolism or signal transduction, which is often

TABLE 4.2 Number of Genes of 8-mer Genes That Are Up-Regulated Against the Other Experimental Conditions in Tiling Array Data

Conditions	Number of 8-mer Genes Down-Regulated	Number of 8-mer Genes Up-Regulated
FL Vs 8-mer	11,103	2,162
SC Vs 8-mer	19,878	1,212
RT Vs 8-mer	2,278	3,353
Control Vs 8-mer	9,986	3,256

revealed in the high-throughput data of protein interaction and gene expression profiles). In our experimental study, we define the function annotation by the GO (gene ontology) biological process [5]. We organized it in a hierarchical structure with nine classes at the top level that are subdivided into more specific classes at subsequent levels. We acquired the GO biological process functional annotation for the known proteins in *A. thaliana* (http://geneontology.org/) and generated a numerical GO index, which represents the hierarchical structure of the classification. The deepest level of hierarchy is 13 (excluding the first level, which always begins with 1, representing a biological process, to distinguish them from the other molecular

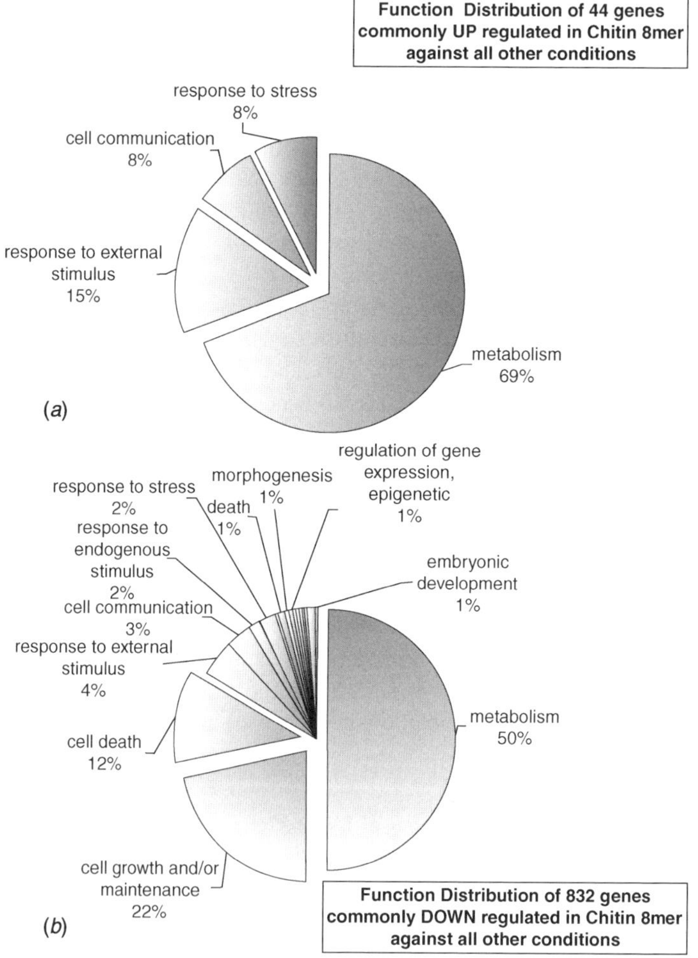

FIGURE 4.5 Distribution of GO biological process functional categories for commonly (*a*) up- and (*b*) down-regulated genes in chitin 8-mer experiments against all other experiments. The category labels in red and blue indicate over- and underrepresentation compared to the GO function distribution for all *A. thaliana* genes already annotated.

function and cellular component categories in the GO annotation). The following shows an example of the GO hierarchy:

1 biological process GO:0000004

1–4 cell growth and/or maintenance GO:0008151

1–4–3 cell cycle GO:0007049

1–4–3–2 DNA replication and chromosome cycle GO:0000067

1–4–3–2–4 DNA replication GO:0006260

1–4–3–2–4–2 DNA-dependent DNA replication GO:0006261

1–4–3–2–4–2–2 DNA ligation GO:0006266

An ORF (open reading frame) can (and usually does) belong to multiple indices at various index levels in the hierarchy, as the proteins may be involved in more than one function in a cell. We analyzed the commonly up- and down-regulated genes in terms of their functional annotations to see which functional categories are over- or underrepresented in comparison to the GO function distribution for all *A. thaliana* genes already annotated (Figure 4.5).

4.3 ANTISENSE REGULATION IDENTIFICATION

4.3.1 Antisense Silencing

The *antisense* of a gene is the noncoding strand in double-stranded DNA. The antisense strand serves as the template for mRNA synthesis. The messenger RNA (mRNA) sequence of nucleotides is called *sense* because it results in a gene product (protein). Normally, its unpaired nucleotides are "read" by transfer RNA anticodons as the ribosome proceeds to translate the message. However, RNA can form duplexes with a second strand of RNA whose sequence of bases is complementary to the first strand, just as DNA does: for example,

$$
\begin{array}{llll}
5' & C\ A\ U\ G & 3' & \text{mRNA} \\
3' & G\ U\ A\ C & 5' & \text{antisense RNA}
\end{array}
$$

When mRNA forms a duplex with a complementary antisense RNA sequence, translation is blocked as shown in Figure 4.6.

4.3.2 Antisense Regulation Identification

Tiling array technology can be applied toward discovery of antisense transcripts. Toward this we have developed a technique using tiling array data as follows:

1. We begin by identifying the down-regulated genes ($\leq$ twofold change) in an experimental condition by comparison with control experiments.

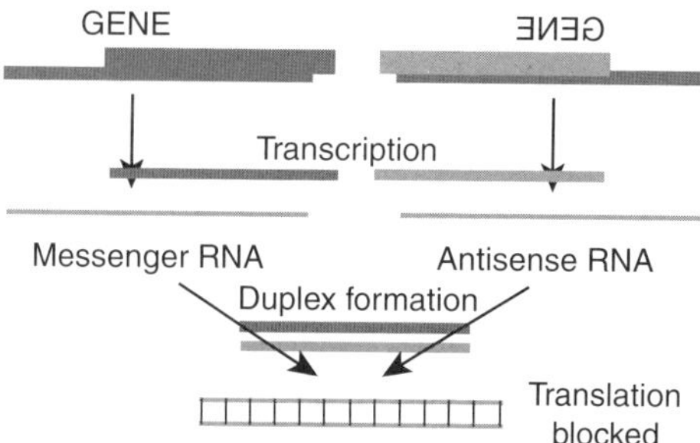

FIGURE 4.6   Antisense RNA mechanism of action.

2. Later we inspect the genomic region corresponding to the opposite strand of the down-regulated genes and calculate the expression value for this identified region on the opposite strand.

3. Then we compare the expression levels of this identified region under experimental conditions against the same genomic location in controlled experiments. If the expression under experimental conditions is up more than two fold compared with the controlled experiments, it is a potential candidate for antisense.

4. Finally, we rule out cross-hybridization by comparing the sequence of the probes tiling the antisense region with all other probes in the tiling array (allowing up to three mismatches).

Table 4.3 shows the total number of antisense candidates identified for different experimental conditions for tiling array data. Figure 4.7 shows examples of two such antisense transcripts, AT1G26290 and AT2G27740, with their expression patterns for the chitin 8-mer tiling array data. Overall analysis of the whole-genome tiling array shows a relatively higher percentage of antisense expression than was expected earlier.

To confirm our antisense transcript prediction, we compared the antisense sequence of AT3G22104 against the nonredundant EST (cDNA) database for *A. thaliana*. The top hits were to mRNA sequences on the opposite strand (plus/

TABLE 4.3   Number of Antisense Candidates Identified for Various Experimental Conditions of Tiling Array Data

Conditions	Number of Antisense Candidates
8-mer vs. control	90
FL vs. control	112
SC vs. control	67
RT vs. control	76

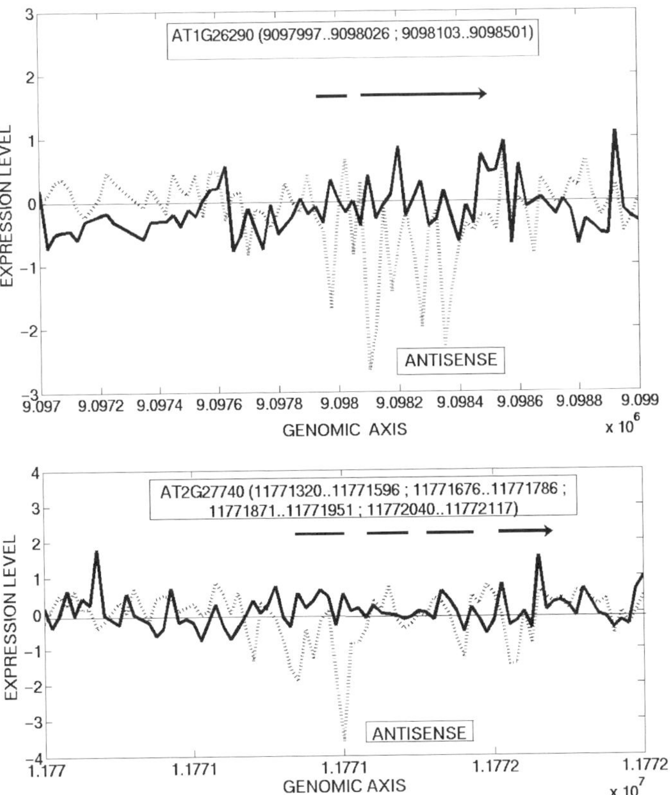

FIGURE 4.7 Example of antisense (AT1G26290 and AT2G27740) candidates for chitin 8-mer experimental conditions.

minus) for the 31–251, 286–374, 1048–1592, and 2347–2412 bases of the antisense query:

1. gi|19869691|dbj|AV827631.1|AV827631 RAFL9 *A. thaliana* cDNA clone RAFL09-17-L02 5′, mRNA sequence.
2. gi|19832809|dbj|AV798826.1|AV798826 RAFL9 *A. thaliana* cDNA clone RAFL09-17-L02 3′, mRNA sequence.

In addition to this, the next hit was to the mRNA sequence on the same strand (plus/plus) for the 1–251 bases of the antisense query:

3. gi|19858281|dbj|AV816427.1|AV816427 RAFL9 *A. thaliana* cDNA clone RAFL09-90-J18 3′, mRNA sequence.

This indicates that the antisense transcript, which hybridizes the first exon of AT3G22104, is expressed (see Figure 4.8).

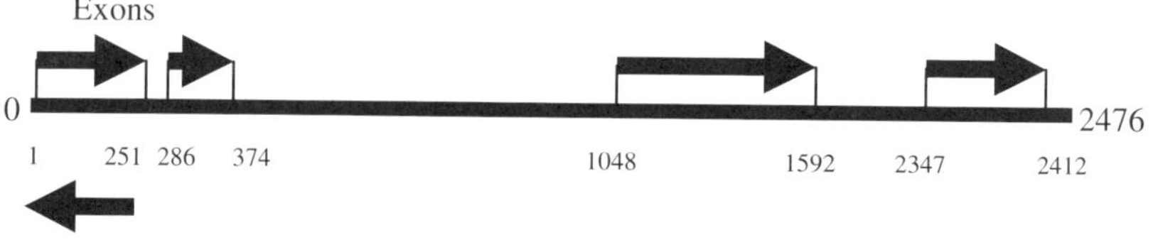

FIGURE 4.8   Antisense transcript expression for AT3G22104 confirmed by sequence comparison. The top arrows represent the exons, and the bottom arrow shows the location of the antisense transcript that matches the EST database.

Furthermore, four such antisense candidates, corresponding to the sense genes AT2G18080, AT3G22104, AT2G42780 and AT4G15030, were also selected for further confirmation by RT–PCR. Briefly, total RNA was isolated from *Arabidopsis* seedlings and turned into cDNA. The cDNA was then used in PCR to amplify *potential* antisense cDNA using gene-specific primer pairs. The gene-specific primers were designed such that they can only amplify the antisense sequence. Out of the four antisense candidates selected, three were shown to be present, suggesting that our technique is very promising in identifying antisense transcripts.

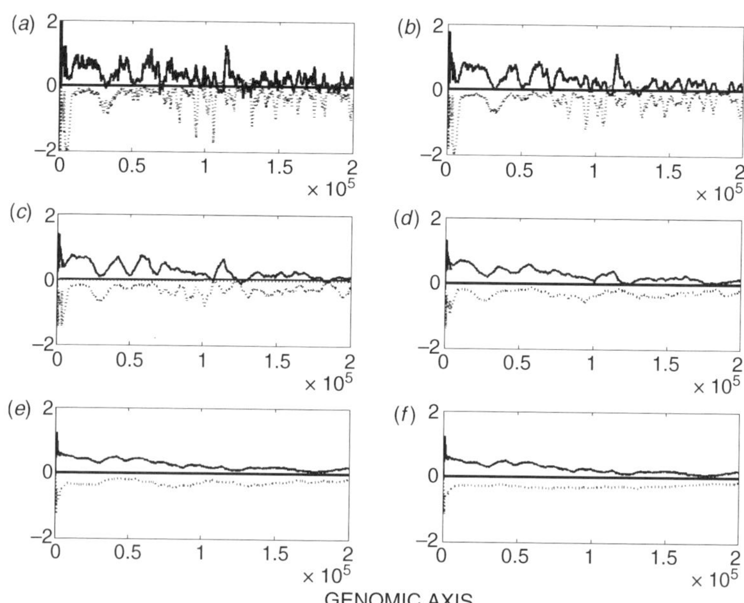

FIGURE 4.9   Detailed view of part of the expression pattern [(*a*) to (*f*)] for window sizes 50, 100, 250, 500, 1000, and 2000, respectively, for partial chip 1–2.

4.4 CORRELATED EXPRESSION BETWEEN TWO DNA STRANDS

Tiling array expression patterns may help identify a gene or the exon–intron boundaries. Toward this, we used the moving-average method to smoothen the expression patterns on both strands of the genomic axis and to calculate the correlation between the data on the two axes. We used a changing window size interval: 50, 100, 250, 500, 1000, and 2000. Based on the expression pattern for all chips for different window sizes (Figure 4.9), the most reliable window size is 250, which is close to the median length of exons and introns in *A. thaliana*.

It was observed from several earlier studies that there is more transcriptional activity in the euchromatin region of the chromosome than in the centromere. We also calculated the correlation of the data for the changing window sizes. Figure 4.10 shows the correlation for the chip 1–2 for various window sizes, and Table 4.4 lists the correlation values for all six chips according to window size. It shows that at a longer scale (with steps of > 1000 bp), the expression patterns of both strands are strongly correlated, which may reflect chromosome-level regulation of transcription (e.g., the expression distribution between the centromeric and euchromatic regions) [6]. At intermediate scales (in steps of hundreds), the correlation still appears to be evident, which may reflect the distribution of gene-rich/gene-poor regions. At shorter scales, the correlation is significant

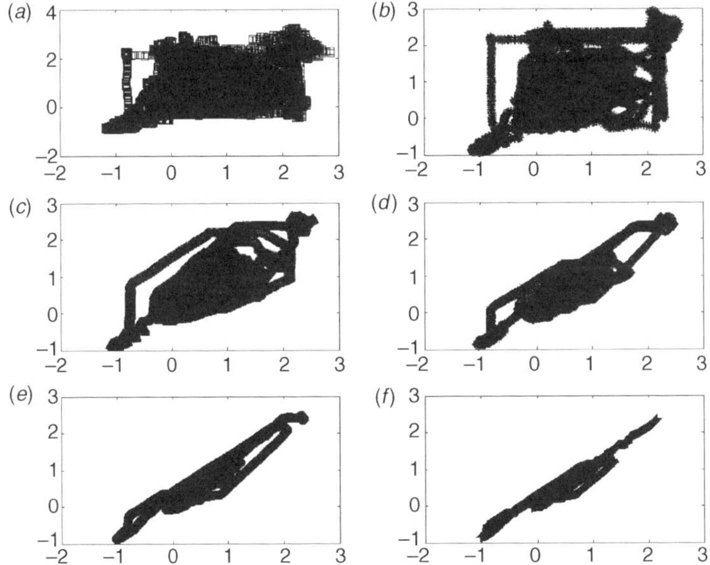

FIGURE 4.10 Correlation between the moving-average expression patterns on both strands with changing window size. The two-dimensional coordinates for a point in a graph shows the averaged expression levels at the sense strand and the antisense strand while moving the window along the genome. Parts (*a*) to (*f*) show window sizes of 50, 100, 250, 500, 1000, and 2000, respectively.

TABLE 4.4 Correlation Between the Expression Pattern on Both Strands of the Genomic Axis for Chitin 8-mer Tiling Array Data with Changing Window Size for Moving Average

	Chip 1–1	Chip 1–2	Chip 1–3	Chip 1–4	Chip 1–5	Chip 1–6
CEL files expression:	0.104	0.139	0.117	0.132	0.082	0.109
Moving-average window size						
50	0.55	0.61	0.54	0.59	0.47	0.54
100	0.68	0.71	0.67	0.71	0.59	0.67
200	0.81	0.82	0.81	0.83	0.74	0.81
500	0.89	0.90	0.89	0.90	0.83	0.9
1000	0.92	0.94	0.93	0.94	0.88	0.93
2000	0.93	0.95	0.94	0.95	0.90	0.94

although it is weak, which may reflect antisense expression and coexpressions of both strands in some regions.

4.5 IDENTIFICATION OF NONPROTEIN CODING mRNA

In eukaryotes, several studies have revealed a new class of mRNAs containing only short open reading frames, termed either sORF-mRNAs, noncoding RNAs, or prlRNAs (protein-lacking RNAs), also referred to as non-protein-coding RNAs (npcRNAs). Their functions may involve the RNA molecule itself and/or short ORF-encoded peptides. To investigate this, we identified transcriptome components outside previously annotated gene models with an ORF no longer than 210 nucleotides

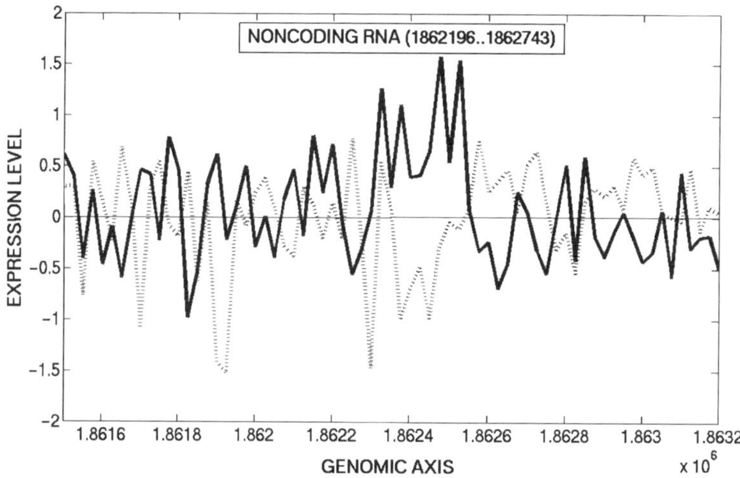

FIGURE 4.11 Expression pattern of a nonprotein coding mRNA.

(70 amino acids). Figure 4.11 shows an example of a non-protein-coding mRNA located on a positive stand of chromosome 1.

4.6 SUMMARY

Until recently, the traditional view of a genome was that genes were the regions of the genome that encoded proteins or structural RNAs; the regions upstream of genes were considered regulatory sequences (promoters), and much of the rest was considered junk. Recent analyses of the transcriptome using unbiased whole-genome tiling arrays have challenged this view. Studies involving arrays containing PCR fragments covering all unique regions of human chromosome 22 [7] or high-density oligonu-cleotide tiling arrays covering essentially all nonrepetitive regions of human chromo-somes 21 and 22 [8–10] or the entire *Arabidopsis* genome [3] have been used to map the sites of transcriptional activity in an unbiased fashion. These studies revealed that up to tenfold more expression than predicted from prior EST or cDNA evidence or gene predictions was evident in the human and plant genomes, although questions remain as to whether these observations reflect transcriptional noise.

Numerous intergenic regions containing no annotated gene models were found to be transcribed, computational gene models were confirmed or corrected, and untrans-lated regions (5- and 3UTRs) of known genes were extended. Novel transcripts were also identified within the genetically defined centromeres. Finally, and most unexpectedly, a strikingly large amount (30%) of antisense transcriptional activity was observed from known genes with sense strand expression (i.e., those with ESTs or cloned full-length cDNAs) [3]. Taken together with natural antisense transcript (NAT) predictions based on tiling array studies [7,10], these observations point to a significant amount of NAT expression in higher eukaryotes that cannot readily be explained as being an artifact of the labeled target–probe preparation because several of these studies involved hybridization of strand-specific labeled targets [3,10] or probes [7] to microarrays. Moreover, these findings are consistent with NAT predic-tions based on transcript sequence evidence, although the functional significance of this phenomenon remains largely unknown. Collectively, these studies show that tiling arrays are a powerful tool for characterizing the transcriptome.

Nevertheless, these recent discoveries made using tiling arrays signal a paradigm shift in how we think about the transcriptome and highlight some limitations of relying on cDNA and EST sequencing projects for gene discovery. An alternative approach that may prove more efficient would be first to attempt to establish the functional relevance of novel transcripts identified using tiling arrays. For example, patterns of coexpression and/or coregulation between novel predicted transcripts and known transcripts could be scored and used to prioritize for further characterization the transcripts predicted.

Acknowledgments

Work in the Xu laboratory was funded by grant CSREES 2004-25604-14708 from the U.S. Department of Agriculture. The Stacey laboratory was funded by grant

DE-FG02-02ER15309 from the Department of Energy, Office of Basic Energy Sciences.

REFERENCES

1. P. Bertone, V. Trifonov, J. S. Rozowsky, F. Schubert, O. Emanuelsson, J. Karro, M. Y. Kao, M. Snyder, and M. Gerstein (2006). Design optimization methods for genomic DNA tiling arrays. *Genome Res.*, 16(2):271–281.

2. T. Mockler and J. Ecker (2005). Applications of DNA tiling arrays for whole-genome analysis. *Genomics.*, 85:1–15.

3. K. Yamada et al. (2003). Empirical analysis of transcriptional activity in the *Arabidopsis* genome. *Science.*, 302:842–846.

4. T. Toyoda and K. Shinozaki (2005). Tiling array-driven elucidation of transcriptional structures based on maximum-likelihood and Markov models. *Plant J.*, 43(4):611–621.

5. The Gene Ontology Consortium (2000). *Nat. Genet.*, 25:25–29.

6. Y. Jiao, P. Jia, X. Wang, N. Su, S. Yu, D. Zhang, L. Ma, et al. (2005). A tiling microarray expression analysis of rice chromosome 4 suggests a chromosome-level regulation of transcription. *Plant Cell*, 17(6):1641–1657.

7. J. L., Rinn et al. (2003). The transcriptional activity of human chromosome 22. *Genes Dev.*, 17:529–540.

8. S. Cawley et al. (2004). Unbiased mapping of transcription factor binding sites along human chromosomes 21 and 22 points to widespread regulation of noncoding RNAs. *Cell*, 116:499–509.

9. P. Kapranov et al. (2002). Large-scale transcriptional activity in chromosomes 21 and 22. *Science*, 296:916–919.

10. D. Kampa et al. (2004). Novel RNAs identified from an in-depth analysis of the transcriptome of human chromosomes 21 and 22. *Genome Res.*, 14:331–342.

5

IDENTIFICATION OF MARKER GENES FROM HIGH-DIMENSIONAL MICROARRAY DATA FOR CANCER CLASSIFICATION

JIEXUN LI, HUA SU, AND HSINCHUN CHEN

Department of Management Information Systems, University of Arizona, Tucson, Arizona

Knowledge discovery in bioinformatics is driven largely by available biological experimental data and knowledge. Cancer research is a major research area in the medical field and accurate classification of tumor types is of great importance in cancer diagnosis and drug discovery (Lu and Han, 2003). Traditional cancer classification has always been based on morphological and clinical measurements, but these conventional methods are limited in their diagnostic ability. To achieve more accurate cancer classification, researchers have proposed approaches based on global genetic data analysis.

Let's first provide some basic background information. Cells are the fundamental working units for every living system. Deoxyribonucleic acid (DNA) contains all the instructions directing cell activities. The DNA sequence is a particular arrangement of base pairs composed of four nitrogen bases: adenine (A), guanine (G), cytosine (C), and thymine (T). This arrangement spells out the exact instructions required to create a particular organism with its own unique characteristics. The entire DNA sequence that codes for a living thing is called its *genome*. A *gene* is a defined section of the entire genomic sequence, with a specific and unique purpose.

DNA acts as a blueprint for the molecule ribonucleic acid (RNA). The process of transcribing a gene's DNA sequence into RNA is called *gene expression*. The

Knowledge Discovery in Bioinformatics: Techniques, Methods, and Applications, Edited by Xiaohua Hu and Yi Pan

expression of a gene provides a measure of gene activity under certain biochemical conditions. It is known that certain diseases, including cancer, are reflected in the changes of the expression values of certain genes.

Due to some practical problems of gene expression, such as the instability of messenger RNA, biomedical researchers also study the relationships between diseases and DNA methylation as an alternative. Methylation is a modification of cytosine that occurs either with or without a methyl group attached. This methylation of cytosine can only appear together with guanine as CpG. The methylated CpG can be seen as a fifth base and is one of the major factors responsible for expression regulation.

The advent of DNA microarray techniques has supported effective identification of a variety of gene functions for cancer diagnosis. Microarray analysis is a relatively new molecular biology methodology that expands on classic probe hybridization methods to provide access to thousands of genes at once, therefore allowing simultaneous measurement of thousands of genes in a cell sample. With this abundance of gene array data, biomedical researchers have been exploring their potential for cancer classification and have seen promising results.

For microarray-based cancer classification, the input variables are measurements of genes and the outcomes are the tumor class. Let $X_1, X_2, \ldots, X_n$, be random variables for genes $g_1, g_2, \ldots, g_n$, respectively. In particular, X_i represents the expression or methylation level of gene g_i. Let C be the random variable for the class labels, and domain $(C) = \{1, \ldots, K\}$, where K denotes the total number of classes. A training set $T = \{(t_1, c_1), (t_2, c_2), \ldots, (t_m, c_m)\}$ of m tuples, where $t_i = \{t_i X_1, t_i X_2, \ldots, t_i X_n\}$ is a set of measurements of n genes in one sample and c_i is its corresponding class label. Microarray-based cancer classification is to construct a classification model from such a training set and to predict the class label of unknown samples in the test set $S = \{t_1, t_2, \ldots, t_l\}$, where l is the size of the test set. The classification *accuracy* is defined as the percentage of correct predictions made by the classifier on the test set.

Various classification algorithms can be applied to cancer classification problems, ranging from the decision tree method, linear discrimination analysis, nearest-neighbor analysis, to the new support vector machines (SVMs). However, the unique nature of microarray data poses some new challenges to cancer classification.

The major problem of all classification algorithms for gene expression and methylation analysis is the high dimensionality of input space compared to the relatively small number of samples available (Model et al., 2001). There are usually thousands to hundreds of thousands in each sample and fewer than 100 samples. Most classification algorithms may suffer from high dimensionality due to overfitting. In addition, the large number of features would increase the computational cost significantly.

Another challenge arises from irrelevant genes. Among the huge number of genes in microarray data, cancer-related genes occupy a small portion only; and most of the genes are irrelevant for cancer distinction. Not only does the presence of these irrelevant genes interfere with the discrimination power of the relevant genes but also introduces difficulty in gaining insights about genes' biological relevance. A critical concern for biomedical researchers is to identify the *marker genes* that can discriminate tumors for cancer diagnosis. Therefore, gene selection is of great importance for accuracy and interpretability of microarray-based cancer classification.

In this chapter we familiarize readers with the research field of gene selection for cancer classification by providing a comprehensive review of the literature and a comparative study. In Section 5.1 we review existing feature selection methods with a focus on the evaluation criterion of features and the generation procedure of candidate features or feature subsets. In Section 5.2 we introduce a framework of gene selection techniques, with emphasis on optimal search-based gene subset selection methods. Under this gene selection framework, in Section 5.3 we present a comparative study of various gene selection methods applied to two real microarray data sets. The experimental results showcase the promises and challenges of gene selection for cancer classification. We conclude by summarizing key insights and future directions.

5.1 FEATURE SELECTION

Identification of marker genes for cancer classification is a typical feature selection problem. In this section we review a variety of feature selection approaches.

5.1.1 Taxonomy of Feature Selection

Feature selection is aimed at identifying a minimal subset of features that are relevant to the target concept (Dash and Liu, 1997). The objective of feature selection is threefold: to improve prediction performance, to provide more cost-effective prediction, and to provide a better understanding of the underlying process that generated the data (Guyon and Elisseeff, 2003).

A feature selection method generates candidates from the feature space and assesses them based on some evaluation criterion to find the best feature subset (Dash and Liu, 1997). According to the *evaluation criterion* and *generation procedure*, we can categorize various feature selection methods into a taxonomy as shown in Table 5.1, which includes examples of the major feature selection methods in each category. We introduce these methods in detail below.

5.1.2 Evaluation Criterion

An evaluation criterion is used to measure the discriminating ability of candidate features. Based on the evaluation criterion, feature selection methods can be divided into filter models and wrapper models (Kohavi and John, 1997). A *filter model* selects good features as a preprocessing step without involving a learning algorithm. In contrast, a *wrapper model* utilizes a learning algorithm as a black box "wrapped" in the feature selection process to score feature subsets according to the prediction accuracy.

Filter Models Filter models select good features based on a certain data intrinsic measure (Dash and Liu, 1997, 2003; Hall, 2000). These measures show the relevance of a feature to the target class. These relevance measures can be grouped further into distance, consistency, and correlation measures (Dash and Liu, 1997).

TABLE 5.1   A Taxonomy of Feature Selection

(a) Evaluation Criterion

Model	Measure	Examples
Filter	Distance: the degree of separation between classes	Fisher criterion (Bishop, 1995), test statistics (Mendenhall and Sincich, 1995), relief (Kira and Rendell, 1992)
	Consistency: finds a minimum number of features that can distinguish classes	Inconsistency rate (Dash and Liu, 2003)
	Correlation: measures the ability to predict one variable from another	Pearson correlation coefficient (Hall, 2000), information gain (Quinlan, 1993)
Wrapper	Classification: the performance of an inductive learning algorithm	Decision tree and naive Bayes (Kohavi and John, 1997)

(b) Generation Procedure

Type	Search	Examples
Individual ranking	Measures the relevance of each feature	Most filters (Guyon and Elisseeff, 2003)
Subset selection	Complete: traverses all the feasible solutions	Branch and bound (Chen, 2003), best-first search (Ginsberg, 1993)
	Heuristic Deterministic: uses a greedy strategy to select features according to local change	Sequential forward selection, sequential backward selection, sequential floating forward selection, sequential floating backward selection (Pudil et al., 1994)
	Nondeterministic: attempts to find the optimal solution in a random fashion	Simulated annealing (Kirkpatrick et al., 1983), Las Vegas filter (Liu and Setiono, 1996), genetic algorithms (Holland, 1975), tabu search (Glover and Laguna, 1999)

Distance Measures   *Distance measures* attempt to quantify the ability of a feature or a feature subset to separate different classes from each other. A classical distance measure to assess the degree of separation between two classes is given by the *Fisher criterion* (Bishop, 1995):

$$J(k) = \frac{(\mu_k^{\mathrm{I}} - \mu_k^{\mathrm{II}})^2}{\sigma_k^{\mathrm{I}^2} + \sigma_k^{\mathrm{II}^2}}$$

where $(\mu_k^{\mathrm{I}}, \mu_k^{\mathrm{II}})$ is the mean and $(\sigma_k^{\mathrm{I}}, \sigma_k^{\mathrm{II}})$ is the standard deviation of all features in instances within classes I and II, respectively. It gives a high score for features where the two classes are far apart compared to the within-class variances. In addition, based on the assumption of normal distribution of feature values within a class, test statistics can also be used as a distance measure. A t-statistic value can measure the significance of the difference between two class means (Mendenhall and Sincich, 1995). Similarly, F- and κ^2-statistics can be used for multiclass problems. A well-known feature selection algorithm, Relief, also uses a distance measure that estimates the ability of a feature to distinguish two instances that are close to each other but of opposite classes (Kira and Rendell, 1992).

Consistency Measures   *Consistency measures* attempt to find a minimum number of features that can distinguish classes as consistently as can a full feature set (Dash and Liu, 2003). An inconsistency is defined as two instances having the same pattern: the same feature values but different class labels. Since the full feature set always has the lowest inconsistency rate, feature selection thus attempts to minimize the number of features in subset S to achieve a certain inconsistency rate. Consistency measures have been shown effective in removing these undesired features. However, these measures are often limited to discrete data to make the patterns of feature values countable. Furthermore, the computational cost of finding the best feature subset based on consistency measures is very high (Dash and Liu, 2003).

Correlation Measures   *Correlation measures* quantify the ability to predict the value of one variable from the value of another (Dash and Liu, 1997). These measures are often based on linear correlation or information theory. Under the first type of linear correlation-based measures, the best known measure is the *Pearson correlation coefficient*. For two continuous variables X and Y, the correlation coefficient is defined as

$$r_{XY} = \frac{\sum_i (x_i - \bar{x})(y_i - \bar{y})}{\sqrt{\sum_i (x_i - \bar{x})^2}\sqrt{\sum_i (y_i - \bar{y})^2}}$$

where $\bar{x}$ (or $\bar{y}$) is the mean of variable X (or Y). This coefficient can also be extended to measure the correlation between a continuous variable and a discrete variable or between two discrete variables (Hall, 2000). Linear correlation measures are simple and easy to understand. However, since they are based on the assumption of linear correlation between features, they may not be able to capture nonlinear correlations in nature.

In information theory, *entropy* is a measure of the uncertainty of a random variable. The amount of entropy reduction of variable X given another variable Y reflects additional information about X provided by Y and is called *information gain* (Quinlan, 1993). Information theory–based correlation measures can capture the different correlation between features in nature. Furthermore, they are applicable to multiclass problems. However, most entropy-based measures require nominal features. They can be applied to continuous features only if the features have been discretized properly in advance, which may lead to information loss.

Wrapper Models None of the evaluation criteria introduced above are dependent on an inductive learning algorithm. By contrast, in wrapper models, classification accuracy is used as an evaluation criterion. Wrapper models (Kohavi and John, 1997) employ—as a subroutine—a statistical resampling technique (such as cross-validation) using an inductive learning algorithm to estimate the accuracy of candidate feature subsets. The one with the highest classification accuracy is identified as the best feature subset.

Since the features selected using the classifier are then used to predict the class labels of unseen instances, accuracy is often very high. However, wrapper models are often criticized for two disadvantages: high computational cost and low generality. First, since the inductive learning algorithm needs to be called as a subroutine repeatedly along the number of iterations, wrappers often suffer from high time complexity. Second, since candidate subsets are assessed based on the error rate of a predetermined classifier, the optimal feature subset is suitable only for this specific classifier. For the same data set, wrappers may get different optimal feature subsets by applying different classifiers.

5.1.3　Generation Procedure

Based on the generation procedure of candidate subsets (i.e., whether features are evaluated individually or collectively), we categorize generation procedures into individual feature ranking and feature subset selection (Blum and Langley, 1997; Guyon and Elisseeff, 2003).

Individual Feature Ranking In individual feature ranking approaches, each feature is measured for its relevance to the class according to a certain criterion. The features are then ranked, and the top ones are selected as a good feature subset. Most filters that aim only at removing irrelevant features belong to this category. Individual feature ranking is commonly used for feature selection because of its simplicity, scalability, and good empirical success (Guyon and Elisseeff, 2003). In particular, it is computationally advantageous because it requires only the computation of N scores and sorting the scores, where N is the total number of features.

However, since no correlations among features are exploited, individual feature rankings have several disadvantages. First, the features selected may be strongly correlated with the class individually, but acting together might not give the best classification performance. Second, some features may contain the same correlation information, thus introducing redundancy. Third, features that are complementary to each other in determining class labels may not be selected if they do not exhibit high individual correlation. Finally, the number of features, m, retained in the feature subset is difficult to determine, often involving human intuition with trial and error.

Feature Subset Selection To overcome these shortcomings of individual feature ranking, feature subset selection attempts to find a set of features that serve together to achieve the best classification performance. The advantage of this approach is that it considers not only the relevance of features and the target concept but also the intercorrelation between features. However, subset selection requires searching a huge number of candidate subsets in the feature space. This search process often requires great computation expense, especially for data with high dimensionality. General approaches were proposed to solve this problem: complete search and heuristic search (Dash and Liu, 1997).

Complete Search Ideally, feature selection should traverse every single feature subset in N-dimensional feature space and try to find the best one among the competing 2^N candidate subsets according to a certain evaluation function. However, this exhaustive search is known to be NP-hard, and the search quickly becomes computationally intractable. For a search to be complete does not mean that it must be exhaustive. Various techniques, such as branch and bound (Narendra and Fukunaga, 1977) and best-first search (Ginsberg, 1993), were developed to reduce the search without jeopardizing the chances of finding the optimal subset. By applying these search techniques, fewer subsets are evaluated, although the order of the search space is $O(2^N)$. Nevertheless, even with these more efficient techniques, complete search is still impractical in cases of high-dimensional data.

Heuristic Search Other search algorithms generate candidate solutions based on certain heuristics. Heuristic searches can be categorized further as deterministic and nondeterministic. *Deterministic heuristic search* methods are basically "hill-climbing" approaches which select or eliminate features in a stepwise manner. At each step of this process, only features yet to be selected (or rejected) are available for selection (or rejection). One considers local changes to the current state to decide on selecting or removing a specific feature. This search strategy comes in two forms: *Sequential forward selection* starts from an empty set and incorporates features progressively; its counterpart, *sequential backward selection*, starts with a full set of features and progressively removes the least promising. These deterministic search methods are particularly computationally advantageous and robust against overfitting (Guyon and Elisseeff, 2003). However, due to the greedy strategy adopted, they often find local optimal solutions, sometimes called the *nesting effect* (Guyon and Elisseeff, 2003). Even their extended versions, such as sequential floating forward selection and sequential floating backward selection (Pudil et al., 1994), cannot overcome this drawback completely.

To jump out of the nesting effect without exploring each of the 2^N feature subsets, *nondeterministic heuristic search* techniques were introduced to select the optimal or suboptimal feature subsets in a random fashion within a prespecified number of iterations. A representative algorithm in this category is the Las Vegas filter (Liu and Setiono, 1996), which searches the space of subsets randomly using a Las Vegas algorithm, which makes probabilistic choices to help guide the search for an optimal set. Simulated annealing (SA) algorithms are based on the analogy between a combinatorial optimization problem and the solid annealing process (Kirkpatrick et al., 1983). SA avoids local optima by allowing backtracking according to Metropolis's criterion based on Boltzmann's probability ($e^{-\Delta E/T}$). A *genetic algorithm* (GA) is an optimal search technique (Holland, 1975) that behaves like the processes of evolution in nature. A GA can find the global (sub)optimal solution in complex multidimensional spaces by applying genetic operators such as selection, crossover, and mutation. A tabu search (TS) algorithm is a metaheuristic method that guides the search for the optimal solution making use of flexible memory, which exploits the history of the search (Glover and Laguna, 1999). Numerous studies have shown that TS can compete and, in many cases, surpass the best-known techniques, such as SA and GA (Glover and Laguna, 1999). These methods are also called *optimal search* because of their ability to find global optimal or suboptimal solutions. In recent years they have been introduced to feature selection and have shown good performance.

5.2 GENE SELECTION

Various feature selection approaches have been applied to gene selection for cancer classification. In this section we introduce two major types of gene selection approaches: individual gene ranking and gene subset selection.

5.2.1 Individual Gene Ranking

Due to its simplicity and scalability, individual gene ranking is the most commonly used approach in gene selection. The most common and simple approach to feature selection is using the correlation between the attribute values and the class labels based on Euclidean distance, Pearson correlation, and so on.

A well-known example is the GS method proposed by Golub et al. (1999). The GS method proposed a correlation metric that measures the relative class separation produced by the values of a gene. It favors genes that have a large between-class mean value and a small within-class variation. For gene j, let $[\mu_+(j),\ \sigma_+(j)]$ and $[\mu_-(j), \sigma_-(j)]$ denote the means and standard deviations of the expression levels of j in the two classes, respectively. A correlation metric $P(j, c)$, called the *signal-to-noise ratio* (SNR), is defined as $[\mu_+(j) - \mu_-(j)]/[\sigma_+(j) + \sigma_-(j)]$. This metric reflects the difference between the two classes relative to the standard deviation within the classes. Larger values of $|P(j, c)|$ indicate a strong correlation between the gene expression and the class distinction. Thus, genes are grouped into positive and negative value groups and ranked according to their absolute values. The top $k/2$ genes from the two groups are selected as the informative genes. Similar distance measures, such as the Fisher criterion, t-statistic, and median vote relevance, have also been applied to identification of marker genes (Chow et al., 2001; Model et al., 2001; Liu et al. 2002; Li et al., 2004a). These measures are often used for binary classification (e.g., distinguishing normal and cancerous tissues).

For multiclass cancer classification, metrics such as the F-statistic and BSS/WSS can be used (Dudoit et al., 2002). BSS/WSS is the ratio of a gene's between-group to within-group sum of squares. For a gene j, the ratio is

$$\frac{\mathrm{BSS}(j)}{\mathrm{WSS}(j)} = \frac{\sum_i \sum_k I(c_i = k)(\bar{x}_{kj} - \bar{x}_{\cdot j})^2}{\sum_i \sum_k I(c_i = k)(x_{ij} - \bar{x}_{kj})^2}$$

where $\bar{x}_{\cdot j}$ denotes the average value of gene j across all samples, $\bar{x}_{kj}$ denotes the average value of gene j across samples belonging to class k, and $I(\cdot)$ denotes the indicator function: 1 if the condition in parentheses is true and 0 otherwise.

5.2.2 Gene Subset Selection

Although the individual feature-ranking methods have been shown to eliminate irrelevant genes effectively, they do not exploit the interaction effects among genes, which are of great importance to cancer classification. In this section we focus on gene subset selection, which takes into account the group performance of genes.

In gene subset selection, for a full set of N genes, each subset can be represented as a string of length N: $[g_1 \, g_2 \cdots g_N]$, where each element takes a Boolean value (0 or 1) to indicate whether or not a gene is selected. For a gene i, g_i equals 1 if this gene is selected in the subset, and equals 0 otherwise. The overall methodology of gene subset selection is as follows: Use a search algorithm to generate candidate gene subsets and assess these subsets by assigning a goodness score based on certain evaluation criteria; then the gene subset with the highest goodness score is regarded as the optimal subset.

Evaluation Criteria for Gene Subset Selection   Gene subset selection approaches treat a gene subset as a group rather than as individuals, to evaluate its predictive power according to a particular criterion. Based on whether a learning algorithm is used as the evaluation criteria, gene subset selection can be categorized into filter models and wrapper models.

Filter Models for Gene Subset Selection   Bø and Jonassen proposed a gene-pair ranking method which evaluates how well a gene pair in combination can separate two classes (Bø and Jonassen, 2002). Each gene pair is evaluated by computing the projected coordinates of each experimental sample on the diagonal linear discriminant axis. The evaluation score is defined as the two-sample t-statistic between the two groups of data points. This method can identify cancer-related genes or gene pairs that are not among the top genes when ranked individually. However, this method can capture only pairwise correlations and is limited to binary classification. A good gene subset contains genes highly relevant with the class, yet uncorrelated with each other. Ding and Peng proposed an approach of minimum redundancy–maximum relevance to remove both irrelevant and redundant genes (Ding and Peng, 2003).

For the objective of maximum relevance, an F-statistic between a gene and the class label could be adopted as a relevance score. The F-statistic value of gene x in K classes denoted by h is denoted as $F(x,h)$. Hence, for a feature set Θ, the objective function of maximum relevance can be written as

$$\max \; V := \frac{1}{|\Theta|} \sum_{x \in \Theta} F(x, h)$$

where V is the average F value of all the features in Θ, and $|\Theta|$ is the number of features. Similarly, mutual information can be adopted to measure the relevance between genes and class distinction for discrete variables.

For the other objective of minimum redundancy, the Pearson correlation coefficient between two genes x and y, denoted as $r(x,y)$, can be used as the score of redundancy. Regarding both high positive and high negative correlation as redundancy, we take the absolute value of correlation. For a feature set Θ, the objective of minimum redundancy can be written as

$$\min \ W := \frac{1}{|\Theta|^2} \sum\nolimits_{x,y \in \Theta} |r(x,y)|$$

where W is the average correlation coefficient between any two features in Θ.

These two objectives can be combined in different ways, such as difference and quotient. For instance, we could choose a quotient of the two objectives due to their good performances, reported by Ding and Peng (2003):

$$\max \ \frac{V}{W} := \frac{\sum_x F(x,h)}{(1/|\Theta|) \sum_{x,y \in \Theta} |r(x,y)|}$$

Wrapper Models for Gene Subset Selection   Unlike these filter models, wrapper models use the estimated accuracy of a specific classifier to evaluate candidate gene subsets. Various inductive learning algorithms can be used to estimate classification accuracy in this step. For example, support vector machine (SVM) classifiers have commonly been used in wrappers due to their good performance and robustness to high-dimensional data (Christianini and Shawe-Taylor, 2000). A standard SVM separates the two classes with a hyperplane in the feature space such that the distance of either class from the hyperplane (i.e., the margin) is maximal. The prediction of an unseen instance $\mathbf{z}$ is either 1 (a positive instance) or -1 (a negative instance), given by the decision function

$$h = f(\mathbf{z}) = \mathrm{sgn}(\mathbf{w} \cdot \mathbf{z} + b)$$

The hyperplane is computed by maximizing a vector of Lagrange multipliers α in

$$L(\alpha) = \sum\nolimits_{i=1}^{n} \alpha_i - \frac{1}{2} \sum\nolimits_{i,j} \alpha_i \alpha_j h_i h_j K(\mathbf{x}_i, \mathbf{x}_j)$$

where $\alpha_1, \alpha_2, \ldots, \alpha_n \geq 0$ and $\sum_{i=1}^{n} \alpha_i h_i = 0$. The function K is a kernel function that maps the features in the input space into a feature space (possibly of a higher dimension) in which a linear class separation is performed. For a linear SVM, the mapping of K is a linear mapping: $K(\mathbf{x}_i, \mathbf{x}_j) = \mathbf{x}_i \cdot \mathbf{x}_j$.

Guyon et al. proposed a support vector machine–based recursive feature elimination (RFE) approach to the selection of genes (Guyon et al. 2002). The

SVM-based RFE gene selection method works as follows: Starting with the full gene set, this approach progressively computes the change in classification error rate for the removal of each gene, then finds and removes the gene with the minimum error rate change and continues to repeat this process until a stopping criterion is satisfied. This process tries to retain the gene subset with the highest discriminating power, which may not necessarily be those with the highest individual relevance.

In addition to the SVM, other classification algorithms, such as k-nearest neighbors (Li et al., 2001), naive Bayes models (Saeys et al., 2003), most likelihood classifier (Ooi and Tan, 2003), linear discriminant analysis (Marchevsky et al., 2004), and linear regression (Xiong et al., 2001), have also been adopted for gene subset selection and have performed well.

Search Algorithms for Gene Subset Selection   Due to the high dimensionality of microarray data, complete searches are often impractical for gene subset selection problems. Greedy searches, such as sequential forward selection and sequential backward selection, are often used to generate candidate gene subsets (Xiong et al., 2001; Bø and Jonassen, 2002; Guyon et al., 2002; Marchevsky et al., 2004). These greedy searches are simple and fast but may result in local optimal solutions.

To achieve global optimal solutions, in recent years some wrapper models replaced greedy search with optimal search in gene selection. Genetic algorithms are optimal search methods that behave like evolution processes in nature (Holland, 1975). Unlike greedy search algorithms, GAs can avoid local optima and provide multi criteria optimization functions. An advantage of a GA is that it tends to retain good features of solutions that are inherited by successive generations. Genetic algorithms have been used successfully in various fields, such as Internet search engines and intelligent information retrieval (Chen et al., 1998). Siedlecki and Sklansky introduced the use of GAs for feature selection and they are now in common use in this field (Siedlecki and Sklansky, 1989).

In GAs, each potential solution to a problem is represented as a chromosome, which in our case is the string representing a gene subset. A pool of chromosomes forms a population. A fitness function is defined to measure the goodness of a solution. A GA seeks an optimal solution by iteratively executing genetic operators to realize evolution. Based on the principle of "survival of the fittest," strings with higher fitness are more likely to be selected and assigned a number of copies into the mating pool. Next, crossovers choose pairs of strings from the pool randomly with probability P_c and produce two offspring strings by exchanging genetic information between the parents. Mutations are performed on each string by changing each element at probability P_m. Each string in the new population is evaluated based on the fitness function. By repeating this process for a number of generations, the string with the best fitness of all generations is regarded as optimum. Following is the principal GA for feature subset selection:

S feature space

k current number of iterations

x solution of feature subset

x^* best solution so far

f fitness/objective function

$f(x)$ fitness/objective value of solution x

V_k current population of solutions

P_c probability of crossover

P probability of mutation

1. Generate an initial feature subset population V_0 from S (population size $=$ *pop_size*). Set $k = 0$.
2. Evaluate each feature subset in V_k with respect to the fitness function.
3. Choose a best solution x in V_k. If $f(x) > f(x^*)$, then set $x^* = x$.
4. Based on the fitness value, choose solutions in V_k to generate a new population V_{k+1}. Set $k = k + 1$.
5. Apply crossover operators on V_k with probability P_c.
6. Apply mutation operators on V_k with probability P_m.
7. If a stopping condition is met, then stop, else go to step 2.

Use of genetic algorithms in gene subset selection can be found in many previous studies. For example, Li et al. (2001) proposed a genetic algorithm/k-nearest neighbors (GA/kNN) method to identify genes that can jointly discriminate normal and tumor samples. The top genes ranked by their frequency of selection through the iterations of GA are selected as the marker genes. Ooi and Tan also used a GA to search the feature space and chose the gene subset with the best fitness among all generations of GAs as the optimal subset (Ooi and Tan, 2003). In addition, Saeys et al. used a estimation of distribution algorithm, a general extension of GA, to select marker genes and reported good performance (Saeys et al., 2003).

5.2.3 Summary of Gene Selection

From the review of related works, we can see that various feature selection methods have been proposed and applied to the identification of marker genes for cancer classification. In summary, we observe that this research domain has the following characteristics. Individual gene-ranking methods are commonly used, due to their simplicity and scalability. However, these univariate methods cannot cover the information regarding gene interactions. In contrast, gene subset selection can capture genes' interaction effects by evaluating the group performance of gene subsets. Greedy searches are commonly used but can only provide local optimal solutions. To achieve global optimal gene subsets, optimal searches such as genetic algorithms can be used for gene subset selection and have shown superior performance.

5.3 COMPARATIVE STUDY OF GENE SELECTION METHODS

In this section we present a comparative study of gene selection methods. In particular, we compare the optimal search-based gene subset selection methods against cancer classification without feature selection and against an individual gene-ranking method. In addition, a comparison of various optimal search-based gene subset methods provides insights regarding their advantages and disadvantages.

5.3.1 Microarray Data Descriptions

We first describe the two microarray data sets involved in our study. The first data set comprises DNA methylation arrays provided by the Arizona Cancer Center. It is derived from the epigenomic analysis of bone marrow specimens from healthy donors and people with myelodysplastic syndrome. This data set contained 678 genes and 55 samples. From this data set we created two test beds to perform a binary and a multiclass classification, respectively. Test bed METH-2 of 55 samples is used to discriminate normal tissues from tumor tissues. Test bed METH-5 of 45 samples is used to discriminate five subtypes of tumors.

The second data set comprises experimental measurements of gene expression with Affymetrix oligonucleotide arrays (Alon et al., 1999). It contains measurements of 2000 human genes in 62 colon tissue samples (40 tumor and 22 normal tissues). This test bed (referred to as COLON) is used to discriminate normal from tumor tissues.

5.3.2 Gene Selection Approaches

In Section 5.2 we introduced a framework of various gene subset selection approaches in detail. In this comparative study, we chose minimum redundancy–maximum relevance (Ding and Peng, 2003) as a representative filter and a SVM classifier as a representative wrapper. In addition, we chose a genetic algorithm as a representative optimal search to explore the feature space. By combining an optimal search algorithm with an evaluation criterion, we developed the following two-gene subset selection methods:

1. *GA/MRMR:* using a genetic algorithm as a search algorithm and minimum redundancy–maximum relevance as an evaluation criterion
2. *GA/SVM:* using a genetic algorithm as a search algorithm and the accuracy of the support vector machine classifier as the evaluation criterion

Two benchmarks were used for comparison in this study. We used cancer classification based on a full set of genes as the first benchmark to demonstrate the power of gene selection. In addition, we chose F-statistic individual ranking as the second benchmark to compare with optimal search-based gene subset selection methods as in the work of Ding and Peng (2003).

5.3.3 Experimental Results

To compare different gene selection approaches, we used the classification accuracy of an SVM classifier as the performance metric. A tenfold cross-validation with an SVM classifier was performed on gene (sub)sets obtained by various methods. Figure 5.1 presents the classification accuracy and number of features for each gene subset on the three test beds. The comparative study of the three test beds showed that gene subsets obtained by different methods all achieved higher classification accuracy than that of a full gene set. Furthermore, optimal search-based gene subset selection often outperformed individual gene ranking. In particular, the optimal search–based wrapper model, GA/SVM, performed best in all three test beds. Interestingly, the marker genes identified by optimal search–based selection methods contain genes that are not among the top genes when ranked individually. This means that gene subset selection can identify marker genes that work collaboratively for cancer distinction. Yet these genes may not be identified by individual ranking. Notably, optimal searches also suffer from high dimensionality, which increases the difficulty and computational cost for a GA to find the optimal solution.

In terms of classification accuracy, wrapper models often outperformed filter models. This is not surprising, because wrappers use classification accuracy as the

```
METH-2
                                   Individual 95% CIs For Mean
                                   Based on Pooled StDev
Gene set      #G    Mean    StDev  -------+---------+---------+---------
Full set     678  92.424   0.689   (--*--)
F-stat        20  94.364   1.809                   (--*--)
GA/MRMR       23  94.424   0.461                   (-*--)
GA/SVM        47  95.697   1.391                              (--*--)
                                   -------+---------+---------+---------
Pooled StDev =     1.229            93.0      94.5      96.0

METH-5
                                   Individual 95% CIs For Mean
                                   Based on Pooled StDev
Gene Set      #G    Mean    StDev  ---------+---------+---------+------
Full set     678  25.891   3.158   (*)
F-stat        40  53.333   2.360                             (*)
GA/MRMR       19  46.124   0.882                    (*)
GA/SVM       156  54.186   3.621                          (*)
                                   ---------+---------+---------+------
Pooled StDev =     2.815             36        48        60

COLON
                                   Individual 95% CIs For Mean
                                   Based on Pooled StDev
Gene set      #G    Mean    StDev  --------+---------+---------+-------
Full set    2000  83.710   2.002   (--*-)
F-stat        70  87.527   0.941                    (-*-)
GA/MRMR        7  86.720   0.694                  (-*-)
GA/SVM       245  90.161   2.133                              (--*-)
                                   --------+---------+---------+-------
Pooled StDev =     1.640            85.0      87.5      90.0
```

FIGURE 5.1 Comparison of gene subsets obtained by various methods. #G, number of genes in the gene set; Mean, mean of classification accuracy; and StDev, standard deviation of classification accuracy.

evaluation criterion, whereas filters do not. However, since wrappers iteratively call a classifier as a subroutine during the feature selection process, they often require even more computational cost than filters require.

5.4 CONCLUSIONS AND DISCUSSION

In this chapter we review the research field of gene selection from microarray data for cancer classification. Based on a framework of gene selection methods, we presented a comparative study of some representative methods (e.g., individual gene ranking, gene subset selection, filter models, and wrapper models). Throughout our literature review and case study we emphasized the high-dimensionality property of microarray data and gene interaction for cancer distinction. Therefore, to achieve global optimal solution, we focused on gene subset selection methods based on optimal search algorithms. The comparative study showcases the advantages of a variety of gene selection methods. In general, individual gene-ranking methods such as GS and F-statistic are commonly used, due to their simplicity and scalability, but cannot capture gene interactions. Gene subset selection methods take into account the genes' group predictive power and can achieve better performance than can be achieved using individual ranking. These methods, especially those based on optimal search algorithms such as GAs, often achieve the better gene subset but require more computational expense.

Studies are going on to identify the optimal subsets of marker genes. Among these gene selection methods for microarray data, many of them are able to identify an effective subset of genes for cancer distinction. For biological researchers, classification accuracy is important in cancer classification but is not the only goal. One critical challenge is to decipher the underlying genetic architecture for cancer distinction (Li et al., 2004b). Studies on analyzing the functions of the marker genes identified will add to the biological significance in the research field of cancer diagnosis and drug discovery.

Acknowledgments

The authors were supported by grant 1 R33 LM07299-01, 2002–2005, from the National Institutes of Health/National Library of Medicine. We thank the Arizona Cancer Center for making the microarray data sets available and for valuable comments.

REFERENCES

Alon, U., N. Barkai, et al. (1999). Broad patterns of gene expression revealed by clustering analysis of tumor and normal colon tissues probed by oligonucleotide arrays. *Proc. Nat. Acad. Sci., USA*, **96**(12): 6745–6750.

Bishop, C. (1995). *Neural Network for Pattern Recognition*. Oxford University Press, New York.

Blum, A. L., and P. Langley (1997). Selection of relevant features and examples in machine learning. *Artif. Intell.*, **97**(1–2): 245–271.

Bø, T. H., and I. Jonassen (2002). New feature subset selection procedures for classification of expression profiles. *Genome Biol.*, **3**(research): 0017.1–0017.11.

Chen, H. C., G. Shankaranarayanan, et al. (1998). A machine learning approach to inductive query by examples: an experiment using relevance feedback, ID3, genetic algorithms, and simulated annealing. *J. Am. Soc. Inf. Sci.*, **49**(8): 693–705.

Chen, X. W. (2003). An improved branch and bound algorithm for feature selection. *Pattern Recognition Lett.*, **24**(12): 1925–1933.

Chow, M. L., E. J. Moler, et al. (2001). Identifying marker genes in transcription profiling data using a mixture of feature relevant experts. *Physiol. Genom.*, **5**: 99–111.

Christianini, N., and J. Shawe-Taylor (2000). *An Introduction to Support Vector Machines.* Cambridge University Press, New York.

Dash, M., and H. Liu (1997). Feature selection for classification. *Intell. Data Anal.*, **1**(3).

Dash, M., and H. Liu (2003). Consistency-based search in feature selection. *Artif. Intell.*, **151**: 155–176.

Ding, C., and H. Peng (2003). Minimum redundancy feature selection from microarray gene expression data. *IEEE Computer Society Conference on Bioinformatics.*

Dudoit, S., J. Fridlyand, et al. (2002). Comparison of discrimination methods for the classification of tumors using gene expression data. *J. Am. Stat. Assoc.*, **97**(457): 77–87.

Ginsberg, M. L. (1993). *Essentials of Artificial Intelligence.* Morgan Kaufmann, San Francisco, CA.

Glover, F., and M. Laguna (1999). *Tabu Search* Kluwer Academic, Norwell, MA.

Golub, T. R., D. K. Slonim, et al. (1999). Molecular classification of cancer: class discovery and class prediction by gene expression monitoring. *Science*, **286**(5439): 531–537.

Guyon, I., and A. Elisseeff (2003). An introduction to variable and feature selection. *J. Mach. Learn. Res.* **3**(Mar.): 1157–1182.

Guyon, I., J. Weston, et al. (2002). Gene selection for cancer classification using support vector machines. *Mach. Learn.*, **46**(1–3): 389–422.

Hall, M. A. (2000). Correlation-based feature selection for discrete and numeric class machine learning. *17th International Conference on Machine Learning.*

Holland, J. H. (1975). *Adaptation in Natural and Artificial Systems.* University of Michigan Press, Ann Arbor, MI.

Kira, K., and L. Rendell (1992). The feature selection problem: traditional methods and a new algorithm. *10th National Conference on Artificial Intelligence.*

Kirkpatrick, S., C. D. J. Gelatt, et al. (1983). Optimization by simulated annealing. *Science*, **220**(4598): 671–680.

Kohavi, R., and G. H. John (1997). Wrappers for feature subset selection. *Artif. Intell.*, **97**(1–2): 273–324.

Li, L. P., C. R. Weinberg, et al. (2001). Gene selection for sample classification based on gene expression data: study of sensitivity to choice of parameters of the GA/KNN method. *Bioinformatics*, **17**(12): 1131–1142.

Li, T., C. L. Zhang, et al. (2004a). A comparative study of feature selection and multiclass classification methods for tissue classification based on gene expression. *Bioinformatics*, **20**(15): 2429–2437.

Li, X., S. Rao, et al. (2004b). Gene mining: a novel and powerful ensemble decision approach to hunting for disease genes using microarray expression profiling. *Nucleic Acids Res.*, **32**(9): 2685–2694.

Liu, H., and R. Setiono (1996). Feature selection and classification: a probabilistic wrapper approach. *9th International Conference on Industrial and Engineering Applications of AI and ES.*

Liu, H., J. Li, et al. (2002). A comparative study of feature selection and classification methods using gene expression profiles and proteomic patterns. *Genome Inf.*, **13**: 51–60.

Lu, Y., and J. W. Han (2003). Cancer classification using gene expression data. *Inf. Syst.*, **28**(4): 243–268.

Marchevsky, A. M., J. A. Tsou, et al. (2004). Classification of individual lung cancer cell lines based on DNA methylation markers: use of linear discriminant analysis and artificial neural networks. *J. Mol. Diagn.*, **6**(1): 28–36.

Mendenhall, W., and T. Sincich (1995). *Statistics for Engineering and the Sciences.* Prentice Hall, Upper Saddle River, NJ.

Model, F., P. Adorján, et al. (2001). Feature selection for DNA methylation based cancer classification. *Bioinformatics*, **17**: 157–164.

Narendra, P. M., and K. Fukunaga (1977). A branch and bound algorithm for feature selection. *IEEE Trans. Comput.*, **C-26**(9): 917–922.

Ooi, C. H., and P. Tan (2003). Genetic algorithms applied to multi-class prediction for the analysis of gene expression data. *Bioinformatics*, **19**(1): 37–44.

Pudil, P., J. Novovicova, et al. (1994). Floating search methods in feature-selection. *Pattern Recognition Lett.*, **15**(11): 1119–1125.

Quinlan, J. (1993). *C4.5: Programs for Machine Learning.* Morgan Kaufmann, San Francisco, CA.

Saeys, Y., S. Degroeve, et al. (2003). Fast feature selection using a simple estimation of distribution algorithm: a case study on splice site prediction. *Bioinformatics*, **19**(Suppl. 2): ii179–ii188.

Siedlecki, W., and J. Sklansky (1989). A note on genetic algorithms for large-scale feature-selection. *Pattern Recognition Lett.*, **10**(5): 335–347.

Xiong, M. M., X. Z. Fang, et al. (2001). Biomarker identification by feature wrappers. *Genome Res.*, **11**(11): 1878–1887.

6

PATIENT SURVIVAL PREDICTION FROM GENE EXPRESSION DATA

HUIQING LIU

Department of Biochemistry and Molecular Biology, University of Georgia, Athens, Georgia

LIMSOON WONG

School of Computing, National University of Singapore, Singapore

YING XU

Department of Biochemistry and Molecular Biology, University of Georgia, Athens, Georgia

A patient outcome study is used to analyze the time required for an event of interest to occur. This event is usually a clinical prognosis; and it can be either unfavorable or favorable, such as death, recurrence of a tumor, cancer distant metastasis, transplant rejection, restoration of renal function, or discharge from a hospital. Among these events, survival status is the most important, especially for cancer patients.

In current practice, medical doctors assess the risk profile of a cancer patient based primarily on various clinical characteristics. For example, the prognostic indicators used to identify the appropriate therapy for acute myeloid leukemia patients include age, cytogenetic findings, white-cell count, and the presence or absence of recurrent cytogenetic aberrations [7]. However, these factors do not fully reflect the molecular heterogeneity of the disease, and treatment stratification is difficult. Another example is diffuse large-B-cell lymphoma (DLBCL), the most common type of adult lymphoma. An international prognostic index (IPI) has been established to predict the outcome of DLBCL patients [22], which is based on five clinical factors: age, tumor stage, performance status, lactate dehydroginase levels, and number of extranodal disease sites. Although these factors are known to be strongly related to patient survival, it is not unusual that patients with DLBCL who have identical IPI values have considerably different survival outcomes [21]. In fact, in many cases, patients with a similar diagnosis respond very differently to the same treatment. Again, taking diffuse large-B-cell lymphoma as an example, only 40% of patients respond well to standard chemotherapy and thus have prolonged survival, while the others die of the disease [1]. With more understanding of molecular variations of cancers, people perceive that at the molecular level, two similar tumors might be completely different diseases. Therefore, traditional cancer diagnosis and treatment based on microscopic analysis are not accurate enough to determine patient outcomes. In other words, prediction models incorporating disease genetic information are desired.

One of the important recent breakthroughs in experimental molecular biology is microarray technology. With this novel and powerful technology, we are able to collect and record the expression levels in cells for thousands of genes simultaneously. Then a question is raised: Is there any relationship between gene expression profiles and patient survival? Here is an excellent story. By analyzing gene expression profiles, Alizadeh and colleagues identified two molecularly distinct subtypes of diffuse large-B-cell lymphoma that possess different gene expression patterns related

FIGURE 6.1 Genes selectly expressed in germinal-center B-like DLBCL and activated B-like DLBCL. (From ref. 1, supplementary material.)

to different stages of B-cell differentiation [1]. One type had high levels of expression of genes, with characteristics of normal germinal-center B cells (termed *germinal-center B-like DLBCL*), while another type expressed genes induced primarily during in vitro activation of peripheral blood B cells (*activated B-like DLBCL*). Most genes that are highly expressed in germinal-center B-like DLBCL were found to have low or undetectable levels in activated B-like DLBCL. Similarly, most genes that defined activated B-like DLBCL were not expressed in normal germinal-center B cells. Figure 6.1 shows the genes expressed selectively in these two genetic subtypes of DLBCL. One significance of this identification is that the overall survival of patients with germinal-center B-like DLBCL was markedly longer than that of patients with activated B-like DLBCL. Therefore, the different genetic profiles of these two different DLBCLs may help explain why current therapy benefits some patients but has no effect on others. In this chapter we introduce a number of methods for predicting cancer patient survival based on gene expression data.

6.1 GENERAL METHODS

To describe an outcome event for each patient, it is necessary to have two types of information available: follow-up time (such as years, months, or days) and status at follow-up (such as alive or dead). Since in an actual clinical trial, it often takes several years to accumulate patient data for the trial, patients being followed for survival will have different start times. Then the patients will have various follow-up times when the results are analyzed at one time. For example, in a study of lung adenocarcinomas [5], the follow-up time of patients varies from 1.5 to 110.6 months. One special characteristic of survival data is that the event will probably not have happened for all patients at the end of the follow-up time. A survival time is said to be *censored* if the event had not occurred on the patient by the end of the observation period. Other possible reasons for censoring include the fact that patients may be lost to follow-up during the study or may have died from other diseases or illness. Although patients in a study may vary as to follow-up time, we assume that the survival prognosis of all the patients stays the same throughout the study.

6.1.1 Kaplan–Meier Survival Analysis

A *Kaplan–Meier plot* estimates survival over time—each point on the plot indicates the probability of being event-free (in other words, still being "at risk") at the corresponding point in time [3,14]. Because of censoring, we cannot simply calculate the probability of the proportion of patients in remission for each time point. For example, an alive patient with two years of follow-up time should contribute to the survival data for the first two years of the curve but not to the portion of the curve that follows. Thus, the data for this patient should be removed from the curve at the end of two years of follow-up time. The Kaplan–Meier survival curve is a type of stairstep plot. On such a curve, when a patient dies, the curve will take a step down at the corresponding time point. The step magnitude gets larger along the time axis due to both dead and censored patients. When a patient is censored, the curve does not take a step down; instead, a tick mark is generally used to indicate where a patient is censored (Figure 6.2*a*), and the proportion of patients who are still "at risk" is thus affected. An alternative way to indicate a censored patient is to show the number of remaining cases at risk at several time points. Patients who have been censored or for whom the outcome event happened before the time point are not counted as at risk.

Let n_t be the number of patients being at risk in the beginning of time point t (e.g., the tth month) and m_t be the number of patients who die (for whom an event has happened) at time point t. The probability of survival for these n_t patients given that

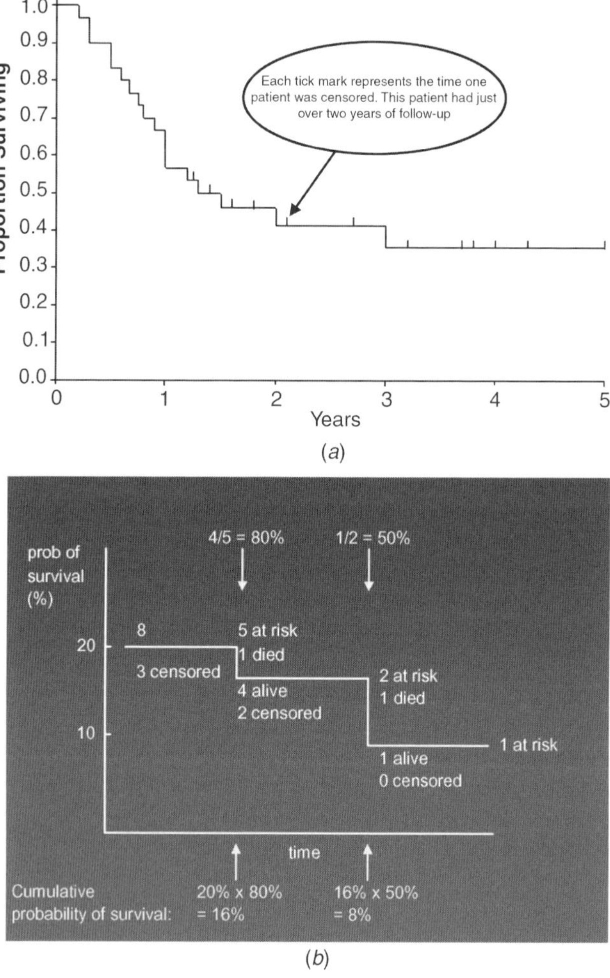

FIGURE 6.2   Samples of Kaplan–Meier survival curves. Part (*a*) is an example of a Kaplan–Meier survival curve. This group of patients has a minimum follow-up of a little over a year. Part (*b*) illustrates how to calculate the fraction of survival at a time.

m_t of them will die can be estimated as

$$p_t = \frac{n_t - m_t}{n_t} \tag{6.1}$$

and the probability of being at risk at time t, known as the *survival function* at t [denoted as $S(t)$], can be estimated as [31]

$$\hat{S}(t) = \prod_{k=1}^{t} p_k \tag{6.2}$$

The standard error in the estimation of the survival function is generally calculated using Greenwood's formula [13]:

$$\delta_{\hat{S}(t)}^2 = \hat{S}(t)^2 \sum_{k=1}^{t} \frac{m_k}{n_k(n_k - m_k)} \tag{6.3}$$

Therefore, a 95% confidence interval for $S(t)$ can be estimated by $S(t) \pm 1.96\delta_{\hat{S}(t)}$.

Shown in Figure 6.2*a* is a complete sample of Kaplan–Meier survival curve with a tick mark representing a censored patient (captured from http://www.cancerguide. org/scurve_km.html), and Figure 6.2*b* illustrates how to calculate the fraction of survival at one time (captured from http://hesweb1.med.virginia.edu/biostat/ teaching/shortcourse/km.lam.pdf).

The Kaplan–Meier plot can also be used to derive an estimation of the median survival time. If there is no censoring, the median survival time can be defined as the time when 50% of the patients studied have experienced the outcome event. In the case of censoring, the median survival time is estimated by the earliest time at which the Kaplan–Meier curve drops below 0.50. Similarly, the qth quantile is estimated as the earliest time at which the curve drops below $1 - q$ [31].

In many cases, Kaplan–Meier curves are used to show the survival characteristics between patient groups for survival prediction studies. The primary statistical tool for this purpose is the *log-rank test*, which is equivalent to the *Mantel–Haenszel test* if there are only two patient groups. The *null hypothesis* for this test is that "the survival distributions being compared are equal at all follow-up times" [31]. This hypothesis implies that Kaplan–Meier curves for different risk groups coincide with each other. The log-rank test calculates a χ^2 statistic with an associated p-value.

6.1.2 Cox Proportional-Hazards Regression

Although the Kaplan–Meier method with the log-rank test is useful for comparing survival curves in two or more groups, people are always eager to know what risk factors are associated with the outcome (i.e., covariates), and how to combine them to make an outcome prediction.

To address this problem, the Cox proportional-hazards regression model [9] is widely used. This regression model for survival data can provide an estimation of the *hazard ratio* and its confidence interval. The hazard ratio is the ratio of the hazard rate of the two groups studied. Here, a hazard rate is the short-term event rate for patients in the group who have not yet experienced the outcome event [31]. This short-term event rate represents an instantaneous rate if the time interval is very short.

The proportional hazards regression model is given by

$$h(t|X) = h_0(t)\exp(x_1\beta_1 + \cdots + x_k\beta_k) \tag{6.4}$$

where $X = x_1, \ldots, x_k$ denotes a vector containing a collection of predictor variables (i.e., risk factors); $\beta = (\beta_1, \ldots, \beta_k)^{\mathrm{T}}$ is the vector of regression coefficients; $h_0(t)$ is the baseline hazard at time t, representing the hazard for a subject (i.e., patient) with the value 0 for all the predictor variables; and $h(t|X)$ is the hazard at time t for a patient given the covariant value X. The model equation (6.4) implies that the contributions of individual predictors to the hazard can be modeled using a multiplicative model, and there is a log-linear relationship between the predictors and the hazard function. $h(t|X)/h_0(t)$ is called the *hazard ratio* and

$$\ln\frac{h(t|X)}{h_0(t)} = x_1\beta_1 + \cdots + x_k\beta_k \tag{6.5}$$

Therefore, given two subjects with different values for the independent predictor variables, the hazard ratio for those two subjects does not depend on time. The Cox model is a semiparametric model, because (1) no assumption is made about the shape of the baseline hazard $h_0(t)$ and it can be estimated nonparametrically; however, (2) the effect of the risk factors on the hazard is measured in a parametric manner. Thus, the Cox model is considered more robust than parametric proportional hazards models [31].

In the Cox model, the estimation of the vector of regression coefficients β can be obtained through maximizing the *partial likelihood* [9]. When there are no ties in the data set (i.e., no two subjects have the same event time), partial likelihood is given by

$$L(\beta) = \prod_{i=1}^{n}\frac{\exp(X_i\beta)}{\sum_{j\in R(t_i)}\exp(X_j\beta)} \tag{6.6}$$

where $R(t_i) = j : t_j \geq t_i, j = 1, \cdots, k$ is the set of subjects who were at risk at time t_i, and n is the total number of subjects in the group. We can easily obtain the log partial likelihood:

$$l(\beta) = \ln L(\beta) = \sum_{i=1}^{n}X_i\beta - \sum_{i=1}^{n}\ln\left[\sum_{j\in R(t_i)}\exp(X_j\beta)\right] \tag{6.7}$$

The estimation of β can therefore be made by calculating the derivatives of $l(\beta)$ (with respect to β). For more details on this estimation and the case with ties in the event time, we refer readers to ref. 9.

With the Cox model and microarray features of each patient, we can measure the correlation between the expression level of genes and numerous clinical parameters, including patient survival. To assess whether an individual microarray feature is associated with a patient's outcome, we use the proportional hazards regression above as a *univariate model* to calculate a regression coefficient for each gene. It should be noted that the vector of risk factors contains only one gene [i.e., $k = 1$ in Eq. (6.4)]. We call the estimated regression coefficient the *Cox score* of this gene. A positive Cox score is associated with a gene for which high values are correlated with a low likelihood of survival; a negative Cox score is associated with a gene for which high values are correlated with a high likelihood of survival. Generally, only genes with an absolute Cox score value that exceeds a certain threshold will be considered relevant to a patient's survival and will be used in later analysis. After identifying a group of genes with the most predictive power to the survival of the patient studied by training data samples, a *risk index* can be defined as a linear combination of the expression levels of the genes selected, weighted by their Cox scores. In the next section we describe some applications of using the Cox model and other methods to predict a patient's survival from the gene expression data.

6.2 APPLICATIONS

In this section we show two applications of feeding patients microarray features into a Cox proportional-hazards regression model to find genes with strong predictive power to survival status and put these genes into a prediction framework.

6.2.1 Diffuse Large-B-Cell Lymphoma

Diffuse large-B-cell lymphoma is a cancer of B-cells (lymphocytes) that normally reside in the lymphatic system. It accounts for 30 to 40% of adult non-Hodgkin's lymphoma, with more than 25,000 cases diagnosed annually in the United States [21]. Current treatments rely primarily on combination chemotherapy, having a success rate lower than 50%.

Recently, Rosenwald and his colleagues studied the effect of molecular features on the survival of patients with diffuse large-B-cell lymphoma after chemotherapy by analyzing the gene expression levels of the tumor-biopsy specimens collected from 240 DLBCL patients [23]. The expression levels were measured using *lymphochip* DNA microarrays, a gene chip composed of genes whose products are expressed preferentially in lymphoid cells and genes believed or known to play a role in cancer or immunal function [1,23]. There were 12,196 cDNA clones on the microarrays by which the expression of mRNA in the tumor can be measured. All patients had received anthracycline-based chemotherapy, and the clinical information shows that [23] (1) the median follow-up times were 2.8 years overall and 7.3 years for survivors; (2) 57% patients died during this period; (3) the median age of the patients was 63 years and 56% were men; and (4) according to the Ann Arbor classification, 15% of patients had stage I disease, 31% stage II, 20% stage III, and 34% stage IV. The entire patient sample set was divided into two groups, a preliminary group containing 160 samples and a validation group containing the remaining 80 samples.

Using the survival information of the patients in the preliminary group, the Cox score for an individual gene was obtained by applying the univariate Cox proportional-hazards model to each microarray feature. Six hundred seventy genes were thus observed to be associated with a good or a bad prognosis at a significant level of 0.01 in a *Wald test*—162 were associated with a favorable outcome and 508 were associated with an unfavorable outcome. To determine the significant of the genes associated with the outcome, a permutation test is designed [23].

1. Within the preliminary group, the association between the gene expression level of samples from individual patients and the overall survival is permuted using a random-number generator.
2. Each gene is fed into a univariate Cox proportional-hazards model to assess the gene's association with either a good or a bad prognosis.
3. Genes are selected at a significant level of $p < 0.01$ in a Wald test.
4. Steps 1 to 3 are repeated 4000 times.

It was noted that only 20 out of 4000 times were as many significant genes found as were found in unpermutated data; hence, the p-value is 0.005 $(= 20/4000)$. The 670 genes identified were further assigned into four gene-expression signatures formed in advance by hierarchical clustering. A gene-expression signature is a group of genes expressed in a specific cell lineage or stage of differentiation or during a particular biological response [23]. By this definition it is likely that genes within the same gene-expression signature are related to similar biological aspects of a tumor. Therefore, to minimize the number of genes in the final outcome predictor, instead of using all 670 genes, only 17 of them were picked as the representatives: three from a *germinal-center B-cell gene-expression signature*, four from an *MHC (major-histocompatibility-complex) class II signature*, six from a *lymph-node signature*, and three from a *proliferation signature*, as well as one individual gene (BMP6) which was not in the four signatures but had a high predictive power on the outcome. The expression values of the genes belonging to the same signature were averaged and the value was put back to the univariate Cox proportional-hazards model to learn a new Cox score for this signature. The final prediction model was constructed as a weighted sum of expression levels of the four gene-expression signatures and BMP6. The formula is (0.241 × average value of the proliferation signature) + (0.310 × value for BMP6) − (0.290 × average value of the germinal − center B-cell signature) − (0.311 × average value of the MHC class II signature) − (0.249 × average value of the lymph-node signature). The higher the score, the poorer the outcome the patient will have.

With the learned prediction model, a risk score was calculated for each DLBCL patient in both the preliminary and validation groups. Within each group, patients were then ranked according to their risk score and divided into quartiles from the highest to the lowest scores. There are 40 and 20 patients in each of the quartiles for

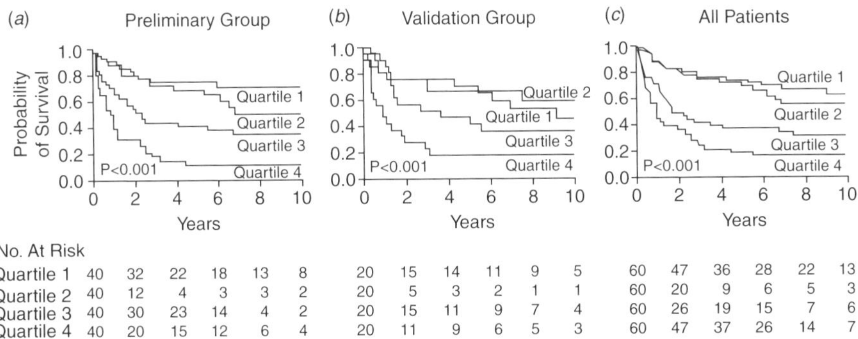

Quartile 1	40	32	22	18	13	8	20	15	14	11	9	5	60	47	36	28	22	13
Quartile 2	40	12	4	3	3	2	20	5	3	2	1	1	60	20	9	6	5	3
Quartile 3	40	30	23	14	4	2	20	15	11	9	7	4	60	26	19	15	7	6
Quartile 4	40	20	15	12	6	4	20	11	9	6	5	3	60	47	37	26	14	7

FIGURE 6.3 Kaplan–Meier survival curves for patients with diffuse large-B-cell lymphoma: (*a*) for 160 patients in the preliminary group; (*b*) for 80 patients in the validation group; (*c*) for all 240 patients. (From ref. 23.)

the preliminary and validation groups, respectively. The Kaplan–Meier plots of overall survival in Figure 6.3 showed distinct differences in five-year survival rates in the various quartiles in both groups (all with $p < 0.001$). Considering all 240 patients [Figure 6.3*c*], the five-year survival rates were 73% in quartile 1, 71% in quartile 2, 34% in quartile 3, and 15% in quartile 4.

In this study, the gene expression–based risk scores were also compared with IPI scores that were calculated on pure clinical factors. The comparison demonstrated that they are two independent predictors. On the other hand, IPI provides little insight into disease biology since it is built solely on the consideration of clinical data [16].

6.2.2 Lung Adenocarcinoma

Most patients with non-small-cell lung cancer present with advanced disease, and the overall 10-year survival rate remains at only 8 to 10% [5]. Even for patients clinically diagnosed with stage I disease who received surgical treatment would have a 35 to 50% chance of relapse within five years. Currently, there is no effective way to identify specific high-risk patients in clinical practice.

Recently, Beer and his colleagues [5] built a model based on gene expression profiles to predict survival for patients with primary lung adenocarcinoma, which is the the major histological subtype of non-small-cell lung cancer. Their lung adeno-carcinoma data set contained 86 primary lung adenocarcinomas, including 67 stage I and 19 stage III tumors. Kaplan–Meier curves and log-rank tests showed poorer survival of patients with stage III adenocarcinomas than of those with stage I.

The microarray data used in this study were generated by oligonucleotide arrays. To identify genes that may be associated with survival, the following two approaches were implemented. The first used an equal training–testing strategy, in which 86 tumor samples were split into a 43-sample training set and a 43-sample testing set with an equal number of randomly assigned stage I and stage III samples. Using the microarray data of training samples, the top 10, 20, 50, or 75 genes identified by the univariate Cox proportional-hazards regression model were selected to construct a risk predictor model as a linear combination of the expression levels of these genes weighted by their Cox score (i.e., coefficient learned from the model). Predictors established were then evaluated by setting different percentile (50th, 60th, 70th) cutoff points to assign patients to high- or low-risk groups. A similar performance was achieved across different cutoff points, and the 50-gene risk predictor was chosen since it had the best overall association with survival in the training set. A p-value of 0.024 was obtained when this predictor was applied to the testing set (using the 60th percentile as the cutoff point). Notably, 11 stage I adenocarcinomas were assigned to the high-risk group.

The second strategy was based on leave-one-out cross-validation. A 50-gene predictor was formed from each training set and evaluated by assigning the held-out test case to either the high- or low-risk group. The result yielded a p-value of 0.0006. Further analysis of the gene-based survival predictor conditionally for stage progression showed that this predictor is independent of disease stage.

6.2.3 Remarks

In both diffuse large-B-cell lymphoma and lung adenocarcinoma analyses, hierarchical clustering was applied to the entire data set of a subgroup of samples. As described earlier, a previous study of DLBCL [1] had shown different prognoses between patients with germinal-center B-cell-like and activated B-cell-like DLBCL. To confirm this, by using the same 100 genes identified in ref. 1, samples in this DLBCL study were clustered into three subgroups: the two known subgroups and a new type 3 subgroup. As expected, overall survival after anthracycline-based chemotherapy predicted by the model using 17 genes differed remarkably among the three subgroups [23]: patients with germinal-center B-cell-like DLBCL had a five-year survival rate of 60%, whereas patients with activated B-cell-like and type 3 DLBCL had rates of 35% and 39%, respectively.

In the lung adenocarcinoma study, the clustering algorithm also identified three tumor subclasses which showed significant differences in terms of tumor stage and tumor differentiation [5]. This finding is consistent with our thinking that tumors with similar histological features of differentiation may have similar gene expression patterns. However, on the other hand, in subclass 3, which contained the largest number of stage III tumors, there were also quite a number of stage I lung adeno-carcinomas. This implied that some stage I tumors were having similar gene expression profiles with tumors at more advanced stages. It is also noteworthy that 10 of the 11 stage I lung adenocarcinomas in subclass 3 were assigned to the high-risk group by the gene expression–based risk prediction model in the leave-one-out cross-validation.

6.3 INCORPORATING DATA MINING TECHNIQUES TO SURVIVAL PREDICTION

All in all, the essential steps of survival prediction using microarray data are *gene identification* and *prediction model construction*. These two problems fit very well into the scope of feature selection and classification in data mining research. Other than the Cox regression model, there are many methods that can be used to address these two issues. Recently, many good results have been achieved by applying various data mining techniques (particularly, clustering, feature selection, and classification) to cancer phenotype discovery and classification from microarray data [2,6,10–12, 17,32]. These successes have greatly encouraged researchers to use similar methodologies on patient survival prediction.

6.3.1 Gene Selection by Statistical Properties

Statistical tests play an important role in the process of feature selection in data mining. The t-test, χ^2 measure, Fisher criterion score, and Wilcoxon rank sum test are general ways to assist the detection of signal features from noises. For an extensive review of these techniques and their applications in the biological domain, we refer readers to ref. 18.

In addition to the description in Section 6.2.1, there have been several other survival studies for diffuse large-B-cell lymphoma on different data sets. For example, Shipp et al. [24] constructed a 13-gene predictive model from a gene expression profile of 58 patients generated by oligonucleotide microarrays. In the study they first categorized the patients into two classes, cured versus fatal/refractory, and then applied a statistical measure called *signal-to-noise* to assess the correlation between genes and patient outcome.

Similar to the classical t-statistic, the signal-to-noise statistic is constructed to test the difference between means of two groups of independent samples. So if samples in different classes are independent, the statistic can be used to find features that have large differences in means between the two classes. These features can then be considered to have the ability to separate samples between classes. Given a data set consisting of samples in two classes, $\mathcal{A}$ and $\mathcal{B}$ (such as *tumor* vs. *normal* or *favorable outcome* vs. *unfavorable outcome*), the signal-to-noise statistic of a feature g, $s(g)$, is constructed as

$$s(g) = \frac{\mu^{\mathcal{A}}(g) - \mu^{\mathcal{B}}(g)}{\delta^{\mathcal{A}}(g) + \delta^{\mathcal{B}}(g)} \tag{6.8}$$

where $\mu^{\mathcal{A}}(g)$ [respectively, $\mu^{\mathcal{B}}(g)$] and $\delta^{\mathcal{A}}(g)$ [respectively, $\delta^{\mathcal{B}}(g)$] are the mean and standard deviation of feature g calculated on the samples in class $\mathcal{A}$ (respectively, $\mathcal{B}$). A large positive value of $s(g)$ indicates strong correlation with class $\mathcal{A}$, whereas a large negative value of $s(g)$ indicates strong correlation with class $\mathcal{B}$. This feature selection method has been proposed to find discriminatory genes that can distinguish tumor cells from normal cells [11,12,25].

To construct a classification model, Shipp and his colleagues [24] employed a weighted-voting algorithm to separate patients with curable disease from those with fatal or refractory disease. This weighted voting scheme is a combination of multiple "univariate" classifiers. They defined in detail $a_g = s(g)$ (reflects the correlation between the expression levels of gene g and the class distinction), and $b_g = [\mu^{\mathcal{A}}(g) + \mu^{\mathcal{B}}(g)]/2$ (the average of the mean expression values in the two classes) [12], $s(g)$ was the signal-to-noise measure of gene g as described above. When doing prediction for a new sample T, let e_g denote the expression value of gene g in the sample. The vote of gene g was determined by $V_g = a_g(e_g - b_g)$, with a positive value representing a vote for class $\mathcal{A}$ and a negative value representing a vote for class $\mathcal{B}$. The total vote for class $\mathcal{A}$ was obtained by adding up the absolute values of the positive votes over the informative genes selected, while the total vote for class $\mathcal{B}$ was obtained by adding up the absolute values of the negative votes. By running a leave-one-out cross-validation test, they constructed survival predictors for each validation iteration by selecting 8 to 16 genes, and finalized a 13-gene model that produced the highest classification accuracy. Although 13 different genes were selected in each validation loop, a large percentage of the genes were in common. By using the classification model established, each of the 58 patients was predicted to have either a curable or a fatal/refractory disease. Kaplan–Meier survival curves drawn for these two groups indicated that the patients predicted to be curable had remarkably improved long-term survival compared with those predicted to have fatal/ refractory disease—the p-value for the log-rank test is 0.00004 and the five-year survival rate was 70% versus 12%. Other learning algorithms, such as *support vector machines* and *k-nearest neighbors*, were also tried and obtained similar good results.

In summary, the strategy used in this study is to (1) categorize patients into several classes according to their survival information, (2) identify survival-related genes by using a statistic test embedded in a cross-validation process, (3) construct a learning model based on the genes selected, and (4) validate a model via Kaplan–Meier survival analysis.

6.3.2 Cancer Subtype Identification via Survival Information

As shown in Section 6.2.3, patients in different cancer subtypes identified through gene expression profiles may have different prognoses. However, when applying unsupervised learning techniques such as hierarchical clustering to identify tumor subclasses purely from microarray data, the cancer subtypes observed may be unrelated to patient survival because we do not use any of the clinical information about the patient. If patients in different subtypes of certain cancers have a similar prognosis, the information regarding such cancer categorization will be very limited [4].

To achieve the goal of discovering subtypes of cancer that are both *clinically relevant* and *biologically meaningful*, we should let the known patient survival information guide the exploration. In practice, Bair and Tibshirani [4] proposed a semisupervised procedure combining the Cox model, clustering, principal components analysis, and some other techniques. Their main idea was implemented through the following steps:

1. Calculate a univariate Cox proportional-hazards score for each gene.
2. Select genes whose Cox score (absolute value) exceeds a certain threshold.
3. Perform clustering algorithms on the genes selected to identify subgroups of patients with similar expression profiles.
4. Use supervised classification methods to classify future patients into the appropriate subgroup identified.

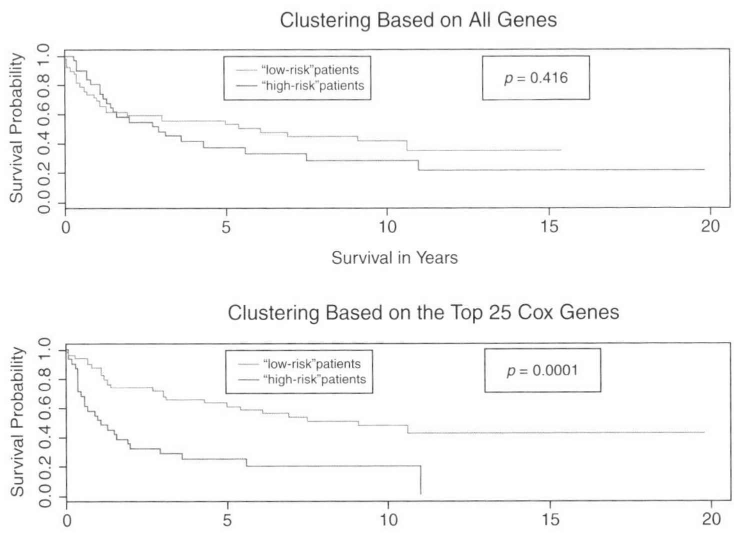

FIGURE 6.4 Kaplan–Meier survival curves for patients with diffuse large-B-cell lymphoma. Plots were drawn on samples in the validation group. (From ref. 4, Figure 3; data from ref. 23.)

When their method was applied to the diffuse large-B-cell lymphoma data described in ref. 23 (see also Section 6.2.1), Bair and Tibshirani selected from 160 training samples the top 25 genes based on their Cox scores and used only these genes to cluster 80 validation samples into two groups. The log-rank statistic comparing the survival times of patients in the two groups was highly significant, with $p = 0.0001$. The Kaplan–Meier curve of this test, as well as the curve obtained by applying clustering to all the genes, are illustrated in Figure 6.4.

To predict survival for future cases, one can train any classifier on samples in the various risk groups formed. At this step, Bair and Tibshirani chose to build a *nearest shrunken centroid model* [4]. The nearest shrunken centroid method computes a standardized centroid for each class—the average gene expression for each gene in each class divided by the within-class standard deviation for that gene. This standardization has the effect of giving higher weight to genes whose expression is stable within samples of the same class [26]. The main feature of the nearest shrunken centroid classification from the nearest standard centroid classification is that it "shrinks" each of the class centroids toward the overall centroid for all classes by an amount called the *threshold.* The selection of the threshold can be determined by the results of cross-validation for a range of candidate values. When classifying a new sample, it follows the usual nearest centroid rule, but using the shrunken class centroids. The idea of shrinkage has two advantages: (1) it achieves better performance by reducing the effect of noisy genes, and (2) it accomplishes automatic gene selection. In particular, if a gene is shrunk to zero for all classes, it will be removed from further consideration. The nearest shrunken centroid method is the core technique of PAM (prediction analysis for microarrays), class prediction software for genomic expression data mining developed at Stanford University (http://www-stat.stanford.edu/~tibs/PAM/). PAM has been

TABLE 6.1   General Information on the Data Sets Used in Ref. 4 (Excluding the DLBCL Data of Ref. 23) to Test a Proposed Semisupervised Method for Patient Survival Prediction

Data Set	No. of Genes[a]	No. of Samples	No. of Training	No. of Testing	Ref.
Breast cancer	4751	97	44	53	[30]
Lung cancer	7129	86	43	43	[5][b]
AML	6283	116	59	57	[7]

[a]Figures in this column do not indicate the number of genes produced by the corresponding microarray chips but the number of genes selected by the paper noted in the last column.
[b]Refer also to the description in Section 6.2.2.

applied to perform classification tasks on several DNA microarray data sets [26,27], such as the small round blue cell tumor data of childhood [15], diffuse large-B-cell lymphoma [1], and acute lymphoblasic and acute myeloid leukemia [12].

When this classification algorithm was tested on the DLBCL data of [24], Bair and Tibshirani trained a nearest shrunken centroid classification model on a clustering based on 343 genes that were observed to have the highest accuracy on cross-validation within 160 training samples. Then they applied the model to assign each of the 80 validation samples to one of the two risk groups. The p-value of the log-rank test comparing the two corresponding survival curves is 0.00777 [4].

Besides the diffuse large-B-cell lymphoma data of ref. 24, survival prediction for other tumors was also studied using the methodology described above [4], including breast cancer, lung cancer, and acute myeloid leukemia (AML). In Table 6.1, general information is given for each data set. To compare the predictive power of the semisupervised method proposed, results of using clinical data alone were also obtained for the same data sets. To generate class labels using only clinical information, Bair and Tibshirani created two classes, high-risk and low-risk, by cutting the survival times at the median survival time of the training samples. A training sample was assigned to the high-risk class if the corresponding patient died before the median survival time; otherwise, it is assigned to the low-risk class. With training samples in these two classes, a nearest shrunken centroid model could be obtained similarly (using cross-validation) and applied to the testing data. This method was named the *median cut*. Table 6.2 displays for each method and each data set the p-value of the

TABLE 6.2 p-Values of the Log-Rank Test Obtained by Various Prediction Methods Proposed in Ref. 4 on DLBCL Data of Ref. 24, Breast Cancer Data of Ref. 30, Lung Cancer Data of Ref. 5, and the Acute Myeloid Leukemia Data of Ref. 7

Method	DLBCL	Breast Cancer	Lung Cancer	AML
Unsupervised[a]	0.286	—	0.33	0.803
Median cut	0.0297	0.00423	0.00162	0.0487
Semisupervised	0.00777	0.000127	0.0499	0.0309

[a]An unsupervised method here is to apply a two-mean clustering algorithm to all genes provided to identify two classes from training samples.

log-rank statistic associated with Kaplan–Meier survival curves drawn on the test samples in the established high-risk versus low-risk group. The data are part of Table 2 of ref. 4.

6.4 SELECTION OF EXTREME PATIENT SAMPLES

So far, all the methods described were proposed to deal with gene selection and predictor modeling. In this section we show how the selection of training samples can help improve the power of prediction.

6.4.1 Short- and Long-Term Survivors

Since the first aim of our survival study is to analyze the relationship between gene expression profiles and patient outcome, two types of extreme cases—*short-term survivors*, who got an unfavorable outcome within a short period, and *long-term survivors*, who maintained a favorable outcome after a long follow-up time—should be more valuable than those in the medium-term range [19]. In other words, a reliable prediction is not expected to emerge from analyzing data of patients who are censored with a short follow-up time. So it is reasonable to include in the training data only short- and long-term survivors.

Formally, for an experimental sample T, if its follow-up time is $F(T)$ and the status at the end of the follow-up time is $E(T)$, then [20]

$$T \text{ is } \begin{cases} \text{short-term survivor} & \text{if } F(T) < c_1 \wedge E(T) = 1 \\ \text{long-term survivor} & \text{if } F(T) > c_2 \\ \text{others} & \text{otherwise} \end{cases} \tag{6.9}$$

where $E(T) = 1$ stands for "dead" or an unfavorable outcome, $E(T) = 0$ stands for "alive" or a favorable outcome, c_1 and c_2 are two thresholds for survival time for selecting short- and long-term survivors, respectively. Note that long-term survivors also include those patients who died after a very long period of time. The two thresholds, c_1 and c_2, can vary from disease to disease and from data set to data set. The basic guideline for the selection of c_1 and c_2 is that the training data selected should contain enough training samples for learning algorithms; generally, it is required that each class have at least 10 samples, and the total number of extreme samples be between one-fourth and one-half of all samples available [20].

6.4.2 SVM-Based Risk Scoring Function

After choosing these two types of extreme informative training samples, one can apply any feature selection algorithm to identify genes whose expression levels can be used to separate two types of samples. With the samples and genes selected, a classification model can be built and used for future prediction. Here we introduce a SVM-based scoring function to estimate the survival risk for patients [20]. In our case, the SVM regression function $G(T)$ is a linear combination of the expression values of the genes identified:

$$G(T) = \sum_i \alpha_i y_i K(T, x(i)) + b \tag{6.10}$$

where the vectors $x(i)$ are the support vectors (samples), y_i are the class labels (1 and -1 used here) of $x(i)$, the vector T represents a test sample, and α_i and b are numerical parameters can be determined from the training data.

We map the class label "short-term survivors" to 1 and "long-term survivors" to -1. If $G(T) > 0$, the sample T is more likely to be a short-term survivor. If $G(T) < 0$, the sample T is more likely to be a long-term survivor. To transform the output of $G(T)$ to probability-like values, a standard sigmoid function $S(T)$ is defined as

$$S(T) = \frac{1}{1 + e^{-G(T)}} \tag{6.11}$$

$S(T)$ falls in the range (0,1). Also note that the smaller the $S(T)$ value, the better outcome the corresponding patient T will have. $S(T)$ is termed the *risk score* of T.

If one categorizes patients only into high- or low-risk groups, the value 0.5 is a natural cutoff for $S(T)$. In other words, if $S(T) > 0.5$, the patient T will be assigned to the high-risk group; otherwise, the patient belongs to the low-risk group. If more than two risk groups are considered (such as high, intermediate, and low), other cutoffs can be used based on the risk scores of the training samples. For example, in a training set, if most short-term survivors have a risk score greater than r_1 and most long-term survivors have a risk score smaller than r_2, then

$$T \text{ is} \begin{cases} \text{high risk} & \text{if } S(T) > r_1 \\ \text{low risk} & \text{if } S(T) < r_2 \\ \text{intermediate risk} & \text{if } r_2 \le S(T) \le r_1 \end{cases} \tag{6.12}$$

In general, we require that $r_1 > 0.5$ and $r_2 < 0.5$; selection of the precise values of r_1 and r_2 can be guided by the risk scores of the training samples. Figure 6.5 shows a diagram of patient survival prediction using the scheme described in this section.

6.4.3 Results

We present some results obtained on the three data sets described in Section 6.4.2 by selecting short- and long-term survivors for training, identifying genes through SAM, and constructing the risk score function using *Weka*'s implementation of SVM (http://www.cs.waikato.ac.nz/ml/weka).

SAM (significance analysis of microarrays), software developed at Stanford University (http://www-stat.stanford.edu/~tibs/SAM/), is designed to find significant genes

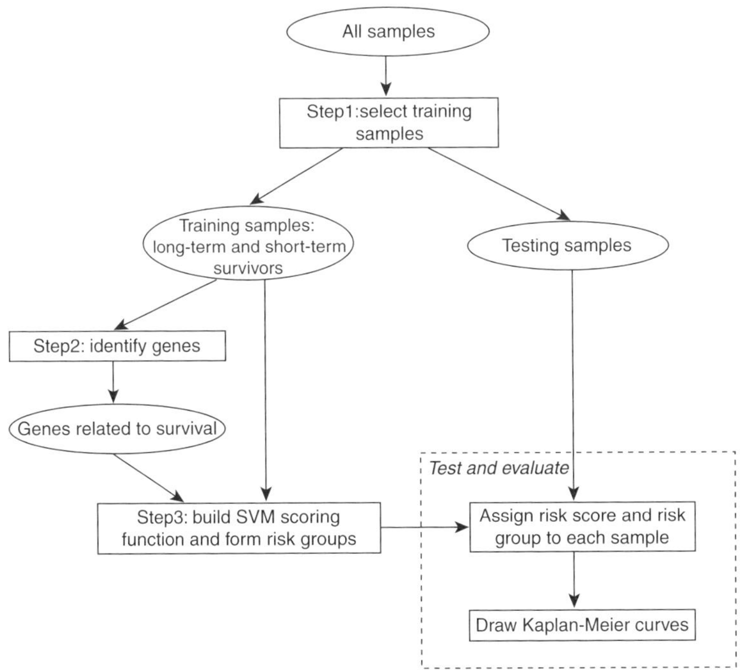

FIGURE 6.5 Process diagram of patient survival study, including three training steps as well as testing and evaluation.

in a set of microarray experiments based on a sophisticated statistical study of genes [28]. SAM first computes a statistic for each gene on the basis of change in gene expression relative to the standard deviation of repeated measurements for the gene. Then for those genes whose statistic is greater than an adjustable threshold, SAM uses permutations of the data to estimate the percentage of genes identified by chance [known as the false discovery rate (FDR)]. The threshold for significance is determined by a *tuning parameter*, chosen by the user based on FDR, or a *fold change parameter* to ensure that the genes selected change by at least a prespecified amount [8]. To identify genes whose expression correlates with survival time, the score assigned by SAM to a gene is defined in terms of Cox's proportional-hazards function. To identify genes whose expression correlates with a quantitative parameter (e.g., a numerical type of class label), such as tumor stage, the score assigned is defined in terms of the Pearson correlation coefficient.

To select extreme samples from the diffuse large-B-cell lymphoma data of ref. 23, we got from 160 samples in the preliminary group of 47 short-term survivors and 26 long-term survivors by setting $c_1 = 1$ year and $c_2 = 8$ years in Eq. (6.9). Therefore, our new training set consists of only 73 samples. SAM picked up 91 genes with a cutoff value at 2.32 (default values used for all the other settings), and a SVM model was trained on the samples and genes selected. The risk score output by the scoring function derived was higher than 0.7 for most of the short-term survivors and lower than 0.3 for most of the long-term survivors. Thus, we categorized patients into three risk groups, defined as

$$
T \text{ is} \begin{cases} \text{high risk} & \text{if } S(T) > 0.7 \\ \text{intermediate risk} & \text{if } 0.3 < S(T) \leq 0.7 \\ \text{low risk} & \text{if } S(T) \leq 0.3 \end{cases} \tag{6.13}
$$

The overall survival Kaplan–Meier curves of the three risk groups are plotted in Figure 6.6 for the validation samples. The software used in this section to generate Kaplan–Meier curves is Prism from GraphPad (http://www.graphpad.com).

For the adult acute myeloid leukemia data of ref. 7, 59 samples were first selected from the total of 116 samples to serve as the candidates of our training set. Among these samples, 26 patients were censored (i.e., with the status "alive"), with the follow-up time ranging from 138 to 1625 days, and 33 were dead, with a follow-up time ranging from 1 to 730 days. Here we can see a big overlap on follow-up time for patients who died and those who were still alive at the end of the follow-up time. By setting $c_1 = 1$ year and $c_2 = 2$ years in Eq. (6.9), 29 short-term survivors and 8 long-term survivors were chosen from the 59 samples to form the training data. With the top 50 genes selected by SAM, $p = 0.0015$ (log-rank test) was obtained for comparing the overall survival of the 57 testing patients in the high- and low-risk groups formed by putting 0.5 as the threshold for the risk score. The p-value would be 0.0045 if the 22 "nonextreme" samples left in the candidate training set were also considered—in this case, 79 validation samples. Similar good results were produced by using the top 100 genes [21]. Note that results presented in this data set should not be compared directly to (1) the one in ref. 7, where 149 genes were selected by SAM and used by a clustering algorithm to estimate the survival for testing samples ($p = 0.006$), and (2) those reported in ref. 4 and Section 6.3.2. The reason is that the test samples might be different due to different partitions on the entire data (to training and validation). For the DLBCL and AML applications above, Table 6.3 summarizes the size change from the original training samples to the training set selected [20].

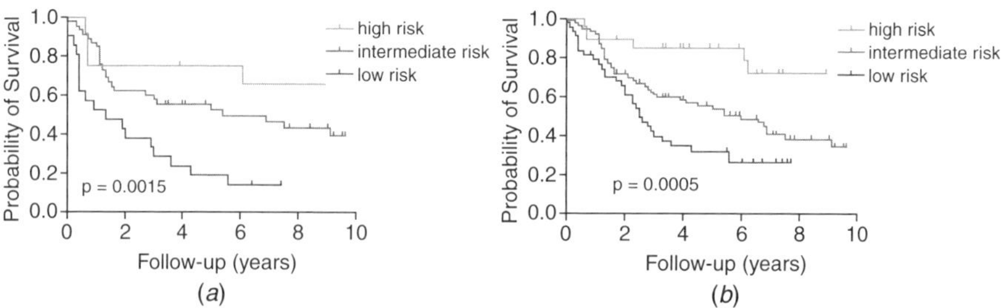

FIGURE 6.6 Kaplan–Meier plots showing the estimation of survival among different risk groups for a DLBCL study: (*a*) for the 80 samples in the validation group, and (*b*) for the 167 samples, including 80 validation samples and 87 nonextreme samples in the preliminary group.

TABLE 6.3 Number of Samples in the Original Training Data and Informative Training Set Selected for the DLBCL and AML Data Sets

Application	Original Training Set			Informative Training Set		
	Alive	Dead	Total	Long Term	Short Term	Total
DLBCL	72	88	160	26	47	73
AML	26	33	59	8	29	37

The last data set tested is the prediction of the time to metastasis of breast cancer patients [29]. In this study, distant metastases are defined as a first event to be a treatment failure, and the outcome data on patients are usually counted from the date of surgery to the time of the first event or the date when the data are censored [29]. There were 295 samples in the data set of ref. 29, each of which had an expression profile with 70 genes. These 70 genes were identified by van't Veer et al. in an earlier study of 78 breast cancer patients [30]. Note that the 295-sample data set contains the 61 lymph-node-negative patients of van't Veer's data set.

Patients involved in this metastasis study include those who had distant metastases as a first event within five years and those who remained free of cancer for at least five years. So there was no overlap on the follow-up times for patients with or without distant metastases; however, we can still select extreme examples to further increase the gap in survival times. From the total of 295 samples, we found that 52 of them had distant metastases within three years, whereas 76 remained free of cancer at least 10 years. Then we selected 40 samples randomly from each of these two types of extreme cases to form our training set, and all the remaining samples (215 samples) are treated as validation data. With all 70 genes or fewer genes ranked by their SAM score, $p < 0.0001$ could be achieved on validation data. The Kaplan–Meier curves in Figure 6.7 illustrate a significant difference in the probability that patients would remain free of distant metastases between the high- and low-risk groups of patients.

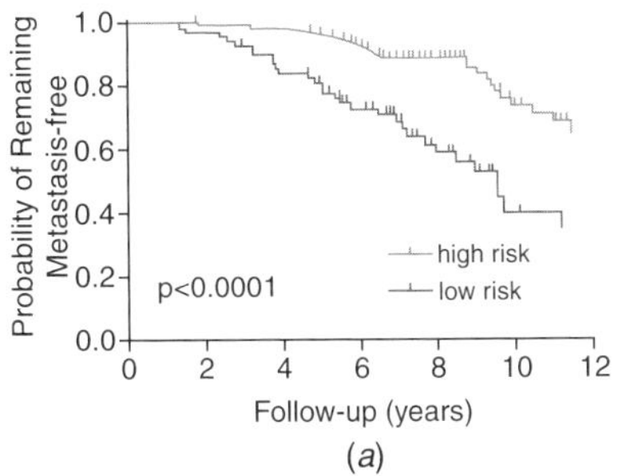

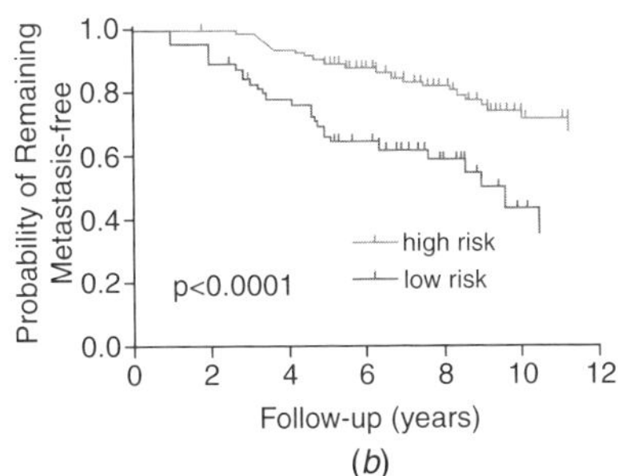

FIGURE 6.7 Kaplan–Meier plots showing the probability that patients would remain metastasis-free among various risk groups for a breast cancer study by using (*a*) all 70 genes provided, and (*b*) 30 genes selected by SAM.

6.5 SUMMARY AND CONCLUDING REMARKS

In this chapter we began with a short introduction to the concept of patient survival prediction followed by some examples to show the urgent needs of more accurate models to predict the clinical outcome based on molecular-level information so that an appropriate treatment can be tailored for individual patients. With the maturity of DNA microarray technology, it becomes possible to develop tools to incorporate gene expression profiles of patients into disease diagnosis and new therapy discovery.

A critical aspect of survival analysis is that the outcome of an interest event may not occur to all patients during the follow-up period. Such patients constitute censored data and we do not know whether or not the event will occur at a later time. To tackle this issue, we have presented the classical Kaplan–Meier survival analysis and Cox proportional-hazards regression model as general approaches. A Kaplan–Meier survival curve aims to estimate over time the probability (or percentage) of patients being event-free. If a patient dies during follow-up, the survival curve will reflect that patient's death at the appropriate point with a step down. When a patient is censored, the curve does not take a step down as it does when a patient dies, but every step down drawn for the later death case will be a little bit larger. Kaplan–Meier plots and the associated p-value of the log-rank test are used throughout the chapter to evaluate survival groups identified by the various methods developed for analysis of gene expression data. To find a correlation between individual microarray features and the clinical outcome, the widely used Cox proportional-hazards regression method can be employed first as a univariate model to estimate the coefficient of each gene associated with survival information and then as a multivariate model to combine the expression level of selected genes to act as a survival predictor.

Two studies, one on diffuse large-B-cell lymphoma after chemotherapy conducted by Rosenwald et al. [23] and another one on lung adenocarcinoma conducted by Beer et al. [5], are then described in detail as examples of applying the Cox proportional-hazards regression model to the framework of patient survival analysis from gene expression data. In both applications, other techniques, such as the cross-validation scheme and hierarchical clustering method, were also employed to make the analyses more practical, more reliable, and the output model simpler.

In fact, two key issues of patient survival analysis based on microarray data, gene identification and prediction model construction, can be very well addressed by feature selection and classification/prediction techniques, two popular topics covered by data mining research. In practice, many data mining methodologies have been used extensively to solve biological and clinical problems [18], and they are playing a more important role in gene expression–related studies. We take the scheme developed by Shipp and his colleagues [24] as an example to illustrate (1) how to categorize patients to form a supervised learning problem, (2) how to use statistical tests in gene selection, and (3) how to build effective models for future prediction. To assist the discovery of new tumor subtypes that are both *clinically relevant* and *biologically meaningful*, a semisupervised idea proposed by Bair and Tibshirani [4] is introduced with an emphasis on clustering patient samples by using only survival-related genes. The effectiveness of the idea is validated by studying four previously analyzed data sets.

In addition to the two main steps in the survival study, a new idea for selecting informative training samples is presented in Section 6.4. Two types of extreme cases, short- and long-term survivors, are picked out to form a new training set in which survival times are remarkably different for different types of patients, so that microarray features with real predictive power might be found easily.

Obviously, identifying significant genes from a huge number of candidates produced by microarray chips is the first and the most difficult problem we are facing. Proper selection not only contributes to an accurate prediction model, but also to potential new drug targets for the disease. Therefore, we must be very careful with genes reported from a single experiment. In other words, more independent validations are always desired. For example, in a recent study on survival prediction in diffuse large-B-cell lymphoma, Lossos et al. [21] showed that using an expression profile of only *six* genes was sufficient to predict overall survival in DLBCL. In their experiment, the gene expression profiles from 66 patients were measured by a quantitative reverse-transcriptase polymerase chain reaction. They also compared the gene list reported by Alizadeh et al. in ref. 1 (71 genes or so), by Rosenwald et al. in ref. 23 (17 genes), and by Shipp et al. in ref. 24 (13 genes), as well as some individual genes reported in the literature as correlating with survival in DLBCL; surprisingly, only *three* genes were present in more than one source. This comparison implies that current gene-based survival prediction models derived from computational methods need to be examined further and verified further by wet laboratory experiments.

Acknowledgments

H.L. and Y.X.'s work is supported in part by a Distinguished Scholar award from the Georgia Cancer Coalition and by the National Science Foundation (NSF/DBI-0354771, NSF/ITR-IIS-0407204).

REFERENCES

1. A. A. Alizadeh, M. B. Eisen, R. E. Davis, C. Ma, I. S. Lossos, A. Rosenwald, J. C. Boldrick, et al. Distinct types of diffuse large B-cell lymphoma identified by gene expression profiling. *Nature*, 403:503–511, 2000.

2. U. Alon, N. Barkai, D. A. Notterman, K. Gish, S. Ybarra, D. Mack, and A. J. Levine. Broad patterns of gene expression revealed by clustering analysis of tumor and normal colon tissues probed by oligonucleotide arrays. *Proc. Natl. Acad. Sci.*, 96:6745–6750, 1999.

3. D. B. Altman. *Practical Statistics for Medical Research.* Chapman & Hall, London, 1991.

4. E. Bair and R. Tibshirani. Semi-supervised methods to predict patient survival from gene expression data. *PLoS Biol.*, 2:0511–0522, Apr. 2004.

5. D. G. Beer, S. L. Kardia, C. C. Huang, T. J. Giordano, A. M. Levin, D. E. Misek, L. Lin, et al. Gene-expression profiles predict survival of patients with lung adenocarcinoma. *Nat. Med.*, 8(8):816–823, 2002.

6. M. P. Brown, W. N. Grundy, D. Lin, N. Cristianini, C. W. Sugnet, T. S. Furey, M. Ares, Jr., and D. Haussler. Knowledge-based analysis of microarray gene expression data using support vector machines. *Proc. Natl. Acad. Sci.*, 97(1):262–267, 2000.

7. L. Bullinger, K. Dohner, E. Bair, S. Frohling, R. F. Schlenk, R. Tibshirani, H. Dohner, and J. R. Pollack. Use of gene-expression profiling to identify prognostic subclasses in adult acute myeloid leukemia. *N. Engl. J. Med.*, 350(16):1605–1616, 2004.

8. G. Chu, B. Narasimhan, R. Tibshirani, and V. G. Tusher. *SAM User Guide and Technical Document.* Stanford University, Stanford, CA, 2004.

9. D. R. Cox. Regression models and life-tables (with discussion). *J. R. Stat. Soc.*, B34:187–220, 1972.

10. S. Dudoit, J. Fridlyand, and T. P. Speed. Comparison of discrimination methods for the classification of tumors using gene expression data. *J. Am. Stat. Assoc.*, 97(457):77–87, 2002.

11. T. S. Furey, N. Cristianini, N. Duffy, D. W. Bednarski, M. Schummer, and D. Haussler. Support vector machine classification and validation of cancer tissue samples using microarray expression data. *Bioinformatics*, 16(10):906–914, 2000.

12. T. R. Golub, D. K. Slonim, P. Tamayo, C. Huard, M. Gaasenbeek, J. P. Mesirov, H. Coller, et al. Molecular classification of cancer: class discovery and class prediction by gene expression monitoring. *Science*, 286:531–537, 1999.

13. J. D. Kalbfleish and R. L. Prentice. *The Statistical Analysis of Failure Time Data.* Wiley, New York, 1980.

14. E. L. Kaplan and P. Meier. Nonparametric estimation from incomplete observations. *J. Am. Stat. Assoc.*, 53:457–481, 1958.

15. J. Khan, J. S. Wei, M. Ringner, L. H. Saal, M. Ladanyi, F. Westermann, F. Berthold, et al. Classification and diagnostic prediction of cancers using gene expression profiling and artificial neural networks. *Nat. Med.*, 7(6):673–679, June 2001.

16. M. LeBlanc, C. Kooperberg, T. M. Grogan, and T. P. Miller. Directed indices for exploring gene expression data. *Bioinformatics*, 19(6):686–693, 2003.

17. J. Li, H. Liu, S.-K. Ng, and L. Wong. Discovery of significant rules for classifying cancer diagnosis data. *Bioinformatics*, 19(Suppl. 2):ii93–ii102, 2003.

18. J. Li, H. Liu, A. Tung, and L. Wong. Data mining techniques for the practical bioinformatician. In L. Wong, Ed., *The Practical Bioinformatician*, pp. 35–70. World Scientific Press, Hackensack, NJ, May 2004.

19. H. Liu, J. Li, and L. Wong. Selection of patient samples and genes for outcome prediction. *Proc. 2004 IEEE Computational Systems Bioinformatics Conference* (CSB'04), pp. 382–392, Stanford University, Stanford, CA, Aug. 2004.

20. H. Liu, J. Li, and L. Wong. Use of extreme patient samples for outcome prediction from gene expression data. *Bioinformatics*, 16(21):3377–3384, 2005.

21. I. S. Lossos, D. K. Czerwinski, A. A. Alizadeh, M. A. Wechser, R. Tibshirani, D. Botstein, and R. Levy. Prediction of survival in diffuse large-B-cell lymphoma based on the expression of six genes. *N. Engl. J. Med.*, 350:1828–1837, 2004.

22. International Non-Hodgkin's Lymphoma Prognostic Factors Project. A predictive model for aggressive non-Hodgkin's lymphoma. *N. Engl. J. Med.*, 329:987–994, 1993.

23. A. Rosenwald, G. Wright, W. C. Chan, J. M. Connors, E. Campo, R. I. Fisher, R. D. Gascoyne, et al. The use of molecular profiling to predict survival after chemotherapy for diffuse large-B-cell lymphoma. *N. Engl. J. Med.*, 346(25):1937–1947, 2002.

24. M. A. Shipp, K. N. Ross, P. Tamayo, A. P. Weng, J. L. Kutok, R. C. Aguiar, M. Gaasenbeek, et al. Diffuse large B-cell lymphoma outcome prediction by gene-expression profiling and supervised machine learning. *Nat. Med.*, 8(1):68–74, 2002.

25. D. K. Slonim, P. Tamayo, T. R. Golub, J. P. Mesirov, and E. S Lander. Class prediction and discovery using gene expression data. *Proc. 4th Annual International Conference on Computational Molecular Biology* (RECOMB), pp. 263–272, Tokyo, 2000.

26. R. Tibshirani, T. Hastie, B. Narasimhan, and G. Chu. Diagnosis of multiple cancer types by shrunken centroids of gene expression. *Proc. Natl. Acad. Sci.*, 99:6567–6572, 2002.

27. R. Tibshirani, T. Hastie, B. Narasimhan, and G. Chu. Class prediction by nearest shrunken centroids, with applications to DNA microarrays. *Stat. Sci.*, 18(1):104–117, 2003.

28. V. G. Tusher, R. Tibshirani, and G. Chu. Significance analysis of microarrays applied to the ionizing radiation response. *Proc. Natl. Acad. Sci.*, 98:5116–5121, 2001.

29. M. J. van de Vijver, Y. D. He, L. J. van't Veer, H. Dai, A. A. Hart, D. W. Voskuil, G. J. Schreiber, et al. A gene-expression signature as a predictor of survival in breast cancer. *N. Engl. J. Med.*, 347(25):1999–2009, 2002.

30. L. J. van't Veer, H. Dai, M. J. van de Vijver, Y. D. He, A. A. Hart, M. Mao, H. L. Peterse, et al. Gene expression profiling predicts clinical outcome of breast cancer. *Nature*, 415(6871):530–536, 2002.

31. E. Vittinghoff, S. C. Shiboski, D. V. Glidden, and C. E. McCulloch. *Regression Methods in Biostatistics: Linear, Logistic, Survival, and Repeated Measures Models.* Springer-Verlag, New York, 2004.

32. E.-J. Yeoh, M. E. Ross, S. A. Shurtleff, W. K. Williams, D. Patel, R. Mahfouz, F. G. Behm, et al. Classification and subtype discovery and prediction of outcome in pediatric acute lymphoblastic leukemia by gene expression profiling. *Cancer Cell*, 1:133–143, Mar. 2002.

7

RNA INTERFERENCE AND microRNA

SHIBIN QIU AND TERRAN LANE

Department of Computer Science, University of New Mexico, Albuquerque, New Mexico

In recent years, RNA interference (RNAi) has surged into the spotlight of pharmacy, genomics, and system biology. Dozens of RNAi-based drugs are entering FDA trials, and as of this writing, at least three companies are based exclusively on RNAi technology. The massive excitement surrounding RNAi stems from its vast potential for therapeutic and genomic applications. Currently, promising RNAi-based therapies include treatments for HIV, Huntington's disease, macular degeneration, hypercholesterolemia, and even prion diseases such as bovine spongiform encephalopathy (BSE; mad cow disease), and potential new therapies are being announced nearly daily.

The fundamental effect of RNAi is *posttranscriptional gene silencing* via targeted knockdown of mRNA. By careful selection of one or more RNAi *initiator molecules*, the RNAi cellular machinery can be coerced into identifying mRNA molecules possessing a specific nucleotide sequence and cleaving those transcripts, thus preventing translation into the corresponding protein or other downstream regulatory effects of the gene. Researchers quickly discovered ways to exploit this effect, including knockdown of viral transcripts to treat viral diseases, knockdown of endogenous genes to treat genetic diseases, gene function study using a loss-of-function method, and probing genetic regulatory networks by knocking down key links.

While a number of laboratories and companies are rushing to turn these potentials into practical therapies, a vital step toward realization is to fully characterize the RNAi mechanism and its effects. RNAi-based therapies are threatened by RNAi characteristics, including nonspecificity, off-target effects, poor silencing efficacy, synergistic pool effects of siRNA/dsRNA, and uncertainties of RNAi kinetics. These negative effects and risky properties of RNAi must be fully characterized before it can be widely applied, especially to human therapies. Meanwhile, investigations into these issues provide opportunities for researchers in biology, bioinformatics, machine learning, and data mining.

In this chapter we describe the biological mechanisms of RNAi, covering both siRNA and microRNA as interfering initiators, followed by in-depth discussions of a few computational topics, including RNAi specificity, microRNA gene and target prediction, and siRNA silencing efficacy estimation. Finally, we point out some open questions related to RNAi research. At the end of the chapter we provide a glossary of terms and acronyms.

7.1 MECHANISMS AND APPLICATIONS OF RNA INTERFERENCE

In this section we first overview the mechanisms, biology, and applications of RNAi and then discuss its design issues and research topics.

7.1.1 Mechanism of RNA Interference

Discovery of the RNAi mechanism is generally credited to Fire and colleagues [19,41], although the first recognition of the effects of RNAi traces back to plant genetics work in the early 1990s [43], where it was labeled as cosuppression. After the initial characterization in *Caenorhabditis elegans* as the result of transfection of double-stranded RNA (dsRNA), similar effects were soon recognized in a number of other organisms under a variety of names: RNA silencing, quelling, posttranscriptional gene silencing (PTGS), and RNA interference. It quickly became apparent that there was a single core mechanism with a couple of major variants and a number of interspecies refinements. Importantly, however, the core mechanism itself appears to be extremely evolutionarily conserved, appearing in fungi [55], insects [31], nematodes [19], plants [68], and mammals [12]. The RNAi mechanism quickly aroused enormous excitement, and by 2002, *Science* had declared it to be the "breakthrough of the year" [14].

RNAi is a cell defense mechanism that represses the expression of viral genes by recognizing and destructing their messenger RNAs (mRNAs), preventing them from being translated into proteins. RNAi also regulates expressions of endogenous genes and suppresses transposable elements. Gene knockdown in RNAi is induced by short interfering RNA (siRNA) of about 21 nucleotides (nt) in length, generated from either a dsRNA by the enzyme dicer, or transfected directly. Target mRNA transcripts that are hybridized with the siRNA are destroyed by the RNA-induced silencing complex (RISC) [23]. We can summarize dsRNA-mediated gene silencing conceptually in the following steps:

1. A dsRNA is introduced into the cell to initiate RNAi.
2. Dicer cleaves the dsRNA into siRNAs of length 21 to 25 nt.

3. RISC unwinds the siRNA to help target the appropriate mRNA.

4. The RISC–siRNA complex recognizes the target via complementarity.

5. RISC cleaves the siRNA–mRNA hybrid and the mRNA is degraded.

A dsRNA, usually a few hundred nucleotides long, is transfected into the cell to initiate RNAi [74]. Dicer, a dimeric enzyme of the RNase III ribonuclease family, contains dual catalytic domains. Since each catalytic domain is a bacterial RNase III protein and has a cleavage length of 9 to 11 nt, one of the active sites of each catalytic domain of dicer remains inactive, resulting in a cleavage length of about 21 nt [9]. Before the target mRNA is silenced, the initiating dsRNA is first cleaved by dicer into siRNAs. A typical siRNA usually has 19 base pairs (bp) and 2-nt 3′ overhangs [21]. Figure 7.1 illustrates the structure of an siRNA and the process of dsRNA-induced RNAi.

In mammalian cells, RNAi cannot be initiated by dsRNA, since dsRNA (>50 bp long) may activate the nonspecific interferon (IFN) pathway [11]. Therefore, siRNAs

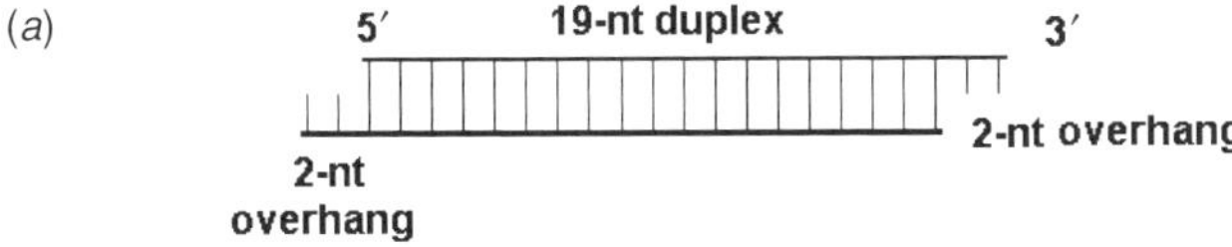

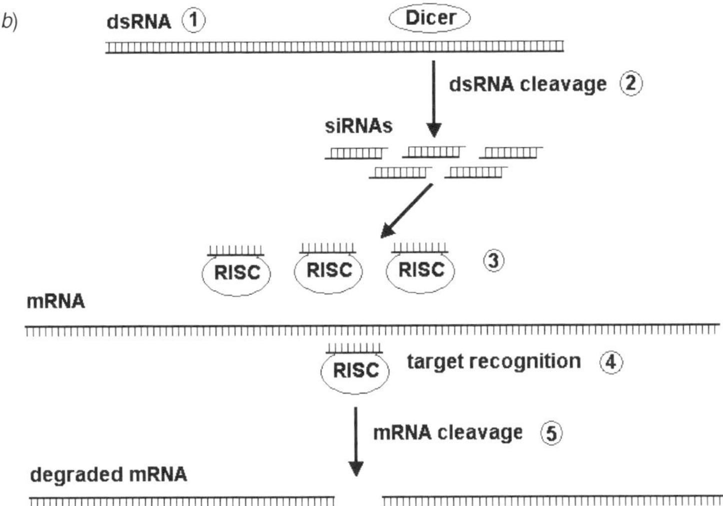

FIGURE 7.1 RNAi mechanism: (*a*) the structure of a 19-nt siRNA; (*b*) the process of dsRNA-initiated RNAi. (1) A dsRNA is introduced to initiate RNAi; (2) dicer cleaves the dsRNA into siRNAs; (3) RISC unwinds the siRNA and takes up the guide strand; (4) the RISC–siRNA complex recognizes the target; (5) the target mRNA is degraded by RISC.

are introduced directly. Alternatively, DNA silencing constructs and short hairpin RNA (shRNA) can be used to generate siRNAs in vivo. In the latter case, DNA templates for the siRNAs desired are transfected into the cells. Polymerase II or III transcribes these templates into shRNAs, which in turn are processed by dicer to generate siRNAs [11].

RISC is a protein–RNA effector nuclease complex [22]. It unwinds the double-strand siRNA, and uptakes one strand, the *guide strand*, to recognize its target [45]. It is generally believed that the guide strand is the antisense strand. Equipped with the guide strand and incubated with ATP, RISC recognizes the target through sequence complementarity (also referred to as *hybridization* or *homology*) and cleaves the target endonucleolytically in the regions homologous to the guide strand. Since target genes are destroyed after they are transcribed, in plants RNAi is also called *posttranscriptional gene silencing* (PTGS). Efficient gene silencing of RNAi implies that RISC has an active search mechanism that scans the transcriptome, consisting of all transcripts for homologous targets in a high-throughput manner. Observations indicated that to make this search mechanism efficient, RISC might be associated with the ribosome that translates mRNA into protein [22].

In plants and worms, siRNAs are amplified and gene silencing is spread by the mechanism of transitive RNAi (tRNAi) [62]. In this mechanism, the RdRP (RNA-directed RNA polymerase) protein produces a dsRNA upstream of the cleavage site hybridized by the initiating siRNA. This novel dsRNA is then cleaved by dicer to generate novel siRNAs (secondary siRNAs), which associate with RISC to silence more targets. Transitive RNAi causes chain reaction due to siRNA proliferation and silencing spreading. It improves degradation efficiency because of siRNA amplification, but also increases the chances of silencing unintentional target mRNAs [15]. RdRP has not been confirmed in humans, and it is not likely that humans have tRNAi. The tRNAi mechanism is illustrated in Figure 7.2, where the targets of the first three stages are displayed.

MicroRNA (miRNA) is a class of 21- to 23-nt single-stranded RNAs processed by dicer from the ∼70-nt precursor of hairpin RNAs formed from endogenous transcripts [35]. Some miRNAs contain homology to 3′ UTR regions of their target mRNA. Other miRNAs target open reading frames (ORFs) of the mRNAs [58]. The class of miRNAs that regulate expressions of genes performing timing and developmental functions are called *short temporal RNAs* (stRNAs) [50]. Most miRNAs do not degrade mRNA but regulate gene expression at the translational level by inhibiting the targeted proteins [47]. Repression of the target protein is effected by an miRNA ribonucleoprotein complex (miRNP) which shares many similarities with RISC. If the miRNA degrades its target mRNA transcripts, it also associates with RISC for target recognition and destruction.

In plants, dsRNA and shRNA also induce genomic methylation at the promoter regions having sequence homologies, resulting in transcriptional gene silencing (TGS) [70]. Inspired by dsRNA's ability to make genomic alteration and initiate TGS, biologists have begun to draw links between PTGS and TGS to unify the functions of RNAi machinery. A three-in-one model was proposed for RISC, where RISC is able to regulate gene expressions at three levels: mRNA knockdown in cytoplasm,

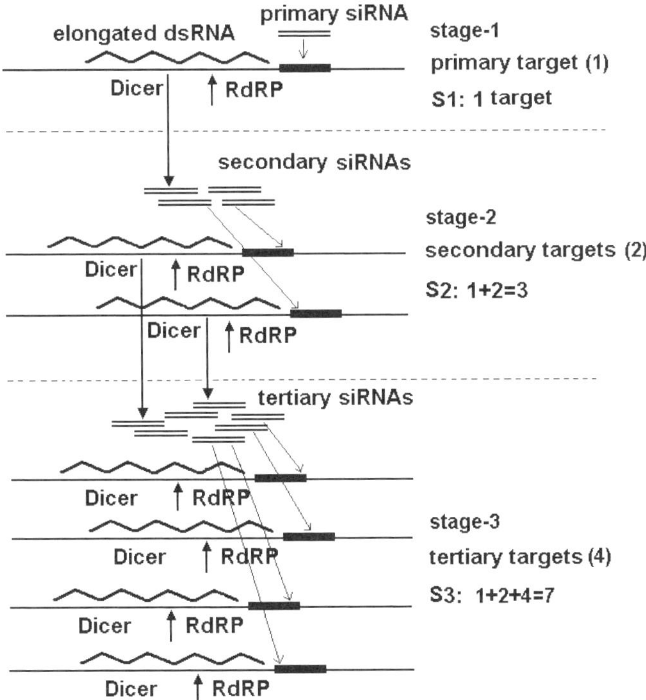

FIGURE 7.2 Transitive RNAi. Seven mRNAs are targeted by a single primary siRNA in the first three stages of the cyclic knockdown.

translation inhibition, and chromatin remodeling in the nucleus. In this model, dsRNAs may be presented in the cell as synthetic RNAs, viral sequences, shRNA, or transcripts of nuclear genes [23]. RISC is associated with the short RNA molecules processed by dicer from these dsRNAs. The activated RISC performs as a flexible platform upon which different regulatory modules may be integrated and functions in three different ways according to three different types of signals. Figure 7.3 shows the functions of this unified RISC.

7.1.2 Applications of RNAi

RNAi is useful for genetic regulation, viral gene knockdown, analysis of gene functions, and therapeutical treatment of diseases. We discuss these applications briefly below.

Genetic Regulation   miRNAs regulate gene expressions in vivo. The 22-nt *lin-4* (lineage-abnormal-4) gene in *C. elegans*, processed by dicer from a ~70-nt hairpin precursor, down-regulates the *lin-14* and *lin-28* genes by targeting to their 3′ UTR regions and inhibiting their translations. Another example of regulatory miRNA is the 21-nt *let-7* (lethal-7), which is a negative regulator of *lin-41*. Since these target genes

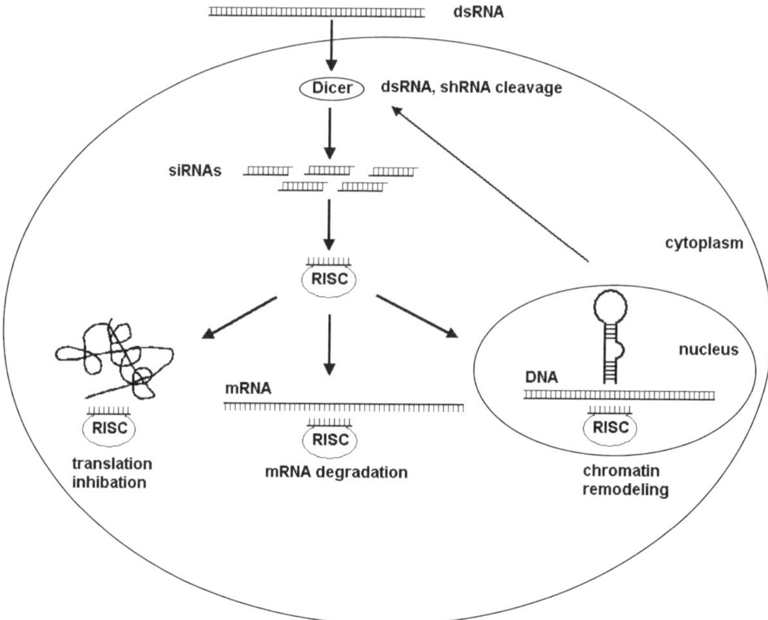

FIGURE 7.3 Flexible functions of RISC. In this three-in-one model, RISC performs three

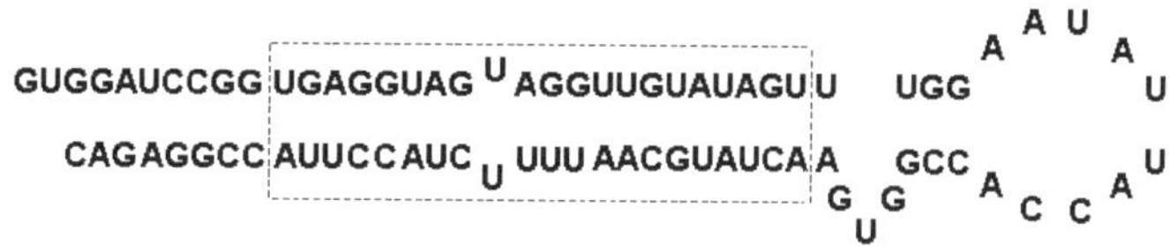

FIGURE 7.4 Hairpin structure of the *C. elegans let-7* gene. (Data from the National Center for
Biotechnology Information.)

are important developmentally, *lin-4* and *let-7* regulate worm growth. Because the *let-7* gene is also found in flies, humans, and other species, the regulatory mechanism of this miRNA is well conserved [50]. Since *let-7* is a well-known and typical miRNA, we show its sequence and hairpin structure in Figure 7.4. To identify miRNA genes and targets, computational methods can be developed to perform reliable predictions.

Viral Gene Knockdown   RNAi can be used to protect host genome against virus infections by suppressing viral expression. An shRNA was constructed to silence the niaprotease (*Pro*) gene of potato virus Y (PVY) in tobacco plants [63]. siRNAs were used successfully to silence the respiratory syncytial virus (RSV), an RNA virus that is responsible for severe respiratory disease in neonates and infants [8]. siRNAs targeting the M-BCR/ABL fusion site were introduced into human K562 cell and specifically silenced leukemic cells [71]. To silence viral genes effectively, we want to silence the target genes maximally and nontarget genes minimally. For these purposes, bioinformatics approaches help optimize siRNA silencing efficacy and specificity.

Gene Function Analysis   The function of a gene can be determined by silencing its mRNA and examining the phenotype changes. This loss-of-function method can be used to study gene functions systematically for all genes in an organism, as has been done for *C. elegans* [30]. RNAi also provides a high-throughput procedure for functional genomics in plants [6]. In addition, loss-of-function of selected genes also helps to build gene regulatory networks and cell signaling pathways [40]. Since loss-of-function studies require maximal target silencing and minimal nontarget silencing, bioinformatics approaches also provide opportunities for optimal solutions with respect to silencing efficacy and specificity.

Disease Treatment   RNAi can specifically knock down genes responsible for human diseases, providing a novel means of therapeutics. Chemically modified siRNAs targeting the apoB gene were injected through the tail of mice and reduced their total level of cholesterol [65]. shRNAs were transfected into mice to investigate the therapeutical treatment of dominant polyglutamine expansion diseases, which include Huntington's disease [72]. These experiments in mice demonstrated the therapeutic potential of RNAi for the treatment of human hypercholesterolaemia and Huntington's disease. HIV-1 RNAs were inhibited by siRNAs and shRNAs in human cell lines [28], showing that HIV virus is amenable to RNAi treatment. In addition to silencing efficacy and specificity, the concentration and temporal changes of the siRNA and the target mRNA should be accessible for efficient transfection. The RNAi kinetics can be modeled by differential equations.

Algorithms Inspired by RNAi   Having discussed some biological applications of RNAi, we point out some possibilities of developing algorithms based on RNAi mechanisms. Potentially, the principles of target recognition and degradation of RNAi can be used to design an algorithm for detecting and destroying computer viruses. Once a binary stream of a virus has been identified, it can be used as a virus identifier and fed into a search–deletion engine that detects and eliminates computer codes that are similar to virus codes. Although it has not yet been realized, this idea holds promise for a large open problem in computer science. We anticipate that control of false positivity is a critical issue, since virus codes might share similarity with normal programs. However, it might be a good strategy in certain situations, such as those where all the working programs have known code structures and virus codes are well distinguishable.

In implementing this RNAi-based search–deletion engine, the principle of siRNA amplification of transitive RNAi can also be incorporated. From the code that has been detected and confirmed to be viral, more virus identifiers can be copied from the code itself. These secondary virus identifiers will increase the degradation efficiency. The siRNA amplification mechanism of tRNAi can also be applied to authorship identification and software forensics. Once a piece of work has been determined to be positively related, its content can be used to produce more identifiers to improve search efficiency.

7.1.3 RNAi Computational and Modeling Issues

Although RNAi is highly effective, attention must be paid to some particular issues when designing an experiment or conducting a bioinformatic research. RNAi silencing components such as dicer and RISC can themselves be silenced, suppressing the RNAi pathway. For example, potyviruses encode a protein, HC-Pro, that potently disables PTGS by preventing the dsRNA cleavage activity of dicer [38]. Therefore, a plant challenged with a potyvirus becomes a battleground in the fight between a defense and a counterdefense strategy. In addition, although RNAi is a widely conserved mechanism, some organisms, such as *Saccharomyces cerevisiae*, are RNAi negative due to the lack of RNAi components. Although the metabolism, protein interaction, and cell cycle behaviors of *S. cerevisiae* have been studied extensively, the lack of RNAi makes it difficult to integrate mRNA regulation into existing networks.

An important issue in RNAi is that when silencing a target gene, it is possible to knock down nontarget genes, causing *off-target effects.* Although RNAi is generally considered specific, off-target effects exist extensively [27,51]. It is conceivable that silencing a large proportion of organismal genes will cause loss of function of too many genes, making the organism unstable. Therefore, off-target silencing must be carefully controlled. Furthermore, siRNA sequences binding to different regions on a target mRNA have variable silencing capabilities. It is important to predict an siRNA's silencing efficacy, measured via the percent remaining level of the target mRNA, before it is synthesized and used. Rational design rules have been developed to predict this silencing efficacy [4,13,26,29,57,59,66]. To use these rules effectively, siRNA efficacy should be predicted using computational approaches. Related to microRNA biology, miRNA genes and their targets are usually identified computationally to determine their regulatory functions.

These issues are critical for the success of RNAi experiments; they also provide opportunities for researchers in bioinformatics and knowledge discovery. To characterize off-target effects thoroughly, all genes to be silenced, with each gene as target, must be examined. It is too expensive, if feasible, to conduct this total investigation experimentally. Computational simulation can, however, simulate the off-target searches and can extend the parameters beyond experimental conditions.

The binding between siRNA and mRNA allows for mismatches, G-U wobbles, and bulges. But it is not clear how flexible nature allows this binding to be. Simulations using graph models can estimate this critical mismatch. In addition, to better solve the siRNA efficacy problem, machine learning and data mining approaches can be used to improve prediction accuracies. Computational methods not only can predict siRNA efficacy using rational design rules, they can also evaluate the relative significance of different efficacy rules. Furthermore, computational approaches have been developed for the prediction of miRNA genes by predicting possible occurrences of the stem-loop structures. The targets regulated by an miRNA are also predictable by searching genes that possess binding regions based on known miRNA–target complementarity. In the following sections we discuss these issues and their computational solutions in detail.

7.2 SPECIFICITY OF RNA INTERFERENCE

RNAi has been regarded as a highly specific means of gene repression [28,30]. But off-target knockdowns are also extensive [27,60,61]. At least two factors contribute to off-target knockdown. First, the length of siRNA matters. If different genes share subsequences of length greater than the siRNA length, off-target silencing occurs. Second, RISC's target recognition process is not perfect and allows for mismatches, G-U wobbles, and bulges [60]. This imperfect matching allows genes that do not share a common subsequence to be silenced at the same time. Since silencing of unwanted or unknown genes has considerable negative implications on RNAi applications, we investigate this specificity issue in detail in this section.

The procedure recommended for target validation is BLAST [3,32]. However, BLAST misses targets in some cases and is not suitable for accurate sequence matching, such as RNAi [46]. The sequence binding between an siRNA and its target allows for mismatches, G-U wobbles, and bulges [60]. Although BLAST allows for deletion, insertion, and substitution, it cannot control the exact patterns of imperfect matches encountered in RNAi. Due to their quadratic complexity, algorithms based on dynamic programming are not feasible for large-scale searching. Alignment algorithms align the input sequences into whatever pattern is needed to get an optimal score based on a cost model, and do not guarantee generating the patterns desired [44,64]. To simulate siRNA–target binding, we describe search algorithms based on string kernels that control matching patterns accurately by adjusting the length and position of the patterns. We first present a formalism using string kernels to represent RNAi computationally. We then present algorithms and results.

7.2.1 Computational Representation of RNAi

As described in Section 7.1, RISC destroys the target mRNA of an siRNA by sequence complementarity between the antisense strand of the siRNA and the target sequence. This sequence complementarity is equivalent to sequence matching between the siRNA sense strand and the target region. Therefore, we can use concepts and techniques in sequence matching. We begin by representing RNAi by considering exact matches only.

We describe each gene by its contiguous subsequences of length n (17 to 28 nt), called n-*mers* or n-*grams*, representing siRNAs. A gene, g_x, represented in the input space, χ, consisting of sequences drawn from the alphabet $\mathcal{A} = \{A, C, G, U\}$, is mapped into an n-gram feature space, $\mathbb{R}^{4^n}$, by the feature map of exact match:

$$\Phi^{ex}(g_x) = (\phi_a(g_x))_{a \in \mathcal{A}^n} \qquad (7.1)$$

where $\phi_a(g_x)$ is the number of times n-gram a occurs in g_x. Therefore, the image of g_x using the exact match is the coordinates in the feature space indexed by the number of occurrences of its constituent n-mers. A gene g_y matches g_x if the condition

$$K(g_x, g_y) = \langle \Phi^{\mathrm{ex}}(g_x), \Phi^{\mathrm{ex}}(g_y) \rangle \geq T \tag{7.2}$$

is met for a threshold T. The similarity measure $K(g_x, g_y) = \langle \Phi^{\mathrm{ex}}(g_x), \Phi^{\mathrm{ex}}(g_y) \rangle$ in Eq. (7.2) defines a kernel as used by a support vector machine classifier [67]. We use this kernel to match an siRNA and its target, instead of classifying. Since any match between an siRNA and an mRNA will silence the gene, we choose $T = 1$.

A simple example helps explain how Eq. (7.2) is used. To compute the similarity measure on short sequences $O_1 = \mathrm{AACGAC}$ and $O_2 = \mathrm{AACGUGG}$ using 3-mer ($n = 3$) exact match, they are mapped into the feature space as $\Phi^{\mathrm{ex}}(O_1) = \{\mathrm{AAC}, \mathrm{ACG}, \mathrm{CGA}, \mathrm{GAC}\}$ and $\Phi^{\mathrm{ex}}(O_2) = \{\mathrm{AAC}, \mathrm{ACG}, \mathrm{CGU}, \mathrm{GUG}, \mathrm{UGG}\}$. Since 3-mers AAC and ACG occur in both of them, $\langle \Phi^{\mathrm{ex}}(O_1), \Phi^{\mathrm{ex}}(O_2) \rangle = 1 + 1 = 2$. Therefore, these two sequences match each other given the parameter and the criterion.

Computing the similarity of (7.2) directly in a vector space requires $O(DF4^n)$ time, where F is the number of n-grams in the genome (40×10^6 for *C. elegans* and 60×10^6 for human) and D (close to F) is the number of n-mers to be compared in the coding sequences. For a genome-wide scan, this computing time is prohibitive and can be improved by taking advantage of the sparsity of the feature space. We use an *inverted file* where the n-mers serve as identifiers and their gene names and positions within the genes serve as attributes. We use the positions for kernels of imperfect matches later. If we ignore n-mers having zero occurrence and allow for duplicate n-mers, a gene g_x can be represented compactly in the feature space as

$$\Phi^{\mathrm{ex}}(g_x) = \{(a_1, p_1), (a_2, p_2), \ldots, (a_{k_x}, p_{k_x})\} \tag{7.3}$$

where a_j, $1 \leq j \leq k_x$, is the jth n-gram of g_x, p_j is its position on g_x, and k_x is the number of n-mers that g_x has. In the inverted file, the records for g_x contain the triples $\langle a_1, g_x, p_1 \rangle, \langle a_2, g_x, p_2 \rangle, \langle a_3, g_x, p_3 \rangle, \ldots, \langle a_{k_x}, g_x, p_{k_x} \rangle$. The inverted file for a genome is a collection of the triples of its genes. To speed up computation, we sort the inverted file on the n-mer field using a binary search tree (BST).

$K(g_x, g_y)$ in (7.2) is computed by searching each n-mer of g_x for g_y in the inverted file. $K(g_x, g_y)$ is the number of occurrences of g_y among the matched genes. Each search in the BST takes $O(\log F)$ time, resulting in a time of $O(k_x \log F)$ for computing $K(g_x, g_y)$.

7.2.2 Definition of Off-Target Error Rates

We define the off-target error using the exact-match kernel, but it is the same for other kernels of imperfect matches defined later. To simulate dicer's cleavage of dsRNA (100 to 400 bp) into siRNAs, we take an oligonucleotide o_x, as dsRNA, from gene g_x

and map it into the feature space using the exact-match kernel. Expressed compactly as in (7.3), $\Phi^{ex}(o_x) = \{(s_1, p_1), (s_2, p_2), \ldots, (s_{l_x}, p_{l_x})\}$, where s_j is the jth n-mer in o_x. To obtain the matched genes based on Eq. (7.2), we compute the kernel $K(o_x, g_y)$ for each gene g_y, for $1 \leq y \leq G$, where G is the total number of genes in a genome. These calculations use BST searches, as described in Section 7.2.1.

Let $C_x = \{g_{x_1}, g_{x_2}, \ldots\}$ be the set of genes whose kernel values with o_x satisfies Eq. (7.2), excluding g_x itself. The *precision* of a search is the proportion of correct documents to the total number of documents matched [20]. Here, only g_x is correct and the number of genes returned is $1 + |C_x|$. So the precision for silencing g_x is $P_x = 1/(1 + |C_x|)$. We define the *off-target error* for repressing g_x as $E_x = 1 - P_x = |C_x|/(1 + |C_x|)$. We define the average error rate as

$$E(\Theta) = \frac{1}{G}\sum_{i=1}^{G} E_i \qquad (7.4)$$

where E_i is the error for silencing gene g_i and Θ is a set of parameters.

The parameters we examine are represented by $\Theta = \langle l, n, m, p, b, q, w, r \rangle$. In Θ, l is dsRNA length. As dsRNA can be synthesized and introduced through a short hairpin experimentally as an alternative to using siRNA directly, its length determines the number of possible siRNAs produced by dicer. By varying l, we can investigate the chances of off-target errors using different dsRNA lengths. n is siRNA length in nucleotides. The length of siRNA produced by dicer appears to vary slightly across organisms [2], so we examine the effect of siRNA length on the off-target error rate. A computational approach is also able to simulate siRNA lengths that do not occur in vivo, allowing us to assess the trade-offs between siRNA lengths and off-target error rates. m is the mismatch length in nucleotides. Experiments have shown that RNAi works despite the existence of mismatches between the siRNA and its target [27,60]. To observe the effect of mismatch on off-target error rate, we use a range of mismatches in our computational procedures. p is the position of the mismatch within the siRNA. Differential silencing efficiency among variable mismatch positions has been reported in biological experiments, demonstrating that mismatches in certain regions within the siRNA are critical for effective knockdown [60]. We examine the positional effect of mismatch by changing p. To control the position of the mismatch, we consider only contiguous mismatches, which are also the most frequently tested. b and w are the lengths of bulge and wobble; and q and r are the positions of bulge and wobble, respectively. Figure 7.5 illustrates some of the parameters. We repeatedly search targets of siRNA from each gene and calculate an average off-target error rate

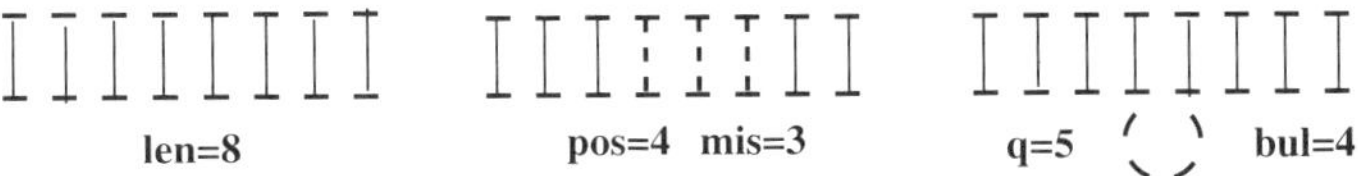

FIGURE 7.5 Matching patterns and parameters between an siRNA and its target. Exact match (left), mismatch and wobble (middle), and bulge (right).

in the organism to investigate the effect of the parameters. After $E(\Theta)$ is evaluated, the average number of incorrect genes targeted by silencing each gene in the genome is $Z(\Theta) = E(\Theta)/[1 - E(\Theta)]$.

7.2.3 Feature Maps of Mismatch, Bulge, and Wobble

The exact match string kernel computes the similarity between two sequences by counting their common n-mers. It does not distinguish the position of the n-mers and does not permit imperfect matches. However, an siRNA is able to silence its target despite the existence of mismatch, bulge, and wobble patterns [27,60]. To study the effect of these patterns, we define the string kernels of imperfect matches through the notion of similarity neighborhoods. Since these binding patterns exist frequently in RNA biology, we call them RNA string kernels. To investigates the positional effects of the imperfect matches, we define contiguous patterns and parameterize their positions.

For an n-mer a from an alphabet $\mathcal{A}$, we define its mismatch neighborhood $N_{m,p}^{\text{mis}}(a)$ as all n-mer γ from $\mathcal{A}$ that differ from a by at most m contiguous mismatches starting at position p in a. And we define the mismatch feature map of a as $\Phi_{m,p}^{\text{mis}}(a) = (\phi_\gamma(a))_{\gamma \in \mathcal{A}^n}$, where $\phi_\gamma(a) = 1$ if $\gamma \in N_{m,p}^{\text{mis}}(a)$, and $\phi_\gamma(a) = 0$ otherwise. The feature map of a gene g_x is defined as the sum of the feature maps of its n-mers:

$$\Phi_{m,p}^{\text{mis}}(g_x) = \sum_{a \in g_x} \Phi_{m,p}^{\text{mis}}(a) \tag{7.5}$$

The bulge neighborhood $N_{b,q}^{\text{bulge}}(a)$ for n-mer a is defined as all $(n+b)$-mers γ from the target that match a exactly everywhere except by a bulge of b nucleotides long starting at position q on γ. The bulge feature map of a is defined as $\Phi_{b,q}^{\text{bulge}}(a) = (\phi_\gamma(a))_{\gamma \in \mathcal{A}^{n+b}}$, where $\phi_\gamma(a) = 1$ if $\gamma \in N_{b,q}^{\text{bulge}}(a)$ and $\phi_\gamma(a) = 0$ otherwise. The feature map of g_x is defined as the sum of the feature maps of its n-mers:

$$\Phi_{b,q}^{\text{bulge}}(g_x) = \sum_{a \in g_x} \Phi_{b,q}^{\text{bulge}}(a) \tag{7.6}$$

The wobble feature map $\Phi_{w,r}^{\text{wobble}}(\cdot)$ is defined similarly to $\Phi_{m,p}^{\text{mis}}(\cdot)$ in Eq. (7.5), except that only G-U wobbles exist in its neighborhood.

Note that the mismatch, bulge, and wobble neighborhoods are supersets of the exact-match regions. By defining the similarity neighborhood for the combination of mismatches, bulges, and wobbles as the union of the separate neighborhoods, we can define the feature map of simultaneous mismatch, bulge, and wobble $\Phi_{m,p,b,q,w,r}^{mbw}(\cdot)$ accordingly. Thus, the RNA string kernel can be computed using inner products based on the feature maps above:

$$K(g_x, g_y) = \langle \Phi_{m,p,b,q,w,r}^{mbw}(g_x), \Phi_{m,p,b,q,w,r}^{mbw}(g_y) \rangle \tag{7.7}$$

7.2.4 Positional Effect

It is possible to control the position of the imperfect matches when designing an RNAi
experiment for a given target. But the exact bindings between the siRNAs and their
nontargets are completely unknown. Therefore, it is reasonable to examine off-target
effects by considering all possible positions of the imperfect matches. Computationally, we average over all positions within the strings as done for the mismatch feature
map below.

$$K^m(g_x, g_y) = \frac{1}{n-m+1} \sum_{p=1}^{n-m+1} \langle \Phi_{m,p}^{\mathrm{mis}}(g_x), \Phi_{m,p}^{\mathrm{mis}}(g_y) \rangle \tag{7.8}$$

However, biological experiments suggested that nucleotides in the region of 2 to
9 nt at the $5'$ end of the guide strand are crucial for gene silencing [32,60]. It therefore
seems that transcripts containing sequence identity within this critical binding region
would have higher chances of being targeted for silencing, and that mismatches
within this region would have a more significant effect on reducing off-target
silencing. To examine this positional effect, we design a weighted scheme where
we assign lower silencing efficiency scores if the mismatches are in the critical
binding region, and higher scores if the mismatches are outside the region. The
silencing efficiency score for silencing a gene is the sum of the scores contributed at
all positions. A gene is considered silenced only when its total efficiency score is
above a threshold. This can be done by using the mismatch kernel and adjusting the
position parameter p:

$$K_p^m(g_x, g_y) = \langle \Phi_{m,p}^{\mathrm{mis}}(g_x), \Phi_{m,p}^{\mathrm{mis}}(g_y) \rangle, \qquad p = 1, 2, \ldots, n-m+1 \tag{7.9}$$

Finding the potential targets for silencing each gene involves scanning the entire
transcriptome. Thus, off-target effect evaluation for an organism requires an immense
amount of processing. To speed up this evaluation, efficient algorithms have been
developed [53,54]. These algorithms use multiple search trees to reduce the search
space by searching different trees when mismatches exist at different positions. In
addition, parallel processing has also been used. Due to space limitations, we omit
implementation of the RNA string kernel.

7.2.5 Results for RNAi Specificity

The average off-target error rate in *C. elegans* has been evaluated. Figure 7.6 *a* shows
the exact-match case, indicating that when the siRNA length increased, the off-target
error decreased. When an siRNA was short, a small length increase improved the
specificity dramatically and the error curve was steep. But when the siRNA was
relatively long, further length increase improved specificity only a little. The siRNA
length observed in biological experiments is between 19 and 25 nt, which is consistent with our finding that increasing siRNA length beyond 25 nt did not improve
specificity significantly. When siRNAs were short, longer dsRNAs yielded larger

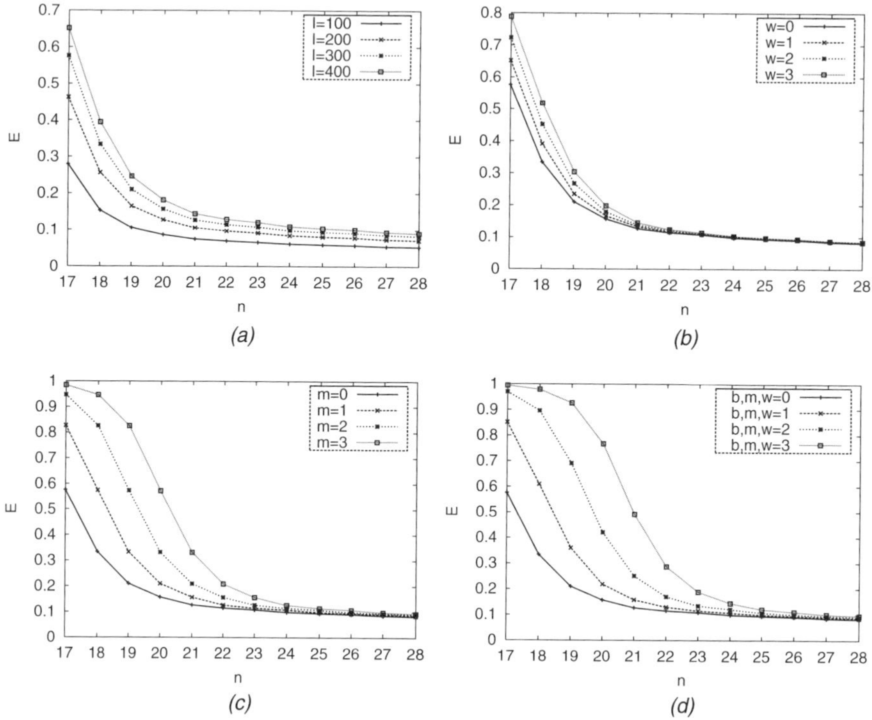

FIGURE 7.6 RNAi off-target effects with exact and imperfect matches in *C. elegans*. n represents siRNA length, E, average off-target error rate. (*a*) Exact match ($l = 100$ to 400, $n = 17$ to 28); (*b*) G-U wobble ($l = 300, w = 0$ to 3); (*c*) mismatch ($l = 300, m = 0$ to 3); (*d*) simultaneous mismatch, bulge, and wobble ($l = 300, b, m, w = 0$ to 3).

errors, and when they were relatively long, longer dsRNAs caused slight increases in the error. The error rates in human and *S. pombe* presented similar patterns (results not shown).

Effects of G-U wobbles are plotted in Figure 7.6*b* for $l = 300$. As shown, G-U wobbles increased off-target error rates slightly, and the increases became insignificant when siRNAs were longer than ~ 21 nt. We found that bulges barely increased off-target errors. This is because the larger neighborhood made by a bulge match was canceled by the reduced chances of target binding caused by a length increase of the subsequence on the target that forms the bulge. Effects of bulges were not plotted separately due to their insignificance, but their contributions are shown in the combined effects. Mismatches caused dramatic increases in error rates, but their effects diminished when siRNAs became longer, as shown in Figure 7.6*c*. The critical siRNA length, which caused the effect of mismatches of up to 3 nt to vanish, is 24 nt. Effects of simultaneous mismatch, bulge, and wobble using combined kernels are shown in Figure 7.6*d*. These figures demonstrate that off-target errors were raised substantially when the imperfect matches existed at the same time compared with the case of exact matches only. The off-target errors caused by mismatch, wobble, and bulge were slightly higher than those caused by mismatch alone, suggesting the dominance of mismatch, as shown in Figure 7.6c and d. Although the increases caused by all the imperfect patterns eventually diminished with the siRNA length, it took much longer siRNA to compensate for their effects. Results also show that when siRNAs were short (e.g., $n = 17$) and simultaneous imperfect patterns existed (e.g., $b, m, w = 3$), the off-target error rates were greater than 90%. However, when siRNAs were long enough, the specificities approached the level of exact matches.

We next extend n to 35 and m to 9. Although these parameter ranges are beyond the scope of those used in biological experiments, they help fully observe the impacts of mismatch and siRNA length on off-target silencing. The results are shown in Figure 7.7, suggesting that off-target chances are increased dramatically by mismatches, but can be reduced by very long enough siRNAs. Presuming that artificial dicer can cleave dsRNA into 35-nt siRNAs in the future, specificity will be improved substantially.

Using the weighted scheme described in Section 7.2.4, we found that off-target error rates corresponding to mismatches within the critical binding region were significantly lower, whereas the error rates corresponding to mismatches outside this region were much higher, consistent with the findings in the literature [27,32,60]. Figure 7.8 shows the positional effect of mismatches in *C. elegans*, *S. pombe*, and human genomes for $p = 1$ to $19, n = 21, l = 100$, and $m = 3$. As demonstrated in the figure, the off-target errors with mismatches in the critical binding region were significantly lower, with small p-values ($< 10^{-4}$) for the three organisms. In the case of *C. elegans*, for example, the average off-target error with mismatches in the critical binding region was 10.2%, whereas the average error with mismatches outside the region was 15.8%, with a standard deviation of 0.41% and a p-value of less

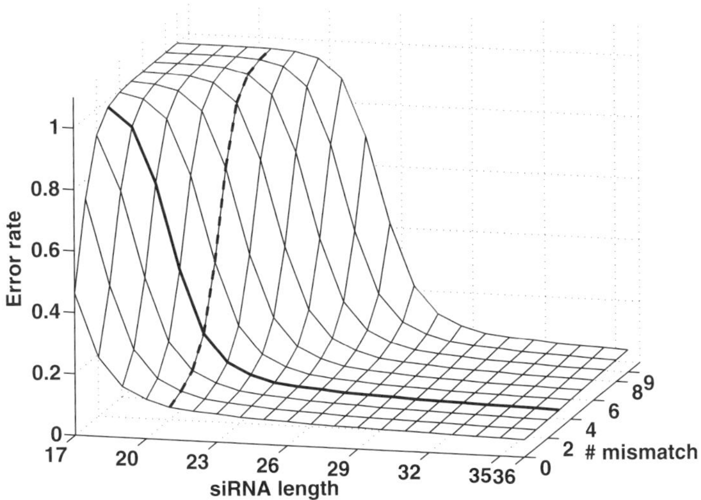

FIGURE 7.7 Effect of longer mismatches and longer siRNAs in *C. elegans*; $l = 200, m = 0$ to 9, and $n = 17$ to 35.

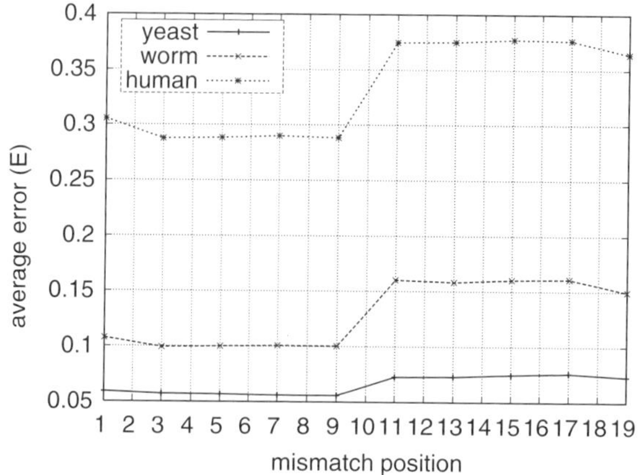

FIGURE 7.8 Effect of mismatch positions in *S. pombe*, *C. elegans*, and *Homo sapiens* ($p = 1$ to 19, $n = 21$, $l = 100$, $m = 3$). Mismatches in the range 2 to 9 nt on the guiding strand reduces off-targeting chances significantly.

than 10^{-4}. Since mismatch dominates off-target errors in the imperfect patterns, the positional effects of wobble should present closely similar behaviors.

The general patterns of off-target error rate demonstrates that increasing the mismatch between siRNA and mRNA raises the chances for off-targeting. If the binding is too flexible, it is possible that silencing one gene will knock down a large number of other genes. If the majority of genes are silenced, the organism cannot function normally. Since RNAi does not cause such instability in reality, there must an upper limit on the flexibility. Previously, we estimated this limit on mismatch allowed by nature using a network model by simulating transitive RNAi in *S. pombe* [52].

In this graph model representing gene knockdown networks, each vertex represented a gene and each edge indicated a knockdown interaction. Connected genes can be knocked down at the same time. Results showed that the degrees followed a power-law-like distribution, as shown in Figure 7.9. This model also demonstrated that increasing m increased the size of the largest cluster of connected genes, making the graph dense. When $m = 6$, a giant component existed, yielding a possibility of silencing the majority of genes in *S. pombe*. Consequently, the critical mismatch is not likely to be greater than 6 nt.

7.2.6 Silencing Multiple Genes

Although most RNAi research focused on targeting a single gene, silencing multiple genes simultaneously has also been studied [75]. Multiple-gene knockdown arises when it is necessary to suppress, simultaneously, the activities of a group of closely related genes with a mutually redundant function or an entire family of genes in a pathway. To silence a family of genes, it is less clear how to design an *optimal set* of siRNAs to target the family of genes. This is further complicated by the need to

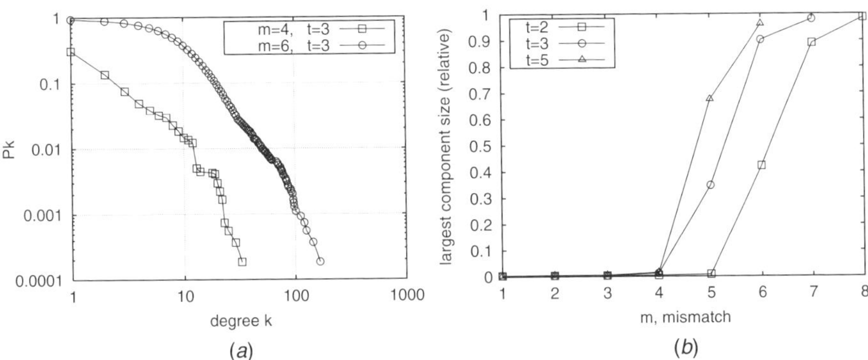

FIGURE 7.9 Properties of gene knockdown networks of tRNAi in *S. pombe*. (*a*) Log-log plots of degree distributions. The *x*-axis is the degree in logarithmic scale; the *y*-axis is the cumulative frequency $P_{\text{cum}}(k) = \sum_{d \geq k} P(d)$ in logarithmic scale. Three stages of tRNAi ($t = 3$) were simulated. (*b*) Phase transitions: *m*, on the x-axis, is the number of mismatches in siRNA–mRNA bindings; the *y*-axis is ρ, the largest component size divided by the number of genes.

account for the variable efficacy of different siRNA molecules and to avoid off-target effects. The goal of *gene family knockdown* is to select a minimal set of siRNAs that (1) cover a targeted gene family or a specified subset of it, (2) do not cover untargeted genes, and (3) are individually highly effective at inducing knockdown. This siRNA covering problem can be reduced to the classical set cover problem, and its exact solution is NP-hard. The problem of silencing multiple genes is an independent topic on its own. But since it is related to RNAi specificity and to space limitations, we discuss it here instead of in a separate section.

A probabilistic greedy algorithm for selecting minimal siRNA covers was proposed. Through randomization, the probabilistic greedy algorithm successfully avoids being trapped in local optima [75]. Results on real biological data showed that the probabilistic greedy algorithm reduces the siRNA cover size by up to 5.3% compared to the standard greedy algorithm and produces siRNAs that cover as well as the exact branch-and-bound algorithm in most cases. Overall, the gene family knockdown approach significantly reduces the number of siRNAs (e.g., up to 52%) required in gene family knockdown experiments compared to knocking down genes independently, and a typical case is shown in Figure 7.10.

7.3 COMPUTATIONAL METHODS FOR microRNAs

There are some similarities and differences between an miRNA and an siRNA. microRNAs are produced from short-hairpin RNAs by dicer. They usually act to degrade translation of their target genes through imperfect hybridization to target sites in the 3′ UTRs [58]. An miRNA usually hybridizes with multiple sites along a single target or different targets. The destruction of the target is effected by an miRNA

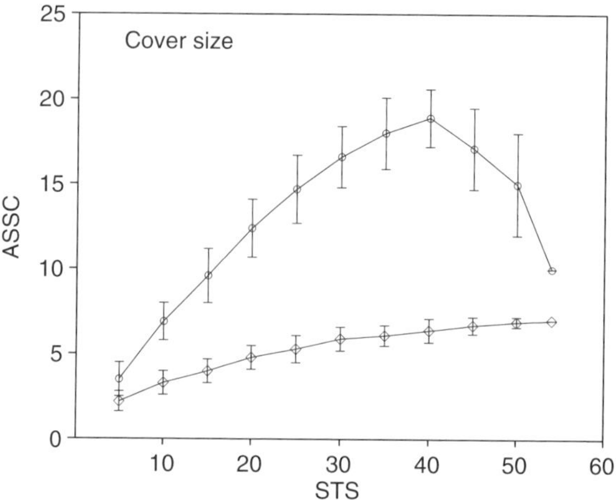

FIGURE 7.10 Average size of siRNA covers for the 54 fibrinogen (FBG) sequences of *B. glabrata.* "ASSC" represents the average size of siRNA covers. "STS" denotes the target size. The upper curve represents the case where the off-target filter is on; in the lower curve, the filter is off. (From ref. 75.)

ribonucleoprotein complex (miRNP), which shares many similarities to RISC in siRNA-initiated RNAi. If an miRNA is perfectly complementary to a 3′ UTR target site, it also results in mRNA degradation, suggesting that an miRNA also enters the RNAi pathway. Since biochemical cloning, which was initially used to find miRNAs, favors detection of abundant miRNA, computational methods are preferred for miRNA prediction. In this section we discuss briefly prediction methods for miRNA genes and target genes.

7.3.1 Prediction of microRNA Genes

Since miRNAs *lin-4* and *let-7* were discovered, some of their orthologs were identified by sequence-similarity searches. However, due to the stem-loop (hairpin) structure from which miRNAs are derived, sequence alignment alone is inadequate for miRNA prediction. Most miRNA prediction approaches are designed according to the principles that miRNA are derived from the precursor transcripts of the stem-loop structures, miRNAs are highly conserved among related species, and miRNAs present evolutionary divergence. The MiRseeker employed these three principles and predicted miRNA genes for *Drosophila melanogaster* [34]. The MiRscan program first predicted secondary structures using RNAfold [24] and then evaluates the stem-loop precursors by assigning a log-likelihood score to each 21-nt sliding window along the precursors of the conserved stem loops [37]. ProMiR used a pair of hidden Markov models (HMMs) to build a probabilistic colearning model and considered the structure and the precursor sequence of the miRNA simultaneously [42]. Table 7.1 summarizes these commonly used methods.

TABLE 7.1   Various Methods for miRNA Gene Prediction

Program/Author	Method[a]	Species	Ref.
MiRseeker	Hairpin, sequence conservation and divergence	*Drosophila*	[34]
MiRscan	Hairpin, likelihood score, sliding window	*C. elegans*	[37]
MIRFINDER	Hairpin, GC, entropy, sequence conservation	*Arabidopsis*	[10]
Wang et al.	Hairpin, GC, loop length, mismatch, conservation	*Arabidopsis*	[69]
ProMiR	Hairpin, hidden Markov model	Humans	[42]

[a]GC denotes G-C content.

7.3.2   Prediction of miRNA Targets

The purpose of miRNA target identification is to determine the regulatory functions of an miRNA. Most prediction methods are based on the rules derived from the characteristics of the imperfect base pairing between known miRNAs and their target sites. Biological observations suggested that the important base-pairing criteria include sequence complementarity between the 3′ UTR of the potential targets and the miRNA, the conservation of target sites in related genomes, and the thermo-dynamics of the association between the miRNAs and their target sites. We discuss some commonly used methods below.

The widely used TargetScan (and its latest version, TargetScanS) requires the sequence identity of six or seven nucleotides and searches for target sites containing sequence conservation across related species [36]. Since plant miRNAs bind to their target sites in a relatively stringent way, predictive rules are straightforward. PatScan, designed for plant miRNA target prediction, employed mainly three rules: 4 nt or less mismatches, no gap, and no G-U wobble in the association between miRNA and its potential targets [16]. The miRanda program used three rules: sequence base pairing using a position-weighted local alignment, free energies of RNA–RNA duplexes, and target site conservation across related species [17]. A kernel method was used for target prediction by employing biologically sensible data and position-based features reflecting the binding between an miRNA and its target [33]. A Bayesian network model was used for target prediction by combining target sequence, miRNA gene, and gene expression data [25]. Methods selected for target prediction are listed in Table 7.2.

TABLE 7.2   Various Methods for miRNA Target Prediction

Program/Author	Method	Species	Ref
PatScan	4-nt mismatch, no gap, no G-U wobble	*A. thaliana*	[16]
TargetScan	Base pairing, conservation	*Drosophila*	[36]
miRanda	Hairpin, cross-species comparison	Human	[17]
RNAhybrid	Energy minimization, dynamic programming	*Drosophila*	[56]
Kim et al.	String kernel, binding features		[33]
GenMiR	Bayesian network, gene expression	Mouse	[25]

7.4 siRNA SILENCING EFFICACY

The goal of an RNAi experiment is to maximally silence the target mRNA and to minimize off-target knockdown. Unfortunately, not only can siRNAs be off-targeting, they can also be defective in that they lead to unsatisfactory or even unobservable target degradations. It has been reported that siRNAs targeting different positions on the same mRNA displayed widely differing effectiveness, and on average less than 30% of random siRNAs are functional [4]. We have studied off-target effects in Section 1.2. In this section we focus on the problem of siRNA silencing efficacy, which is defined as the percentage of mRNA knockdown. Most biologists believe that efficacy is dependent on siRNA sequence rather than on the accessibility of the target site as determined by its secondary structure [4,57], although the impacts of the target site have been reported [73]. We focus on the design rules derived from siRNA sequence descriptors.

7.4.1 siRNA Design Rules

We briefly discuss commonly used siRNA design rules.

Ui-Tei's Rule Ui-Tei et al. [66] investigated the relationship between siRNA sequence and RNAi effectiveness in human, Chinese hamster, and *Drosophila* cells, and presented four rules for describing functional siRNAs: (1) an A/U residue at the 5$'$ end of the antisense strand; (2) G/C at the 5$'$ end of the sense strand; (3) at least five A/U residues in the 5$'$ terminal one-third of the antisense strand; and (4) the absence of any G/C stretch of more than 9 nt in length. This work also proposed a siRNA model, which suggests that the A/U richness at the 5$'$ termini on the antisense strand facilitates RISC's unwinding of the siRNA.

Amarzguioui's Rule Amarzguioui and Prydz [4] studied siRNA silencing efficacy in human and mouse cells and found the properties characterizing functional and nonfunctional siRNA sequences (70% activity). Their rules are: (1) $\Delta T_3 = T_3 - T_5$, the difference between the number of A/U residues in the three terminal positions at the 3$'$ and 5$'$ ends (relative to the sense strand of the siRNA) [$\Delta T_3 > 1$ is positively correlated (correlated with functional siRNA)]; (2) S_1, a G or C residue at position 1, positively correlated; (3) U_1, a U residue at position 1, negatively correlated; (4) A_6, an A residue at position 6, positively correlated; (5) W_{19}, A or U at position 19, positively correlated; and (6) G_{19}, G at position 19, negatively correlated. They also found that a G/C content of 31.6 to 57.9% was preferable. If a positive determinant is satisfied by an siRNA, its score is increased by 1; if it satisfies a negative determinant, its score is decremented. A total score, including ΔT_3, of 3 or higher is indicative of a functional siRNA.

Reynolds' Rule Reynolds et al. [57] examined 180 siRNAs targeting the firefly luciferase and human cyclophilin genes and discovered eight rules describing functional siRNA sequences. These rules are summarized in Table 7.3. If an siRNA

TABLE 7.3 Reynolds's Rules for Functional siRNA Sequences of 19 nt

No.	Conditions[a]
1	G/C content: 30–50%
2	At least 3 A/U bases at positions 15–19
3	Absence of internal repeats ($T_m < 20°$)
4	An A base at position 19
5	An A base at position 3
6	A U base at position 10
7	A base other than G or C at 19
8	A base other than G at position 13

[a]Positions refer to the sense strand.

TABLE 7.4 Jagla's Rule Sets for Functional siRNA Sequences of 19 nt

Set	Criterion[a]
1	position 19 = (A/U), position 13–19 > 3(A/U), position 1 = (G/C), position 10 = (A/U)
2	position 19 = (A/U), position 13–19 > 3(A/U), position 1 = (G/C), position 10 = (G/C)
3	position 19 = (G/C), position 1 = (G/C), position 5–19 > 6(A/U), position 11 = (G/C)
4	position 19 = (A/U), position 13–19 > 3(A/U), position 1 = (A/U)

[a]Positions refer to the sense strand.

satisfies a criterion, its score is incremented. A score of 6 (or 5) or higher signifies an effective sequence.

Jagla's Rules   Jagla and co-workers [29] uncovered four rule sets using decision tree algorithms on 600 chemically modified 21-mer duplexes targeting human genes. These rule sets are listed in Table 7.4. A sequence satisfying the criteria of one of the rule sets is considered functional.

Huesken's Motifs   Huesken and colleagues trained a neural network using more than 2400 siRNAs targeting human and rodent genes, and discovered a collection of important motifs characterizing the most active and most inactive siRNA sequences [26]. They consider single nucleotide motifs of statistical significance (p-value < 0.05) as strong contributions to siRNA efficacy. The p-values were computed by comparing the background frequencies with the overrepresented frequencies using a binomial distribution. Table 7.5 lists motifs for the first two positions using the IUPAC ambiguous nucleotide code, taken from the table describing all the motifs [26]. Although motifs listed here for two positions out of the entirety of 21 positions are far from sufficient, they show the flavor of this rule set. These motifs serve as sequence features and can be used by machine learning algorithms such as neural networks and support vector machines for efficacy prediction.

These rules describe different aspects of the siRNA sequences and have demonstrated variable success in efficacy prediction. It can be seen that one common

**TABLE 7.5 Huesken's Motifs on the First Two Positions for
Functional and Nonfunctional 21-mer siRNA Sequences**

Most Active			Most Inactive		
Position	Motif	p-Value	Position	Motif	p-Value
1	A	4.5×10^{-6}	1	B	9.7×10^{-6}
1	D	8.8×10^{-7}	1	C	1.5×10^{-5}
1	H	8.9×10^{-16}	1	G	1.6×10^{-10}
1	T	4.9×10^{-11}	1	R	1.3×10^{-2}
1	W	2.9×10^{-27}	1	S	4.3×10^{-25}
1	Y	1.9×10^{-2}	1	V	2.7×10^{-16}
2	H	5.3×10^{-3}	2	C	9×10^{-5}
2	T	8×10^{-4}	2	M	2.2×10^{-2}

[a]Positions refer to the antisense strand. Nucleotides are denoted by the IUPAC
ambiguous nucleotide code: B = no A; D = no C; H = no G; K = G or U/T; M = A
or C; R = A or G; S = G or C; V = no U/T; W = A or U/T; Y = C or U/T.

descriptor of these rules is the requirement for the A/U richness at the $5'$ end of the
antisense strand, suggesting that RISC's unwinding is fundamental. We also notice
that the rules of Ui-Tei, Amarzguioui, Reynolds, and Jagla contained some rational
criteria and motif descriptors, whereas Huesken's rule is totally motif based. The
criteria in the rules of Ui-Tei, Amarzguioui, and Reynolds can be used to construct
filters according to the scores, whereas Jagla's and Huesken's rules are mainly for use
by computer programs. Next, we discuss efficacy prediction using these rules and
machine learning algorithms.

7.4.2 Efficacy Prediction with Support Vector Regression

Using a threshold on silencing efficacy, 70% for example, an siRNA can be categorized into two classes: functional, if its silencing efficacy is at least 70%, and
nonfunctional, otherwise. Then classification algorithms such as support vector
machines (SVMs) can be trained and used for efficacy prediction [1,59]. A decision
tree algorithm was used for classification by Jagla et al. [29]. These applications
classify siRNA sequences without evaluating their actual efficacy and are acceptable
for applications that do not require accurate estimations. But predicting the exact
silencing efficacy is necessary for advanced designs. For instance, some siRNAs are
more off-targeting within the functional group, and detailed efficacy estimation helps
trade off between silencing efficiency and off-target knockdown. However, accurate
efficacy prediction using methods such as regression is a much more difficult problem
than classification. Huesken et al. constructed a neural network to estimate the actual
efficacy for each siRNA and reported the Pearson correlation coefficient between the
real efficacy and its estimate [26], but the prediction error rate was not shown. In this
section we use support vector (SV) regression to predict siRNA efficacy.

We used three data sets: (1) 73 examples originally used to verify Reynolds'
rules (we call this data set KR); (2) 600 data points used to develop Jagla's rules

(we call it JA); and (3) 2400 samples used by Huesken et al. to extract the motifs (we call it HU). We applied three rule sets: Reynolds' rule, Amarzguioui's rule, and Huesken's motifs. We used two binary attributes to code each criterion of Reynolds' rule, resulting in an input dimension of 16. We employed 76 attributes to describe Huesken's motifs. We coded Amarzguioui's rules using a nine-dimensional input space, the three extra descriptors being T_3, T_5, and G/C content.

To predict the siRNA silencing efficacy, we trained a kernel SV regressor on each of the three input sets generated by the three rules on the three data sets. The SV regressor is in the form

$$f(x) = \sum_{i=1}^{l} (\alpha_i^+ - \alpha_i^-) k(x, x_i) + b \tag{7.10}$$

where $k(\cdot, \cdot)$ is a kernel function, x_i are the training points, x is an unseen data point whose efficacy is to be predicted, b is a threshold constant, and $\alpha_i^+ - \alpha_i^-$ are the representing coefficients of the support vectors [67]. The kernel functions we employed are Gaussian, $k(x, y) = \exp(-\|x - y\|^2 / 2\sigma^2)$; exponential, $k(x, y) = \exp(-\gamma \|x - y\|)$; linear, $k(x, y) = x^T y$; and polynomial, $k(x, y) = (1 + x^T y)^d$. On the KR data set, 10-fold cross-validation demonstrated that the best kernels are Gaussian, with $\sigma = 10$ for Reynolds' rule, $\sigma = 22.4$ for Amarzguioui's rule, and $\sigma = 17.1$ for features generated by Huesken's motifs. On the JA data set, we found that Gaussian kernels generated the best performance for the rules of Reynolds and Amarzguioui, but an exponential kernel produced the best cross-validation accuracy for Huesken's descriptors. Their parameters are $\sigma = 8.77$ (Reynolds' rule), $\sigma = 3.33$ (Amarzguioui's rule), and $\gamma = 0.0085$ (Huesken's motifs). Finally, on the HU data set, Gaussian kernels also led to the best accuracies, with parameters of $\sigma = 5$ (Reynolds' rule), $\sigma = 3.33$ (Amarzguioui's rule), and $\sigma = 10.5$ (Huskens's motifs).

The mean squared errors (MSE) and Pearson correlation coefficients (R) of the predictions are listed in Table 7.6. These correlation coefficients generated by SV regression are comparable with those by neural networks [26]. As shown, Reynolds' rules produced the best performance on the KR data set, which was used to derive this rule set. But it consistently generated the lowest performance among the three rule sets, except that on the KR data set it yielded better performance than Amarzguioui's rules. Overall, Husken's motifs produced the best performance. The rules of Amarzguioui

TABLE 7.6 siRNA Efficacy Prediction Using Kernel SV Regression[a]

Data Set (Size)	Reynolds' Rule		Amarzguioui's Rule		Huesken's Motif	
	MSE	R	MSE	R	MSE	R
KR (73)	0.0928	0.44	0.119	0.42	0.0892	0.52
JA (600)	0.0728	0.35	0.0618	0.52	0.0566	0.57
HU (2400)	0.0359	0.33	0.0268	0.59	0.0257	0.60

[a]MSE denotes mean squared error, and R, the Pearson correlation coefficient.

and Huesken both performed better on larger data sets than on smaller ones. On the KR data set, which was the smallest, Huesken's motifs yielded much better performance than Amarzguioui's rules, improving MSE by 25% and correlation coefficient by 24%. But on the HU data, which is large enough, Huesken's motifs produced only slightly better performance than Amarzguioui's rules, improving MSE by 4% and correlation coefficient by 2%. This observation might suggest that sequence motifs are more effective descriptors on smaller data sets.

To summarize, our empirical study of siRNA silencing efficacy on the three rule sets suggested that Huesken's motifs are the best descriptors and Amarzguioui's rules are close to Huesken's motifs on large data sets. It also demonstrated that SV regression and the kernel method are applicable to the problem of siRNA efficacy prediction using the descriptors generated by the design rules.

7.5 SUMMARY AND OPEN QUESTIONS

In this chapter we have described the mechanism and biology of RNA interference, followed by detailed discussions on RNAi specificity, gene family knockdown, microRNA and target prediction, and siRNA silencing efficacy. We noted that the techniques used in these computational studies are novel and diverse. For example, to investigate RNAi specificity, the concept of precision in information retrieval was used to define a quantitative measure for off-target error. String kernels as used by SVMs were applied to detect siRNA targets, and RNA behaviors were simulated to develop a similarity model that gave rise to the RNA string kernels. The problem of gene family knockdown was an NP-hard problem whose solution required approximation. To predict miRNA genes, the concepts of entropy in information theory and of hidden Markov models were employed. For effective miRNA target estimation, dynamic programming, the kernel, and Bayesian network methods were utilized. We also noted that many different algorithms and data structures are used extensively in RNAi design and research. For example, algorithms for siRNA target detection and miRNA prediction employed search trees, including BST, trie, B-tree, suffix tree, and graph algorithms such as connected component search. In addition, greedy approximation was developed for gene family knockdown. With respect to applications of machine learning and data mining to RNAi and miRNA, we found that decision tree, neural network, SVM classifier, SV regressor, HMM, and Bayesian network methods have been used successfully. We believe that more algorithms will be applied to solve problems in RNAi. Problems in this area also motivate novel computational models, methodologies, and algorithms for their efficient representations and solutions.

As computational biology and bioinformatics advance rapidly, recent publications are good sources for the newest developments. Furthermore, some problems are not agreed on in the bioinformatics community. For example, some results showed that siRNA silencing efficacy depends on the secondary structure of target mRNA, but most reports claim that target accessibility does not affect efficacy. There is also a need to develop more accurate and high-quality models. We address these debatable and open questions next.

7.5.1 siRNA Efficacy and Target mRNA Secondary Structures

Most rules, such as those of Ui-Tei, Amarzguioui, and Reynolds, suggested that siRNA silencing efficacy does not depend on target secondary structure. Sætrom et al. also claimed that efficacy is independent of target structure [59]. Experiments have shown that the nonessential nuclear lamina protein lamin A/C expressed in human and mouse cells have very different predicted secondary structures but have the same silencing efficacy using the siRNAs [4]. It has also been reported that shifting the siRNA target sequence by only a few nucleotides (keeping the accessibility on the secondary structure unchanged) resulted in quite different efficacy [4]. This evidence supported the independence of efficacy of target structures.

On the other hand, silencing efficacy has been reported to be strongly dependent on target structure [18,39,48,73]. In a computational model for siRNA efficacy prediction using a Eulerian graph representation, target context was also considered an important factor [49]. Therefore, it is debatable whether siRNA efficacy depends on target structure. Although this issue can only be solved through biological experiments and will be clear in the future, we can foresee its computational consequences. The independence of target structure makes easier siRNA design and off-target control. But if efficacy really depends on target accessibility, mRNA secondary structure prediction must be incorporated into siRNA efficacy rules. Since mRNA structure prediction is no simple task, efficacy prediction will become more complicated.

7.5.2 Dynamics of Target mRNA and siRNA

Most RNAi experiments measure target mRNA concentration after a period of time (e.g., 48 hours) after transfection. Therefore, the exact dynamics of target and siRNA concentration is not totally clear. For therapeutic applications, accurate estimation for the levels of target mRNA and siRNA helps transfect in time and avoid waste of siRNA duplexes. Accurate models for mRNA and siRNA concentration can also better predict the pool effects of siRNA, the synergistic effects of efficacy, and off-target effects. Differential equations were used to model the kinetics for transitive RNAi [7]. But the accuracy of this model needs further improvement to be practically useful. Along with the increasing availability of observatory data, more models will be introduced for RNAi dynamics.

7.5.3 Integration of RNAi into Network Models

Biological interactions such as metabolic systems, protein interactions, and cell signaling networks are complex systems, and network (graphs) models have been used to represent these systems [5]. Although RNAi plays a role in systems biology [40], most network models do not model the regulatory function of RNAi directly. In the future, more work could be done in this area. In a metabolic network, for example, the regulatory role of RNAi can be incorporated into the network so that quantities of the metabolites can be leveraged by the miRNAs or siRNAs. Integrating RNAi into the graphs is an important step in building the network of networks to totally model complex biological systems.

APPENDIX: GLOSSARY

dicer

A dimeric enzyme of the RNase III ribonuclease family that contains dual catalytic domains, which act in concert to cleave dsRNA or shRNA into siRNA or miRNA of length $\sim$21 nt.

dsRNA

(double-stranded RNA). Usually a few hundred nucleotides long in RNAi.

gene family knockdown

Knockdown of a group of closely related genes with mutually redundant functions or an entire family of genes in a pathway.

gene knockdown network

A graph representing knockdown interactions due to silencing different genes in an organism, where each gene is represented as a node and each interaction is represented as an edge.

guide strand

One strand, generally believed to be the antisense strand, of double-stranded siRNA; used to guide RISC's target recognition.

miRNA

(microRNA). A class of 21- to 23-nt single-stranded RNAs processed by Dicer from a $\sim$70-nt precursor of hairpin RNAs formed from endogenous transcripts. miRNAs usually inhibit target protein translation by hybridizing with $3'$ UTR regions of their target mRNA.

miRNP

The miRNA ribonucleoprotein complex that represses the target of miRNAs and shares many similarities with RISC.

mRNA

(messenger RNA). Transcribed by RNA polymerase from the genomic DNA. Ribosomes use mRNA as an information prototype and translate it into a protein that performs particular functions.

off-target effect

The effect of possible knockdown of unintentional genes when silencing a true target gene.

phase transition

A change in size of the largest component in a graph to a level covering a substantial portion of the graph, due to the change of parameters affecting the density of the graph. In the knockdown network, increasing the number of mismatched nucleotides creates a giant component and generates a phase transition.

PTGS — Posttranscriptional gene silencing. *See* RNAi.

RdRP — (RNA-directed RNA polymerase). A protein that produces a dsRNA upstream of the cleavage site hybridized by the initiating siRNA. This dsRNA is cleaved by Dicer to generate secondary siRNAs.

RISC — A protein–RNA effector nuclease complex that unwinds double-stranded siRNA and uptakes one strand (the guide strand) to recognize and destroy its target.

RNAi — (RNA interference). A cell defense mechanism that represses the expression of viral genes by recognizing and destroying their messenger RNAs (mRNAs), preventing them from being translated into proteins. RNAi also regulates expressions of endogenous genes and suppresses transposable elements. It is called *RNA silencing, quelling*, and *posttranscriptional gene silencing* (PTGS) in some situations.

RNAi kinetics — Concentrations and their dynamic changes of siRNA and mRNA in RNAi, usually modeled with differential equations.

RNA string kernels — String kernels that model common RNA behaviors, including mismatches, G-U wobbles, and bulges.

silencing efficacy — (siRNA silencing efficacy). Measured by the percentage level of target mRNA concentration.

siRNA — (Short interfering RNA). Usually 21 to 25 nucleotides long, cleaved by Dicer from dsRNA or shRNA.

siRNA design rules — Criteria describing siRNA silencing efficacy based on siRNA thermodynamic properties and sequence motifs. If target accessibility affects efficacy, these criteria also include target site descriptions.

siRNA efficacy prediction — The use of statistical and machine learning methods to estimate the silencing efficacy of a given siRNA.

shRNA — (short hairpin RNA). Formed endogenously or artificially synthesized from the precursor of a stem-loop structure.

stRNA — (short temporal RNA). A class of miRNAs that regulate expressions of genes performing timing and developmental functions.

tRNAi — (transitive RNAi). It causes silencing to spread and induces chain reactions of gene knockdown due to siRNA proliferation.

REFERENCES

1. S. Abubucker. Machine learning approaches to siRNA efficacy prediction. Technical Report TR-CS-2005-26. Computer Science Department, University of New Mexico, Albuquerque NM, 2005.

2. N. Agrawal, P. V. N. Dasaradhi, A. Mohmmed, P. Malhotra, R. K. Bhatnagar and S. K. Mukherjee. RNA interference: biology, mechanism and applications. *Microbiol. Mol. Biol. Rev.*, 67(4):657–685, 2003.

3. S. F. Altschul, W. Gish, W. Miller, E. W. Myers, and D. J. Lipman. Basic local alignment search tool. *J. Mol. Biol.*, 215:403–410, 1990.

4. M. Amarzguioui and H. Prydz. An algorithm for selection of functional siRNA sequences. *Biochem. Biophys. Res. Commun.*, 316:1050–1058, 2004.

5. A.-L. Barabási and Z. N. Oltvai. Network biology: understanding the cell's functional organization. *Nat. Rev. Genet.*, pp. 101–113, 2004.

6. D. C. Baulcombe. Fast forward genetics based on virus induced gene silencing. *Curr. Opin. Plant Biol.*, 2:109–113, 2001.

7. C. T. Bergstrom, E. McKittrick, and R. Antia. Mathematical models of RNA silencing: unidirectional amplification limits accidental self-directed reactions. *Proc. Natl. Acad. Sci. USA*, 100:11511–11516, 2003.

8. V. Bitko and S. Barik. Phenotypic silencing of cytoplasmic genes using sequence-specific double-stranded short interfering RNA and its application in the reverse genetics of wild-type negative-strand RNA viruses. *BMC Microbiol.*, 1, 2001.

9. J. Blaszczyk, J. E. Tropea, M. Bubunenko, K. M. Routzahn, D. S. Waugh, D. L. Court, and X. Ji. Crystallographic and modeling studies of RNase III suggest a mechanism for double-stranded RNA cleavage. *Structure (Cambridge)*, 9:1225–1236, 2001.

10. E. Bonnet, J. Wuyts, P. Rouze, and Y. Van de Peer. Detection of 91 potential conserved plant microRNAs in *Arabidopsis thaliana* and *Oryza sativa* identifies important target genes. *Proc. Natl. Acad. Sci. USA*, 101:11511–11516, 2004.

11. T. R. Brummelkamp, R. Bernards, and R. Agami. A system for stable expression of short interfering RNAs in mammalian cells. *Science*, 296:550–553, 2002.

12. N. J. Caplen, S. Parrish, F. Imani, A. Fire, and R. A. Morgan. Specific inhibition of gene expression by small double-stranded RNAs in invertebrate and vertebrate systems. 98(17):9742–9747, Aug. 2001.

13. A. M. Chalk, C. Wahlestedt, and E.L.L. Sonnhammer. Improved and automated prediction of effective siRNA. *Biochem. Biophys. Res. Commun.*, 319:264–274, 2004.

14. J. Couzin. Breakthrough of the year: small RNAs make big splash. *Science*, 298(5602):2296–2297, Dec. 2002.

15. A. Dillin. The specifics of small interfering RNA specificity. *Proc. Natl. Acad. Sci. USA*, 100:6289–6291, 2003.

16. M. Dsouza, N. Larsen, and R. Overbeek. Searching for patterns in genomic data. *Trends Genet.*, 13:497–498, 1997.

17. A. J. Enright, B. John, U. Gaul, T. Tuschl, C. Sander, and D. S. Marks. MicroRNA targets in *Drosophila*. *Genome Biol.*, 5(1):R1, 2003.

18. R. K.-K. Far and G. Sczakiel. The activity of siRNA in mammalian cells is related to structural target accessibility: a comparison with antisense oligonucleotides. *Nucleic Acids Res.*, 31:4417–4424, 2003.

19. A. Fire, S. Q. Xu, M. K. Montgomery, S. A. Kostas, S. E. Driver, and C. C. Mello. Potent and specific genetic interference by double-stranded RNA in *Caenorhabditis elegans. Nature*, 391(6669):806–811, 1998.

20. W. B. Frakes and R. Baeza-Yates, Eds. *Information Retrieval Data Structures and Algorithms.* Prentice Hall, Englewood Cliffs, NJ, 1992.

21. A. J. Hamilton and D. C. Baulcombe. A species of small antisense RNA in posttranscriptional gene silencing in plants. *Science*, 286:950–952, 1999.

22. S. M. Hammond, E. Bernstein, D. Beach, and G. J. Hannon. An RNA-directed nuclease mediates post-transcriptional gene silencing in *Drosophila* cells. *Nature*, 404:293–296, 2000.

23. G. J. Hannon. RNA interference. *Nature*, 418:244–251, July 2002.

24. I. L. Hofacker, W. Fontana, P. F. Stadler, S. Bonhöffer, M. Tacker, and P. Schuster. Fast folding and comparison of RNA secondary structures. *Monatsh. Chem.*, 125:167–188, 1994.

25. J. C. Huang, Q. D. Morris, and B. J. Frey. Detecting microRNA targets by linking sequence, microRNA and gene expression data. *Proc. 10th Annual International Conference on Research in Computational Molecular Biology* (RECOMB), Venice, Italy, 2006.

26. D. Huesken, J. Lange, C. Mickanin, J. Weiler, F. Asselbergs, J. Warner, B. Meloon, et al. Design of a genome-wide siRNA library using an artificial neural network. *Nat. Biotechnol.*, 23(8):995–1001, 2005.

27. A. L. Jackson, S. R. Bartz, J. Schelter, S. V. Kobayashi, J. Burchard, M. Mao, B. Li, G. Cavet, and P. S. Linsley. Expression profiling reveals off-target gene regulation by RNAi. *Nat. Biotechnol.*, 21(6):635–637, 2003.

28. J. M. Jacque, K. Triques, and M. Stevenson. Modulation of HIV-1 replication by RNA interference. *Nature*, 418:435–438, July 2002.

29. B. Jagla, N. Aulner, P. D. Kelly, D. Song, A. Volchuk, A. Zatorski, D. Shum, et al. Sequence characteristics of functional siRNAs. *RNA*, pp. 864–872, June 2005.

30. R. S. Kamath, A. G. Fraser, Y. Dong, G. Poulin, R. Durbin, M. Gotta, A. Kanapin, et al. Systematic function analysis of the *C. elegans* genome using RNAi. *Nature*, 421:231–237, 2003.

31. J. R. Kennerdell and R. W. Carthew. Use of dsRNA-mediated genetic interference to demonstrate that *frizzled* and *frizzled 2* act in the wingless pathway. *Cell*, 95(7):1017–1026, Dec. 1998.

32. A. Khvorova, A. Reynolds, and S. D. Jayasena1. Functional siRNAs and miRNAs exhibit strand bias. *Cell*, 115:209–216, 2003.

33. S.-K. Kim, J.-W. Nam, W.-J. Lee, and B.-T. Zhang. A kernel method for microRNA target prediction using sensible data and position-based features. *Proc. 2005 IEEE Symposium on Computational Intelligence in Bioinformatics and Computational Biology* (CIBCB'05), pp. 1–7, 2005.

34. E. C. Lai, P. Tomancak, R. W. Williams, and G. M. Rubin. Computational identification of *Drosophila* microRNA genes. *Genome Biol.*, 4(7):R:42, 2003.

35. R. C. Lee, R. L. Feinbaum, and V. Ambros. The *C. elegans* heterochronic gene *lin-4* encodes small RNAs with antisense complementarity to *lin-14. Cell*, 75:843–854, 1993.

36. B. P. Lewis, I-hung Shih, M. W. W. Jones-Rhoades, D. P. Bartel, and C. B. Burge. Prediction of mammalian microRNA targets. *Cell*, 115:787–798, 2003.

37. L. P. Lim, M. E. Glasner, S. Yekta, C. B. Burge, and D. P. Bartel. Vertebrate microRNA genes. *Science*, 299:1540, 2003.

38. C. Llave, K. D. Kasschau, and J. C. Carrington. Virus-encoded suppressor of posttranscriptional gene silencing targets a maintenance step in the silencing pathway. *Proc. Natl. Acad. Sci. USA*, 97:13401–13406, 2000.

39. K. Q. Luo and D. C. Chang. The gene-silencing efficiency of siRNA is strongly dependent on the local structure of mRNA at the targeted region. *Biochem. Biophys. Res. Commun.*, 318:303–310, 2004.

40. J. Moffat and D. M. Sabatini. Building mammalian signalling pathways with RNAi screens. *Nat. Rev. Mol. Cell Biol.*, 7:177–187, 2006.

41. M. K. Montgomery, S. Xu, and A. Fire. RNA as a target of double-stranded RNA-mediated genetic interference in *Caenorhabditis elegans*. *Proc. Nat. Acad. Sci. USA*, 95:15502–15507, Dec. 1998.

42. J.-W. Nam, K.-R. Shin, J. Han, Y. Lee, V. N. Kim, and B.-T. Zhang. Human microRNA prediction through a probabilistic co-learning model of sequence and structure. *Nucleic Acids Res.*, 33:3570–3581, 2005.

43. C. Napoli, C. Lemieux, and R. Jorgensen. Introduction of a chimeric chalcone synthase gene into *Petunia* results in reversible co-suppression of homologous genes in trans. *Plant Cell*, 2(4):279–289, Apr. 1990.

44. S. B. Needleman and C. D. Wunsch. *J. Mol. Biol.*, 48:443–453, 1970.

45. A. Nykanen, B. Haley, and P. D. Zamore. ATP requirements and small interfering RNA structure in the RNA interference pathway. *Cell*, 107:309–321, 2001.

46. J. O. Snøve and T. Holen. Many commonly used siRNA risks off-target activity. *Biochem. Biophys. Res. Commun.*, 319:256–263, 2004.

47. P. H. Olsen and V. Ambros. The *lin-4* regulatory RNA controls developmental timing in *C. elegans* by blocking *lin-14* protein synthesis after the initiation of translation. *Dev. Biol.*, 216:671–680, 1999.

48. M. Overhoff, M. Alken, R. K. Far, M. Lemaitre, B. Lebleu, G. Sczakiel, and I. Robbins. Local RNA target structure influences siRNA efficacy: a systematic global analysis. *J. Mol. Biol.*, 348:871–881, 2005.

49. P. Pancoska, Z. Moravek, and U.M. Moll. Effcient RNA interference depends on global context of the target sequence: quantitative analysis of silencing effciency using Eulerian graph representation of siRNA. *Nucleic Acids Res.*, 32(4):1469–1479, 2004.

50. A. E. Pasquinelli, B. J. Reinhart, F. Slack, M. Q. Martindale, M. I. Kuroda, B. Maller, D. C. Hayward, et al. Conservation of the sequence and temporal expression of *let-7* heterochronic regulatory RNA. *Nature*, 408:86–89, 2000.

51. S. Qiu, C. M. Adema, and T. Lane. A computational study of off-target effects of RNA interference. *Nucleic Acids Res.*, 33:1834–1847, 2005.

52. S. Qiu and T. Lane. Phase transitions in gene knockdown networks of transitive RNAi. *Proc. 6th International Conference on Computer Science* (ICCS'06), pp. 895–903, Reading, UK, 2006. Springer-Verlag LNCS 3992.

53. S. Qiu and T. Lane. RNA string kernels for RNAi off-target evaluation. *Int. J. Bioinf. Res. and Appl.*, 2(2):132–146, 2006.

54. S. Qiu, C. Yang, and T. Lane. Efficient target detection for RNA interference. *Proc. 2006 Grid and Pervasive Computing* (GPC'06), pp. 22–31, Taichung, Taiwan, 2006. Springer-Verlag LNCS 3947.

55. M. Raponi and G. M. Arndt. Double-stranded RNA-mediated gene silencing in fission yeast. 31(15):4481–4489, Aug. 2003.

56. M. Rehmsmeier, P. Steffen, M. Hochsmann, and R. Giegerich. Fast and effective prediction of microRNA/target duplexes. *RNA*, 10:1507–1517, 2004.

57. A. Reynolds, D. Leake, Q. Boese, S. Scaringe, W. S. Marshall, and A. Khovorova. Rational siRNA design for RNA interference. *Nat. Biotechnol.*, 22:326–330, 2004.

58. M. W. Rhoades, B. J. Reinhart, L. P. Lim, C. B. Burge, B. Bartel, and D. P. Bartel. Prediction of plant microRNA targets. *Cell*, 110:513–520, 2002.

59. P. Sætrom and J. O. Snøve. A comparison of siRNA efficacy predictors. *Biochem. Biophys. Res. Commun.*, 321:247–253, 2004.

60. S. Saxena, Z. O. Jonsson, and A. Dutta. Small RNAs with imperfect match to endogenous mRNA repress translation. *J. Biol. Chem.*, 278(45):44312–44319, 2003.

61. P. C. Scacheri, O. Rozenblatt-Rosen, N. J. Caplen, T. G. Wolfsberg, L. Umayam, J. C. Lee, C. M. Hughes, K. S. Shanmugam, A. Bhattacharjee, M. Meyerson, and F. S. Collins. Short interfering RNAs can induce unexpected and divergent changes in the levels of untargeted proteins in mammalian cells. *Proc. Natl. Acad. Sci. USA*, 101(7):1892–1897, 2004.

62. T. Sijen, J. Fleenor, F. Simmer, K. L. Thijssen, S. Parrish, L. Timmons, R. H. A. Plasterk, and A. Fire. On the role of RNA amplification in dsRNA-triggered gene silencing. *Cell*, 107:465–476, Nov. 2001.

63. N. A. Smith, S. P. Singh, M.-B. Wang, P. A. Stoutjesdijk, A. G. Green, and P. M. Waterhouse. Total silencing by intronspliced hairpin RNAs. *Nature*, 407:319–320, 2000.

64. T. F. Smith and M. S. Waterman. *J. Mol. Biol.*, 147(1):195–197, 1981.

65. J. Soutschek, A. Akinc, B. Bramlage, K. Charisse, R. Constien, M. Donoghue, S. Elbashir, et al. Therapeutic silencing of an endogenous gene by systemic administration of modified siRNAs. *Nature*, 432:173–178, 2004.

66. K. Ui-Tei, Y. Naito, F. Takahashi, T. Haraguchi, H. Ohki-Hamazaki, A. Juni, R. Ueda, and K. Saigo. Guidelines for the selection of highly effective siRNA sequences for mammalian and chick RNA interference. *Nucleic Acids Res.*, 32:936–948, 2004.

67. V. N. Vapnik. *Statistical Learning Theory.* Wiley, New York, 1998.

68. H. Vaucheret and M. Fagard. Transcriptional gene silencing in plants: targets, inducers and regulators. *Trends Genet.*, 17(1):29–35, Jan. 2001.

69. X.-J. Wang, J. L. Reyes, N.-H. Chua, and T. Gaasterland. Prediction and identification of *Arabidopsis thaliana* microRNAs and their mRNA targets. *Genome Biol.*, 5:R65, 2004.

70. P. M. Waterhouse, M.-B. Wang, and T. Lough. Gene silencing as an adaptive defence against viruses. *Nature*, 411:834–842, 2001.

71. M. Wilda, U. Fuchs, W. Wossmann, and A. Borkhardt. Killing of leukemic cells with BCR/ABL fusion gene by RNA interference (rnai). *Oncogene*, 21:5716–5724, 2002.

72. H. Xia, Q. Mao, S. L. Eliason, S. Q. Harper, I. H. Martins, H. T. Orr, H. L. Paulson, L. Yang, R. M. Kotin, and B. L. Davidson. RNAi suppresses polyglutamine-induced neurodegeneration in a model of spinocerebellar ataxia. *Nat. Med.*, 10:816–820, July 2004.

73. K. Yoshinari, M. Miyagishi, and K. Taira. Effects on RNAi of the tight structure, sequence and position of the targeted region. *Nucleic Acids Res.*, 32:691–699, 2004.

74. P. D. Zamore, T. Tuschl, P. A. Sharp, and D. P. Bartel. RNAi: double-stranded RNA directs the ATP-dependent cleavage of mRNA at 21 to 23 nucleotide intervals. *Cell*, 101:25–33, 2000.

75. W. Zhao, T. Lane, and M. Fanning. Efficient RNAi-based gene family knockdown via set cover optimization. *Artif. Intell. Med.*, 35:61–73, 2005.

8

PROTEIN STRUCTURE PREDICTION USING STRING KERNELS

HUZEFA RANGWALA, KEVIN DERONNE, AND GEORGE KARYPIS

Department of Computer Science and Engineering, University of Minnesota–Twin Cities, Minneapolis, Minnesota

With recent advances in large-scale sequencing technologies, we have seen exponential growth in protein sequence information. Currently, our ability to produce sequence information far outpaces the rate at which we can produce structural and functional information. Consequently, researchers rely increasingly on computational techniques to extract useful information from known structures contained in large databases, although such approaches remain incomplete. As such, unraveling the relationship between pure sequence information and three-dimensional structure remains one of the great fundamental problems in molecular biology.

The motivation behind the structural determination of proteins is based on the belief that structural information will ultimately result in a better understanding of intricate biological processes. Many methods exist to predict protein structure at different levels of granularity. Due to the interest in this subject from a wide range of research communities, a biennial competition, the Critical Assessment for Structure Prediction (CASP; http://predictioncenter.org/) assesses the performance of current structure prediction methods. In this chapter we show several ways in which researchers try to characterize the structural, functional, and evolutionary nature of proteins.

Knowledge Discovery in Bioinformatics: Techniques, Methods, and Applications, Edited by Xiaohua Hu and Yi Pan

8.1 PROTEIN STRUCTURE: GRANULARITIES

Within each structural entity called a protein there lies a set of recurring substructures, and within these substructures are smaller substructures still. As an example, consider hemoglobin, the oxygen-carrying molecule in human blood. Hemoglobin has four domains that come together to form its quaternary structure. Each domain assembles (i.e., folds) itself independently to form a tertiary structure. These tertiary structures are comprised of multiple secondary structure elements—in hemoglobin's case, α helixes. Alpha-helixes (and their counterpart, β-sheets) have elegant repeating patterns, that depend on sequences of amino acids. These sequences form the primary structure of a protein, the smallest structural division aside from atoms. Hence, the linear ordering of amino acids forms secondary structure, arranging secondary structures yields tertiary structure, and the arrangement of tertiary structures forms quaternary structure (Figure 8.1). Research in computational structure prediction concerns itself mainly with predicting secondary and tertiary structure from known experimentally determined primary structure. This is due to the relative ease of determining primary structure and the complexity involved in quaternary structure. In this chapter we provide an overview of current secondary-structure prediction techniques, followed by a breakdown of the tertiary-structure prediction problem and descriptions of algorithms for each of several more restricted problems.

8.1.1 Secondary-Structure Prediction

A sequence of characters representing the secondary structure of a protein describes the general three-dimensional form of local regions. These regions organize themselves into patterns of repeatedly occurring structural fragments independent of the rest of the protein. The most dominant local conformations of polypeptide chains are α helixes and β sheets. These local structures have a certain regularity in their form, attributed to the hydrogen-bond interactions between various residues. An α-helix has a coil-like structure, whereas a β-sheet consists of parallel strands of residues (see Figure 8.1). In addition to regular secondary structure elements, irregular shapes form an important part of the structure and function of proteins. These elements are typically termed *coil regions*.

Secondary structure can be divided into several types, although usually at least three classes (α-helixes, coils, and β-sheets) are used. No unique method of assigning residues to a particular secondary-structure state from atomic coordinates exists, although the most widely accepted protocol is based on the DSSP algorithm [25]. DSSP uses the following structural classes: H (α-helix), G (3_{10}-helix), I (π-helix), E (β-strand), B (isolated β-bridge), T (turn), S (bend), and − (other). Several other secondary-structure assignment algorithms use a reduction scheme that converts this eight-state assignment to three states by assigning H and G to the helix state (H), E and B to a the strand state (E), and the rest (I, T, S, and −) to a coil state (C). This is the format generally used in structure databases. Within the secondary-structure prediction problem, the task is to learn a model that assigns a secondary-structure state to each residue of an input sequence in the absence of atomic coordinates.

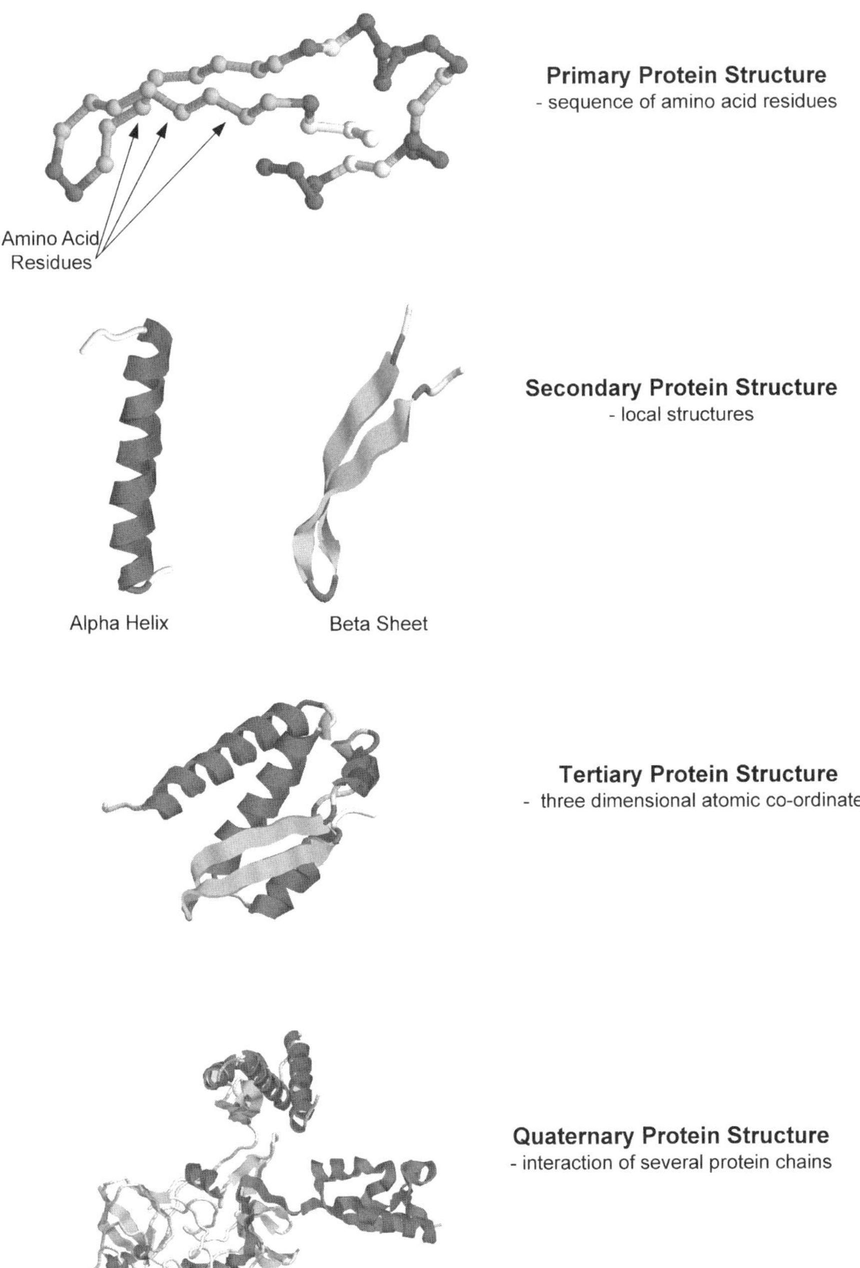

FIGURE 8.1   Overview of the protein structure prediction problem.

8.1.2 Protein Tertiary Structure

One of the biggest goals in structural bioinformatics is the prediction of the three-dimensional structure of a protein from its one-dimensional protein sequence. The goal is to be able to determine the shape (known as a *fold*) that a given amino acid sequence will adopt. The problem is divided further based on whether the sequence will adopt a new fold or resemble an existing fold (template) in a protein structure database. Fold recognition is easy when the sequence in question has a high degree of sequence similarity to a sequence with known structure [7]. If the two sequences share evolutionary ancestry, they are said to be homologous. For such sequence pairs we can build a structure for the query protein by choosing the structure of the known homologous sequence as a template. This is known as *comparative modeling*.

When the query lacks a good template structure, one must attempt to build a protein tertiary structure from scratch. These methods are usually called *ab initio methods*. In a third fold-prediction scenario, there may not necessarily be good sequence similarity with a known structure, but a structural template may still exist for the given sequence. To clarify this case, a person aware of the target structure could extract the template using structure–structure alignments of the target against the entire structural database. It is important to note that the target and template need not be homologous. These two cases define the fold prediction (homologous) and fold prediction (analogous) problems during CASP competition.

Comparative Modeling Comparative or homology modeling is used when there exists a clear relationship between the sequence of a query protein (unknown structure) to that of a sequence of a known structure. The most basic approach to structure prediction for such (query) proteins is to perform a pairwise sequence alignment against each sequence in protein sequence databases. This can be accomplished using sequence alignment algorithms such as Smith–Waterman [55] or sequence search algorithms (e.g., BLAST [3]). With a good sequence alignment in hand, the challenge in comparative modeling becomes how best to build a three-dimensional protein structure for a query protein using the template structure. The heart of the process is the selection of a suitable structural template based on sequence pair similarity. This is followed by the alignment of query sequence to the template structure selected to build the backbone of the query protein. Finally the entire structure modeled is refined by loop construction and side-chain modeling. Several comparative modeling methods, more commonly known as *modeler programs*, focusing on various parts of the problem have been developed over the past several years [6,13].

Fold Prediction (Homologous) Although satisfactory methods exist to detect homologs (proteins that share similar evolutionary ancestry) with high levels of similarity, accurately detecting homologs at low levels of sequence similarity (remote homology detection) remains a challenging problem. Some of the most popular approaches for remote homology prediction compare a protein with a collection of related proteins using methods such as PSI-BLAST [2], protein family profiles [15], hidden Markov models (HMMs) [5,30], and SAM [26]. These schemes produce models that are generative, in the sense that they build a model for a set of related proteins and then check to see how well this model explains a candidate protein.

In recent years, the performance of remote homology detection has been improved further through the use of methods that explicitly model the differences between the various protein families (classes) by building discriminative models. In particular, a number of different methods have been developed that use support vector machines (SVMs) [56] to produce results that are generally superior to those produced by either pairwise sequence comparisons or to approaches based on generative models— provided that there are sufficient training data. [17–19,31,33–35,52].

Fold Prediction (Analogous)   Occasionally, a query sequence will have a native fold similar to another known fold in a database but the two sequences will have no detectable similarity. In many cases the two proteins will also lack an evolutionary relationship. As the definition of this problem relies on the inability of current methods to detect sequential similarity, the set of proteins falling into this category remains in flux. As new methods continue to improve at finding sequential similarities as a result of increasing database size and better techniques, the number of proteins in question decreases. Techniques to find structures for such query sequences revolve around mounting the query sequence on a series of template structures in a process known as *threading* [18,20,21]. An objective energy function provides a score for each alignment, and the highest-scoring template is chosen. Obviously, if the correct template does not exist in the series, the method will not produce an accurate prediction. As a result of this limitation, predicting the structure of proteins in this category usually falls to new fold prediction techniques.

New Fold   Techniques to predict novel protein structure have come a long way in recent years, although a definitive solution to the problem remains elusive. Research in this area can be divided roughly into fragment assembly [24,28,32] and first-principle-based approaches, although occasionally the two are combined [9]. The former attempt to assign a fragment with known structure to a section of the unknown query sequence. The latter start with an unfolded conformation, usually surrounded by solvent, and allow simulated physical forces to fold the protein as would normally happen in vivo. Usually, algorithms from either class will use reduced representations of query proteins during initial stages to reduce the overall complexity of the problem.

8.2   LEARNING FROM DATA

Supervised learning is the task of creating a function that maps a set of inputs to a particular set of outputs by examining labeled training data. This form of learning plays a vital role in several bioinformatic applications, including protein structure prediction. Several books [10,11,56] cover the foundations of supervised learning in detail. The general framework of a supervised learning problem is as follows. Given an input domain $\mathcal{X}$ and output domain $\mathcal{Y}$, learn a function mapping each element of $\mathcal{X}$ to an element in domain $\mathcal{Y}$. In formal terms, given some training data $(X_1, Y_1) \cdots (X_n, Y_n)$, we need to learn a function $h\colon \mathcal{X} \to \mathcal{Y}$ mapping each object $X_i \in \mathcal{X}$ to a classification label $Y_i \in \mathcal{Y}$.

It is assumed that there exists an underlying probability distribution $D(X, Y)$ over $\mathcal{X} \times \mathcal{Y}$. This distribution remains unchanged for the training and test samples, but this distribution is unknown. The training and test samples are assumed to be drawn independently, identically distributed from $D(X, Y)$.

Classifiers can be categorized as parametric models and distribution-free models. Parametric models attempt to solve the supervised learning problem by explicitly modeling the joint distribution $D(X, Y)$ or conditional distribution $D(Y|X)$ for all X. Bayesian and hidden Markov models are examples of parametric models. Distribution-free models make no attempt to learn the distribution, but rather, choose a function in a selected hypothesis space for classification purposes. Margin-based learners (e.g., SVMs) are distribution-free classifiers.

8.2.1 Kernel Methods

Given a set of positive training examples $\mathcal{S}^+$ and a set of negative training examples $\mathcal{S}^-$, a support vector machine (SVM) learns a classification function $f(X)$ of the form

$$f(X) = \sum_{X_i \in \mathcal{S}^+} \lambda_i^+ \mathcal{K}(X, X_i) - \sum_{X_i \in \mathcal{S}^-} \lambda_i^- \mathcal{K}(X, X_i) \tag{8.1}$$

where λ_i^+ and λ_i^- are nonnegative weights that are computed during training by maximizing a quadratic objective function, and $\mathcal{K}(\cdot, \cdot)$ called the *kernel function*, is computed over all training-set and test-set instances. Given this function, a new instance X is predicted to be positive or negative depending on whether $f(X)$ is positive or negative. In addition, the value of $f(X)$ can be used to obtain a meaningful ranking of a set of instances, as it represents the strength by which they are members of the positive or negative class.

When computed over all pairs of training instances, the kernel function produces a symmetric matrix. To ensure the validity of a kernel, it is necessary to ensure that it satisfies Mercer's conditions, which require the pairwise matrix generated by the kernel function to be positive semidefinite. Formally, any function can be used as a kernel as long as for any number n and any possible set of distinct instances $\{X_1, \ldots, X_n\}$, the $n \times n$ Gram matrix defined by $K_{i,j} = \mathcal{K}(X_i, X_j)$ is symmetric positive semidefinite.

A symmetric function defined on the training set instances can be converted into a positive definite by adding to the diagonal of the training Gram matrix a sufficiently large nonnegative constant [52]. For example, the constant shift embedding kernelizing approach proposes use of the smallest negative eigenvalue subtracted from the main diagonal [58].

8.3 STRUCTURE PREDICTION: CAPTURING THE RIGHT SIGNALS

Thus far we have looked at several problems within the larger context of protein structure prediction. An ideal solution to the structure prediction problem would correctly predict, using only sequence information, the complete native conformation of a protein in three-dimensional space. Due to the difficulty of developing such a grand solution, decomposing the problem has led to good solutions to smaller parts of the problem.

In the remainder of this chapter we focus on three common prediction problems: secondary structure prediction, remote homology, and fold prediction. We also describe a class of methods that employs large margin classifiers with novel kernel functions for solving these problems.

One of the fundamental steps in building good classification models is selecting features that fit the classification task well. The input domain X for the protein structure prediction problems comprises the amino acid residues and their properties.

A protein sequence X of length n is represented by a sequence of characters $X = \langle a_1, a_2, \ldots, a_n \rangle$ such that each character corresponds to one of the 20 standard amino acids. Quite often, the learning and prediction algorithms segment the sequence into short contiguous segments called w-mers. Specifically, given a sequence X of length n and a user-supplied parameter w, the w-mer at position i of X ($w < i \leq n - w$) is defined to be the $(2w + 1)$-length subsequence of X centered at position i. That is, the w-mer contains a_i, the w amino acids before a_i, and the w amino acids after a_i. We denote this subsequence w-mer$_X(i)$.

It is widely believed that a sequence of amino acids encodes a structural signal [4], and this belief forms the underlying premise of the protein structure prediction problem. Working under this assumption, researchers have tried to encapsulate protein sequence information in various forms for structure analysis. One common way to incorporate more information about the structure of a sequence is to consider similar (and hopefully, therefore, related) sequences. Using multiple sequence alignments one can infer structural information about conserved regions. Many classifiers take as input profiles constructed from such alignments.

The profile of a sequence X of length n can be represented by two $n \times 20$ matrices. The first is its position-specific *scoring matrix* PSSM$_X$, which is computed directly by PSI-BLAST using the scheme described in ref. 2. The rows of this matrix correspond to the various positions in X, and the columns correspond to the 20 distinct amino acids. The second matrix is its position-specific *frequency matrix* PSFM$_X$, which contains the frequencies used by PSI-BLAST to derive PSSM$_X$. These frequencies (also referred to as *target frequencies* [38]) contain both the sequence-weighted observed frequencies (also referred to as *effective frequencies* [38]) and the BLOSUM62 [16]-derived pseudocounts [2].

We use the notation defined above to illustrate the machine learning methods used for secondary structure prediction, remote homology detection, and fold recognition.

8.4 SECONDARY-STRUCTURE PREDICTION

A large number of secondary-structure prediction algorithms have been developed, and since their inception, prediction accuracy has been improved continuously. Currently, many algorithms can achieve a sustained three-state prediction accuracy in the range 77 to 78%, and combinations of them can sometimes improve accuracy further by one to two percentage points. These improvements have been well documented [51] and are attributed to an ever-expanding set of experimentally determined tertiary structures, to the use of evolutionary information, and to algorithmic advances.

The secondary-structure prediction approaches in use today can be broadly categorized into three groups: neighbor-based, model-based, and metapredictor-based. The neighbor-based approaches [14,23,53] predict the secondary structure by identifying a set of similar sequence fragments with known secondary structure; the model-based approaches [22,42,44,49], employ sophisticated machine learning techniques to learn a predictive model trained on sequences of known structure, whereas the metapredictor-based approaches [12,41] predict based on a combination of the results of various neighbor and/or model-based techniques. The nearly real-time evaluation of many of these methods performed by the EVA server [48] shows that model-based approaches tend to produce statistically better results than neighbor-based schemes, which are further improved by some of the more recently developed metapredictor-based approaches [41].

Historically, the most successful model-based approaches, such as PHD [49], PSIPRED [22], and SSPro [42], were based on neural network (NN) learning techniques. However, in recent years, a number of researchers have also developed secondary structure prediction algorithms based on support vector machines. In the remainder of this section we present one such SVM-based secondary structure prediction algorithm called YASSPP, which shows exemplary performance [29].

8.4.1 YASSPP Overview

The overall structure of YASSPP is similar to that used by many existing secondary-structure prediction algorithms such as PHD and PSIPRED. The approach is illustrated in Figure 8.2. It consists of two models, referred to as L_1 and L_2, that are connected together in cascaded fashion. The L_1 model assigns to each position a weight for each of the three secondary structure elements $\{C, E, H\}$, which are provided as input to the L_2 model to predict the actual secondary structure class of each position. The L_1 model treats each position of the sequence as an independent prediction problem, and the purpose of the L_2 model is to determine the structure of a position by taking into account the structure predicted for adjacent positions. YASSPP splits the training set equally between L_1 and L_2 models.

Both the L_1 and L_2 models consist of three binary SVM classifiers ($\{M_1^{C/\bar{C}}, M_1^{E/\bar{E}}, M_1^{H/\bar{H}}\}$ and $\{M_2^{C/\bar{C}}, M_2^{E/\bar{E}}, M_2^{H/\bar{H}}\}$, respectively) trained to predict whether or not a position belongs to a particular secondary structure state (i.e., one-vs. the rest models). The output values of the L_1 model are the raw functional outputs of these binary classifiers (i.e., $M_1^{C/\bar{C}}$, $M_1^{E/\bar{E}}$, and $M_1^{H/\bar{H}}$), whereas the secondary state predicted for the L_2 model corresponds to the state whose corresponding binary classifier achieves the maximum value. That is,

$$\text{predicted state} = \operatorname*{argmax}_{x \in \{C,E,H\}} (M_2^{x/\bar{x}}) \tag{8.2}$$

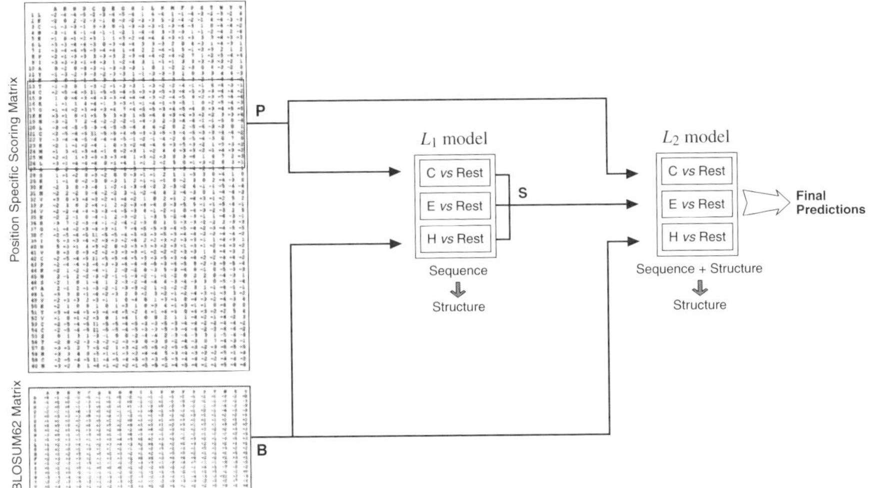

FIGURE 8.2 General architecture of YASSPP's learning framework.

During training, for each position i that belongs to one of the three secondary structure states (i.e., classes) of a sequence X, the input to the SVM is a $(2w + 1)$-length subsequence w-mer of X. The proper value for the parameter w is determined experimentally. During secondary structure prediction, a similar approach is used to construct a w-mer around each position i of a *query* sequence X with unknown secondary structure.

8.4.2 Input Sequence Coding

For the input sequence coding there are two different approaches for the L_1 model and two different approaches for the L_2 model. L_1's first coding scheme represents each w-mer x as a $(2w + 1) \times 20$ matrix P_x, whose rows are obtained directly from the rows of the PSSM for each position. The second coding scheme augments this PSSM-based representation by adding another $(2w + 1) \times 20$ matrix B_x, whose rows are the rows of the BLOSUM62 matrix corresponding to each position's amino acid. These schemes are referred as the P and PB coding schemes, respectively.

By augmenting the w-mer coding scheme to contain both PSSM- and BLOSUM62-based information, the SVM can learn a model that is also partially based on the non-position-specific information. This information will remain valid even in cases in which PSI-BLAST could not or failed to generate correct alignments.

The two coding schemes for the L_2 model are derived from the corresponding coding schemes of L_1 by including the predictions computed by L_1's three binary classifiers. This is done by adding another $(2w + 1) \times 3$ matrix S_x, whose columns store the raw functional predictions of the $M_1^{C/\bar{C}}$, $M_1^{E/\bar{E}}$, and $M_1^{H/\bar{H}}$ models, respectively. Thus, the first coding scheme consists of matrices P_x and S_x, and the second coding scheme consists of matrices P_x, B_x, and S_x. These coding schemes are novel compared to the existing methods.

8.4.3 Profile-Based Kernel Functions

YASSPP shows a methodology for designing and evaluation various kernel functions for use by binary SVM classifiers of the L_1 and L_2 models. It develops kernel functions that are derived by combining a normalized second-order kernel in which the contribution of each position decreases based on how far away it is from the central residue, along with an exponential function.

The general structure of the kernel functions used in YASSPP is given by

$$\mathcal{K}(x,y) = \exp\left(1.0 + \frac{\mathcal{K}_1(x,y)}{\sqrt{\mathcal{K}_1(x,x)\mathcal{K}_1(y,y)}}\right) \tag{8.3}$$

where x and y are two w-mers, $\mathcal{K}_1(x,y)$ is given by

$$\mathcal{K}_1(x,y) = \overset{cs}{\mathcal{K}_2}(x,y) + [\overset{cs}{\mathcal{K}_2}(x,y)]^2 \tag{8.4}$$

and $\mathcal{K}_2^{cs}(x,y)$ is a kernel function that depends on the choice of the particular input coding scheme cs, and for each one of the P, PB, PS, and PBS coding schemes is defined as follows:

$$\overset{P}{\mathcal{K}_2}(x,y) = \sum_{j=-w}^{j=w} \frac{P_x(j,:)P_y^t(j,:)}{1+|j|} \tag{8.5}$$

$$\overset{PB}{\mathcal{K}_2}(x,y) = \mathcal{K}_2^P(x,y) + \sum_{j=-w}^{j=w} \frac{B_x(j,:)B_y^t(j,:)}{1+|j|} \tag{8.6}$$

$$\overset{PS}{\mathcal{K}_2}(x,y) = \mathcal{K}_2^P(x,y) + \gamma \sum_{j=-w}^{j=w} \frac{S_x(j,:)S_y^t(j,:)}{1+|j|} \tag{8.7}$$

$$\overset{PBS}{\mathcal{K}_2}(x,y) = \mathcal{K}_2^{PB}(x,y) + \gamma \sum_{j=-w}^{j=w} \frac{S_x(j,:)S_y^t(j,:)}{1+|j|} \tag{8.8}$$

The various terms involving the rows of the P, B, and S matrices [e.g., $P_x(j,:)P_y^t(j,:)$] correspond to the dot products of the rows corresponding to the jth positions of the w-mers (indexed from $-w$ to $+w$). We do not delve into the various characteristics that are coded in the constructions of the kernel functions but direct the reader to ref. 29 for further details.

8.4.4 Performance Evaluation

A wide variety of data sets were used to assess the performance of YASSPP. A thorough parameter study was done to study the impact of the various coding schemes, the kernel choices, and the best parameters. We show some of the comparative performance study results for YASSPP.

The prediction accuracy is assessed using four widely used performance measures. These are the three-state per-residue accuracy (Q_3), the segment overlap measure (SOV), the per-state Matthews correlation coefficients (C_C, C_E, C_H), and the information index (Info). Q_3 is a measure of the overall three-state prediction accuracy and is defined as the percentage of residues whose structural class is predicted correctly [49]. The SOV is a segment-level measure of the overall prediction accuracy. This measure was introduced in ref. 50 and subsequently refined [54]. Matthews correlation coefficients [37] provide a per-state measure of prediction performance, and for a particular state $i \in \{C, E, H\}$ it is given by

$$C_i = \frac{p_i n_i - u_i o_i}{\sqrt{(p_i + u_i)(p_i + o_i)(n_i + u_i)(n_i + o_i)}} \tag{8.9}$$

TABLE 8.1 Performance on the EVAc4 Data Set

Scheme[a]	Q_3	SOV	Info	C_C	C_E	C_H
PHDpsi	74.52	70.69	0.346	0.529	0.685	0.665
PSIPRED	77.62	76.05	0.375	0.561	0.735	0.696
SAM-T99sec	77.64	75.05	0.385	0.578	0.721	0.675
PROFsec	76.54	75.39	0.378	0.562	0.714	0.677
[1]YASSPP$_{P+PS}$	78.35	77.20	0.407	0.589	0.746	0.708
ErrSig	0.86	1.21	0.015	0.015	0.021	0.017
[1]YASSPP$_{PB+PBS}$	79.34	78.65	0.419	0.608	0.747	0.722
ErrSig	0.82	1.16	0.015	0.015	0.021	0.016
SCRATCH	75.75	71.38	0.357	0.545	0.690	0.659
[2]YASSPP$_{P+PS}$	78.39	77.69	0.406	0.586	0.750	0.711
ErrSig	0.97	1.36	0.016	0.017	0.023	0.018
[2]YASSPP$_{PB+PBS}$	79.31	78.75	0.416	0.602	0.751	0.722
ErrSig	0.94	1.29	0.016	0.017	0.023	0.018
SSPro4	77.96	72.73	0.385	0.559	0.711	0.696
[3]YASSPP$_{P+PS}$	79.21	78.60	0.418	0.590	0.749	0.723
ErrSig	1.19	1.67	0.021	0.023	0.030	0.022
[3]YASSPP$_{PB+PBS}$	80.03	79.00	0.430	0.605	0.751	0.736
ErrSig	1.18	1.68	0.022	0.024	0.030	0.022
SABLE2	76.85	73.55	0.376	0.546	0.725	0.682
[4]YASSPP$_{P+PS}$	78.70	78.09	0.417	0.596	0.766	0.715
ErrSig	1.00	1.42	0.018	0.018	0.025	0.019
[4]YASSPP$_{PB+PBS}$	79.85	79.71	0.432	0.615	0.768	0.730
ErrSig	0.97	1.39	0.018	0.019	0.025	0.019

[a]YASSPP$_{P+PS}$ uses the $P + PS$ input coding and the YASSPP$_{PB+PBS}$ uses the $PB + PBS$ input coding and were obtained using $w = 7$ (i.e., w-mers of size 15). The [1]YASSPP are the averages over the set of sequences in common with PHDpsi, PSIPRED, SAM-T99sec, and PROFsec. The [2]YASSPP are the averages over the set of sequences in common with SCRATCH. The [3]YASSPP are the averages over the set of sequences in common with SSPro4. The [4]YASSPP are the averages over the set of sequences in common with SABLE2.

where p_i is the number of correctly predicted residues in state i, n_i the number of residues that were rejected correctly (true negatives), u_i the number of residues that were rejected incorrectly (false negatives), and o_i the number of residues that were predicted incorrectly to be in state i (false positives). Finally, the information index [49] is an entropy-related measure that merges the observed and predicted state-specific accuracy measures into a single number, with all these elements contributing equally.

Table 8.1 compares the performance achieved by YASSPP against that achieved by PHDpsi[44], PSIPRED[22], SAM-T99sec[27], PROFsec[47], SCRATCH[42], SSPro4[42], and SABLE2[43]. These schemes represent some of the best-performing schemes currently evaluated by the EVA server, and their results were obtained directly from EVA. Since EVA did not use all the methods to predict all the sequences of EVAc4, Table 8.1 presents four different sets of results for $YASSPP_{P+PS}$ and $YASSPP_{PB+PBS}$ (indicated by the superscripts 1 to 4), each obtained by averaging the various performance assessment methods over the common subset. These common subsets contained 165, 134, 86, and 115 sequences, respectively.

These results show that both $YASSPP_{P+PS}$ and $YASSPP_{PB+PBS}$ achieve better prediction performance than that achieved by any of the other schemes across all the

TABLE 8.2 Comparative Performance of YASSPP Against Other Secondary-Structure Prediction Servers

Scheme[a]	Q_3	SOV	Info	C_C	C_E	C_H
			RS126 Data Set			
PSIPRED	81.01	76.24	0.45	0.65	0.70	0.77
PHD	76.92	72.57	0.38	0.57	0.63	0.73
Prof	76.95	71.70	0.38	0.58	0.63	0.73
SSPro	77.01	70.24	0.38	0.58	0.61	0.72
$YASSPP_{P+PS}$	79.81	74.41	0.42	0.61	0.70	0.76
ErrSig	0.80	1.28	0.02	0.02	0.02	0.02
$YASSPP_{PB+PBS}$	80.29	75.65	0.43	0.61	0.70	0.75
ErrSig	0.79	1.25	0.02	0.02	0.02	0.02
			CB513 Data Set			
PSIPRED	79.95	76.48	0.43	0.63	0.68	0.76
PHD	77.61	74.98	0.39	0.59	0.65	0.73
Prof	77.13	73.74	0.39	0.58	0.64	0.73
SSPro	79.07	74.39	0.42	0.61	0.65	0.76
$YASSPP_{P+PS}$	80.52	77.39	0.45	0.62	0.70	0.74
ErrSig	0.40	0.60	0.01	0.01	0.01	0.01
$YASSPP_{PB+PBS}$	80.99	77.86	0.45	0.63	0.70	0.75
ErrSig	0.39	0.60	0.01	0.01	0.01	0.01

[a]$YASSPP_{P+PS}$ uses the $P + PS$ input coding and the $YASSPP_{PB+PBS}$ uses the $PB + PBS$ input coding. Both schemes use w-mers of length $15(w = 7)$. The results for PSIPRED, PHD, Prof, and SSPro were obtained from ref. 46. ErrSig is the significant difference margin for each score (to distinguish between two methods) and is defined as the standard deviation divided by the square root of the number of proteins $(\sigma/\sqrt{N})$.

different performance assessment measures. In particular, for the entire data set, YASSPP$_{PB+PBS}$ achieves a Q_3 score of 79.34%, which is 1.7 percentage points higher than the second-best-performing scheme in terms of Q_3 (SAM-T99sec), and an SOV score of 78.65%, which is 2.6 percentage points higher than the second-best-performing scheme in terms of SOV (PSIPRED).

Table 8.2 compares the performance achieved by YASSPP's production server with that achieved by other model-based servers, such as PSIPRED, PHD, Prof, and SSPro [46]. These results show that the performance achieved by YASSPP$_{P+PS}$ and YASSPP$_{PB+PBS}$ is in general higher than that achieved by the other servers. YASSPP$_{PB+PBS}$'s performance is one to four percentage points higher in terms of Q_3 and SOV. The only exception is the RS126 data set, for which PSIPRED achieves somewhat better prediction performance than either YASSPP$_{P+PS}$ or YASSPP$_{PB+PBS}$ (PSIPRED achieves a Q_3 score of 81.01 vs. 80.29 for YASSPP$_{PB+PBS}$). However, as measured by ErrSig, this performance difference is not statistically significant. Also, as was the case with the previous results, YASSPP$_{PB+PBS}$ achieves better prediction performance than that achieved by YASSPP$_{P+PS}$.

8.5 REMOTE HOMOLOGY AND FOLD PREDICTION

Both remote homology detection and fold recognition are central problems in computational biology and bioinformatics, with the aim of classifying protein sequences into structural and functional groups or classes. Pairwise sequence comparison methods (e.g., sequence alignment algorithms such as Smith–Waterman [55] and sequence database search tools such as BLAST [1]) are able to detect homologous sequences with a high percentage sequence identity. However, as the percent identity between sequence pairs decreases, the problem of finding the correct homologous pairs becomes increasingly difficult.

Some of the better-performing schemes in this domain use profile information to compare a query protein with a collection of related proteins. Profiles for a sequence can be defined in terms of a multiple-sequence alignment of a query sequence with its statistically significant homologs (as computed by PSI-BLAST [2]) or in the form of hidden Markov model (HMM) states [5,30]. The models built in this fashion are examples of generative models.

The current state-of-the-art methods employ discriminative-based modeling techniques and have a distinct advantage over generative models in this domain. Support vector machines have been the popular choice of discriminative learners.

One of the early attempts at using a feature-space-based approach is the SVM–Fisher method [19], in which a profile HMM model is estimated on a set of proteins belonging to the positive class. This HMM is then used to extract a vector representation for each protein. Another approach is the SVM-pairwise scheme [35], which represents each sequence as a vector of pairwise similarities between all sequences in a training set. A relatively simpler feature space that contains all possible short subsequences ranging from three to eight amino acids (k-mers) is explored in a series of papers (Spectrum kernel [33], Mismatch kernel [34], and Profile kernel [31]).

All three of these methods represent a sequence X as a vector in this simpler feature space, but differ in the scheme they employ to actually determine if a particular dimension u (i.e., k-mer) has a nonzero weight in X's vector. The Spectrum kernel considers u to be present if X contains u as a substring, the Mismatch kernel considers u to be present if X contains a substring that differs with u in at most a predefined number of positions (i.e., mismatches), whereas the Profile kernel considers u to be present if X contains a substring whose PSSM-based ungapped alignment score with u is above a user-supplied threshold. An entirely different feature space is explored by the SVM-Isites [17] and SVM-HMMSTR[18] methods, which take advantage of a set of local structural motifs (SVM-Isites) and their relationships (SVM-HMMSTR).

An alternative to measuring pairwise similarity through a dot product of vector representations is to calculate an explicit protein similarity measure. The recently developed LA-Kernel method [52] represents one such example of a *direct kernel function*. This scheme measures the similarity between a pair of protein sequences by taking into account all the optimal gapped local alignment scores between all possible subsequences of the pair. The experiments presented in ref. 52 show that this kernel is superior to schemes developed previously that do not take into account sequence profiles and that the overall classification performance improves by taking all possible local alignments into account.

8.5.1 Profile-Based Kernel Functions

Recently, a set of direct profile-based kernel functions were developed and tested to show very good performance [45]. The first class, referred to as *window-based*, determines the similarity between a pair of sequences by combining ungapped alignment scores of fixed-length subsequences. The second, referred to as *local alignment-based*, determines the similarity between a pair of sequences using Smith–Waterman alignments and a position independent affine gap model, optimized for the characteristics of the scoring system. Both kernel classes utilize profiles constructed automatically via PSI-BLAST and employ a profile-to-profile scoring scheme that extends a recently introduced profile alignment method [38].

One way of computing the profile-to-profile scores would be to take the dot product between the profile columns for the two positions:

$$S_{X,Y}(i,j) = \sum_{k=1}^{20} \text{PSSM}_X(i,k)\text{PSSM}_Y(j,k) \tag{8.10}$$

Another example of such a scoring function [45] is given by

$$S_{X,Y}(i,j) = \sum_{k=1}^{20} \text{PSFM}_X(i,k)\text{PSSM}_Y(j,k)$$
$$+ \sum_{k=1}^{20} \text{PSFM}_Y(j,k)\text{PSSM}_X(i,k) \tag{8.11}$$

This particular scoring function captures the similarity between the two profile positions using both position-specific scoring matrices and position-specific frequency matrices.

Smith–Waterman-Based Kernel Functions   As explained in Section 8.2.1, the choice of kernel function plays a critical role in the performance of a classifier. A simple Smith–Waterman-based alignment scoring scheme can be used as a kernel function provided that steps are followed to ensure its validity: specifically, that it follows Mercer's conditions. The Smith–Waterman kernel computes the similarity between a pair of sequences X and Y by finding an optimal alignment between them that optimizes a particular scoring function. Given two sequences X and Y of length n and m, respectively, the SW-PSSM kernel computes their similarity as the score of the optimal local alignment. In this alignment, the similarity between two sequence positions is determined using the profile-to-profile scoring scheme of Eq. (8.11) and a position-independent affine gap model.

Within this local alignment framework, the similarity score between a pair of sequences depends on the gap-opening (go) and gap-extension (ge) costs and the intrinsic characteristics of the profile-to-profile scoring scheme. A scoring system whose average score is positive will tend to produce very long alignments, potentially covering segments of low biologically relevant similarity. On the other hand, if the scoring system cannot easily produce alignments with positive scores, it may fail to identify any nonempty similar subsequences. To obtain meaningful local alignments, the scoring scheme that is used should produce alignments whose score must on average be negative, with the maximum score being positive [55].

To ensure that the SW-PSSM kernel can account correctly for the characteristics of the scoring system, the profile-to-profile scores calculated from equation (8.11) are modified by adding a constant value. This scheme, commonly referred to as *zero shifting* [57], ensures that the resulting alignments have scores that are negative on the average, while allowing for positive maximum scores.

Window-Based Kernel Functions   The local alignment-based kernels capture the similarity between sequence pairs by combining the ungapped alignment scores of w-mer subsequences between the various positions of the sequences. Based on the combination of fixed- and variable-length w-mers for different pair positions between sequences, [45] introduces three novel window-based kernel functions.

The ungapped alignment score between two w-mers is computed using the profile-to-profile scoring method of Eq. (8.11) as follows:

$$w\text{-score}_{X,Y}(i,j) = \sum_{k=-w}^{w} S_{X,Y}(i+k,j+k) \tag{8.12}$$

The all fixed-width w-mer (AF-PSSM) kernel computes the similarity between a pair of sequences X and Y by adding up the alignment scores of all possible w-mers between X and Y that have a positive ungapped alignment score. Specifically, if the ungapped alignment score between two w-mers at positions i and j of X and Y, respectively, is denoted by $w\text{-score}_{X,Y}(i,j)$, n and m are the lengths of X and Y, respectively, and $\mathcal{P}_w$ is the set of all possible w-mer pairs of X and Y with a positive ungapped alignment score, that is,

$$\mathcal{P}_w = \{(w\text{-mer}_X(i), w\text{-mer}_Y(j)) | w\text{-score}_{X,Y}(i,j) > 0\} \tag{8.13}$$

for $w + 1 \leq i \leq n - w$ and $w + 1 \leq j \leq m - w$, the AF-PSSM kernel computes the similarity between X and Y as

$$\text{AF-PSSM}_{X,Y}(w) = \sum_{\substack{(w\text{-mer}_X(i), \\ w\text{-mer}_Y(j)) \in \mathcal{P}_w}} w\text{-score}_{X,Y}(i,j) \tag{8.14}$$

The best fixed-width w-mer (BF-PSSM) kernel improves on the AF-PSSM kernel by selecting a subset $\mathcal{P}'_w$ of $\mathcal{P}_w$ [as defined in Eq. (8.13)] such that (1) each position of X and each position of Y is present in at most one w-mer pair and (2) the sum of the w-scores of the pairs selected is maximized. Given $\mathcal{P}'_w$, the similarity between the pair of sequences is then computed as follows:

$$\text{BF-PSSM}_{X,Y}(w) = \sum_{\substack{(w\text{-mer}(X,i), \\ w\text{-mer}(Y,j)) \in \mathcal{P}'_w}} w\text{-score}_{X,Y}(i,j) \tag{8.15}$$

The relation between $\mathcal{P}'_w$ and $\mathcal{P}_w$ can be better understood if the possible w-mer pairs in $\mathcal{P}_w$ are viewed as forming an $n \times m$ matrix whose rows correspond to the positions of X, columns to the positions of Y, and values correspond to their respective w-scores. Within this context, $\mathcal{P}'_w$ corresponds to a matching of the rows and columns [40] whose weight is high (bipartite graph matching problem). Since the selection forms a matching, each position of X (or Y) contributes a single w-mer in Eq. (8.15) and as such eliminates the multiplicity present in the AF-PSSM kernel. At the same time, the BF-PSSM kernel attempts to select the *best* w-mers for each position.

In fixed-width w-mer-based kernels, the width of the w-mers is fixed for all pairs of sequences and throughout the entire sequence. As a result, if w is set to a relatively high value, it may fail to identify positive scoring subsequences whose length is smaller than $2w + 1$, whereas if it is set too low, it may fail to reward sequence pairs that have relatively long similar subsequences.

The best variable-width w-mer (BV-PSSM) kernel overcomes this problem by using variable-length w-mers. It is derived from the BF-PSSM kernel, where for a given user-supplied width w, the BV-PSSM kernel considers the set of all possible w-mer pairs whose length ranges from 1 to a maximum w:

$$\mathcal{P}_{1\cdots w} = \mathcal{P}_1 \cup \cdots \cup \mathcal{P}_w \tag{8.16}$$

From this set $\mathcal{P}_{1\cdots w}$ the BV-PSSM kernel uses the greedy scheme employed by BF-PSSM to select a subset $\mathcal{P}'_{1\cdots w}$ of w-mer pairs that form a high weight matching. The similarity between the pair of sequences is then computed as follows:

$$\text{BV-PSSM}_{X,Y}(w) = \sum_{\substack{(w\text{-mer}(X,i), \\ w\text{-mer}(Y,j)) \in \mathcal{P}'_{1\cdots w}}} w\text{-score}_{X,Y}(i,j) \qquad (8.17)$$

Since for each position of X (and Y), $\mathcal{P}'_{1\cdots w}$ is constructed by including the highest-scoring w-mer for i that does not conflict with the previous selections, this scheme can automatically select the highest-scoring w-mer, whose length can vary from 1 up to w, thus achieving the desired effect.

8.5.2 Performance Evaluation

The fold prediction algorithms can be evaluated using the sets of sequences obtained from the SCOP database [39]. The SCOP database is a manually curated protein structure database assigning proteins into hierarchically defined classes. The fold prediction problem in the context of SCOP can be defined as assigning a protein sequence to its correct fold. On a similar basis, the remote homology problem can be defined as predicting the correct superfamily for a protein.

To evaluate the techniques described above, remote homology detection is simulated by formulating it as a superfamily classification problem within the context of the SCOP database. The same data set and classification problems (definitions of which are available at http://www.cs.columbia.edu/compbio/svm-pairwise) have been used in a number of earlier studies [18,35,52], allowing for direct comparison of the relative performance of the various schemes. The data consist of 4352 sequences from SCOP version 1.53 extracted from the Astral database, grouped into families and superfamilies. The data set is processed so that it does not contain any sequence pairs with an E-value threshold smaller than 10^{-25}. For each family, the protein domains within the family are considered positive test examples, and protein domains within the superfamily but outside the family are considered positive training examples. This yields 54 families with at least 10 positive training examples and five positive test examples. Negative examples for the family are chosen from outside the positive sequences' fold and are randomly split into training and test sets in the same ratio as that used for positive examples.

Employing the same data set and overall methodology as in remote homology detection, we simulate fold detection by formulating it as a fold classification problem within the context of SCOP's hierarchical classification scheme. In this setting, protein domains within the same superfamily are considered positive test examples, and protein domains within the same fold but outside the superfamily are considered positive training examples. This yields 23 superfamilies with at least 10 positive training and five positive test examples. Negative examples for the superfamily are chosen from outside the positive sequences' fold and split equally into test and training sets (the classification problem definitions are available at http://bioinfo.cs.umn.edu/supplements/remote-homology/). Since the positive test and training

TABLE 8.3 Comparison of Various Schemes for the Superfamily-Level Classification Problem

Kernel[a]	ROC	ROC50	mRFP
SVM-Fisher	0.773	0.250	0.204
SVM-Pairwise	0.896	0.464	0.084
LA-eig($\beta = 0.2$)	0.923	0.661	0.064
LA-eig($\beta = 0.5$)	0.925	0.649	0.054
LA-ekm($\beta = 0.5$)	0.929	0.600	0.052
SVM-HMMSTR-Ave	–	0.640	0.038
SVM-HMMSTR-Max	–	0.618	0.043
SVM-HMMSTR-Hybrid	–	0.617	0.048
Mismatch	0.872	0.400	0.084
Profile(4,6)	0.974	0.756	0.013
Profile(5,7.5)	0.980	0.794	0.010
AF-PSSM(2)	0.978	0.816	0.013
BF-PSSM(2)	0.980	0.854	0.015
BV-PSSM(2)	0.973	0.855	0.018
SW-PSSM(3.0,0.750,1.50)	0.982	**0.904**	0.015
AF-GSM(6)	0.926	0.549	0.048
BF-GSM(6)	0.934	0.669	0.053
BV-GSM(6)	0.930	0.666	0.052
SW-GSM(B62,5.0,1,0.5)	0.948	0.711	0.039

[a]The SVM-Fisher, SVM-Pairwise, LA-Kernel, and Mismatch results were obtained from ref. 52. The SVM-HMMSTR results were obtained from ref. 18 and correspond to the best-performing scheme (the authors did not report ROC values). The Profile results were obtained locally by running the publicly available implementation of the scheme obtained from the authors. The ROC50 value of the best-performing scheme is shown in bold.

instances are members of different superfamilies within the same fold, this new problem is significantly more difficult than remote homology detection, as the sequences in the various superfamilies do not have any apparent sequence similarity [39]. The quality of these methods is evaluated by using the receiver operating characteristic (ROC) scores, the ROC50 scores, and the median rate of false positives (mRFPs).

Tables 8.3 and 8.4 compare the performance of various kernel functions developed in this chapter against that achieved by a number of schemes developed previously for the superfamily- and fold-level classification problems, respectively. In the case of the superfamily-level classification problem, the performance is compared against SVM-Fisher [19], SVM-Pairwise [35], and various instances of LA-Kernel [52], SVM-HMMSTR [18], Mismatch [34], and Profile [31].

The results in these tables show that both the window- and local alignment-based kernels derived from sequence profiles (i.e., AF-PSSM, BF-PSSM, BV-PSSM, and SW-PSSM) lead to results that are in general better than those obtained by existing schemes. The performance advantage of these direct kernels is greater over existing schemes that rely on sequence information alone (e.g., SVM-Pairwise, LA-Kernel), but still remains significant compared against schemes that either take profile

TABLE 8.4 Comparison of Various Schemes for the Fold-Level Classification Problem

Kernel[a]	ROC	ROC50	mRFP
LA-eig($\beta = 0.2$)	0.847	0.212	0.129
LA-eig($\beta = 0.5$)	0.771	0.172	0.193
Profile(4,6)	0.912	0.305	0.071
Profile(5,7.5)	0.924	0.314	0.069
AF-PSSM(4)	0.911	0.374	0.067
BF-PSSM(4)	0.918	0.414	0.060
BV-PSSM(4)	0.941	0.481	0.043
SW-PSSM(3.0,0.750,2.0)	0.936	**0.571**	0.054
AF-GSM(6)	0.770	0.197	0.217
BF-GSM(6)	0.822	0.240	0.157
BV-GSM(7)	0.845	0.244	0.133
SW-GSM(B62,5,1.0,0.5)	0.826	0.223	0.176

[a]The results for LA-Kernel were obtained using kernel matrices available publically at the author's Web site. The Profile results were obtained locally by running the publicly available implementation of the scheme obtained from the authors. The ROC50 value of the best-performing scheme is shown in bold.

information into account directly (e.g., SVM-Fisher, Profile) or utilize higher-level features derived by analyzing sequence-structure information (e.g., SVM-HMMSTR). Also, the relative advantage of profile-based methods over existing schemes is greater for the much harder fold-level classification problem over the superfamily-level classification problem. For example, the SW-PSSM scheme achieves ROC50 values that are 13.8% and 81.8% better than the best values achieved by existing schemes for superfamily- and fold-level classification problems, respectively.

To get a better understanding of the relative performance of the various schemes across classes, Figures 8.3 and 8.4. plot the number of classes whose ROC50 are greater than a given threshold that ranges from 0 to 1. Specifically, Figure 8.3 shows the results for the remote homology detection problem, and Figure 8.4 shows the results for the fold detection problem. (Note that these figures contain only results for the schemes that we are able to run locally.) These results show that our profile-based methods lead to higher ROC50 values for a greater number of classes than either Profile or LA-Kernel, especially for larger ROC50 values (e.g., in the range 0.6 to 0.95). Also, SW-PSSM tends consistently to outperform the rest of the profile-based direct kernel methods.

In addition, the results for the BF-GSM, BV-GSM, and SW-GSM kernels that rely on the BLOSUM scoring matrices show that these kernel functions are capable of producing results that are superior to all of the existing non-profile-based schemes. In particular, the properly optimized SW-GSM scheme is able to achieve significant improvements over the best LA-Kernel-based scheme (7.6% higher ROC50 value) and the best SVM-HMMSTR-based scheme (15.1% higher ROC50 value).

From the evaluation of direct profile-based kernels for fold classification, three major observations can be made. First, as was the case with a number of studies on the

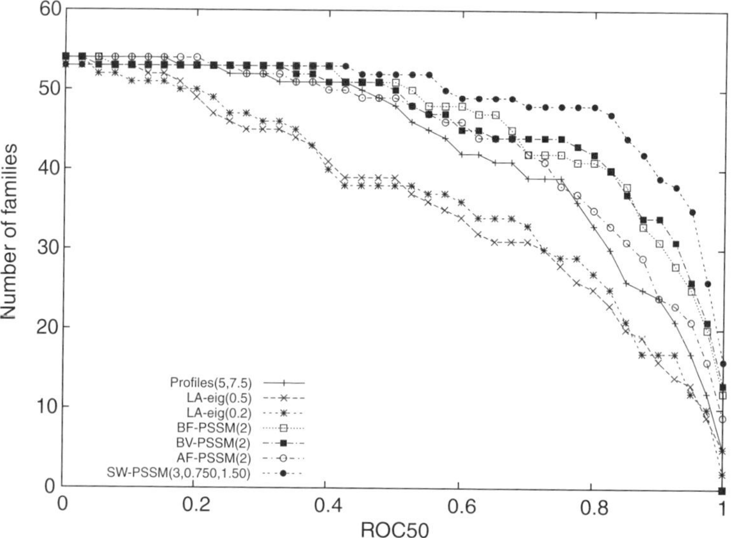

FIGURE 8.3 Comparison of various SVM-based methods for remote homology detection on the SCOP 1.53 benchmark data set. The graph plots the total number of families for which a given method exceeds the ROC-50 score threshold along the *x*-axis.

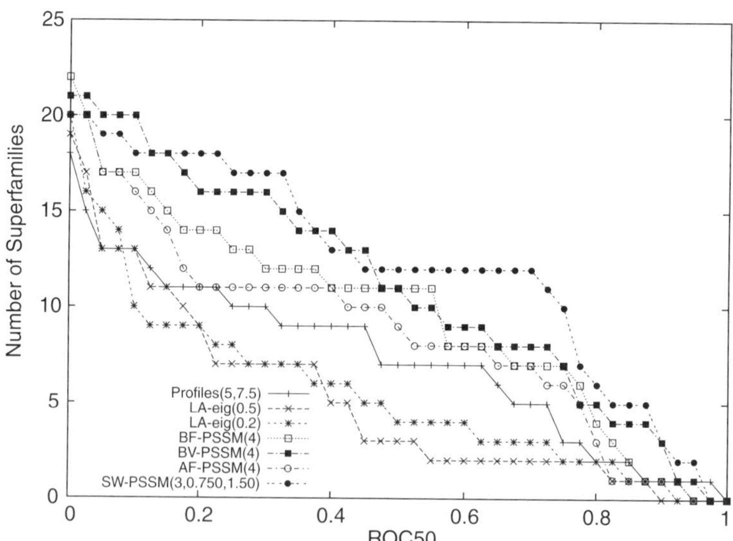

FIGURE 8.4 Comparison of various SVM-based methods for fold detection on the SCOP 1.53 benchmark data set. The graph plots the total number of superfamilies for which a given method exceeds the ROC-50 score threshold along the *x*-axis.

accuracy of protein sequence alignment [36,38,57], proper use of sequence profiles leads to dramatic improvements in the overall ability to detect remote homologs and to identify proteins that share the same structural fold. Second, kernel functions that are constructed by taking into account directly the similarity between the various protein sequences tend to outperform schemes that are based on a feature-space representation [where each dimension of the space is constructed as one of k-possibilities in a k-residue-long subsequence or using structural motifs (Isites) in the case of SVM-HMMSTR]. This is especially evident by comparing the relative advantage of the window-based kernels over the Profile kernel. Third, time-tested methods for comparing protein sequences based on optimal local alignments (as well as global and local-global alignments), when properly optimized for the classification problem at hand, lead to kernel functions that are in general superior to those based on either short subsequences (e.g., Spectrum, Mismatch, Profile, or window-based kernel functions) or local structural motifs (e.g., SVM-HMMSTR). The fact that these widely used methods produce good results in the context of SVM-based classification is reassuring as to the validity of these approaches and their ability to capture biologically relevant information.

8.6 CONCLUDING REMARKS

Predicting protein structure from primary sequence information is a challenging problem that has attracted and continues to attract attention from several fields of research. The current challenges within this problem stem from two factors. First, we still do not have a complete understanding of the basic physical principles that govern protein folding. Second, the number of experimentally resolved three-dimensional protein structures remains small compared to the number of known proteins. Despite these obstacles, recent advances in applying machine learning to evolutionary analysis have improved the quality of current structural predictions significantly.

In this chapter we provided a brief overview of some of these machine learning techniques. Specifically, we examined the design of state-of-the-art kernel functions within a discriminative learning framework for secondary structure prediction, remote homology detection, and fold recognition. We have given a flavor of string kernels along with the use of evolutionary information in our methods. Hopefully, increasingly better solutions to subproblems within complete structure prediction will lead to an accurate method for native fold prediction from sequence.

REFERENCES

1. S. F. Altschul, W. Gish, E. W. Miller, and D. J. Lipman. Basic local alignment search tool. *J. Mol. Biol.*, 215:403–410, 1990.

2. S. F. Altschul, L. T. Madden, A. A. Schäffer, J. Zhang, Z. Zhang, W. Miller, and D. J. Lipman. Gapped BLAST and PSI-BLAST: a new generation of protein database search programs. *Nucleic Acids Res.*, 25(17):3389–3402, 1997.

3. S. Altschul, W. Gish, W. Miller, E. Myers, and D. Lipman. Basic local alignment search tool. *J. Mol. Biol.*, 215:403–410, 1990.

4. C. Anfinsen. Principles that govern the folding of protein chains. *Science*, 181:223–230, 1973.

5. P. Baldi, Y. Chauvin, T. Hunkapiller, and M. McClure. Hidden Markov models of biological primary sequence information. *Proc. Natl. Acad. Sci. USA*, 91:1053–1063, 1994.

6. P. A. Bates and M. J. E. Sternberg. Model building by comparison at *casp3*: Using expert knowledge and computer automation. *Proteins: Struct. Funct. Genet.*, 3:47–54, 1999.

7. P. Bourne and H. Weissig. *Structural Bioinformatics.* Wiley, Hoboken, NJ, 2003.

8. J. U. Bowie, R. Luethy, and D. Eisenberg, A method to identify protein sequences that fold into a known three-dimensional structure. *Science*, 253:797–815, 1991.

9. K. M. S. Misura, C. A. Rohl, and C. E. M. Strauss, and D. Baker. Protein structure prediction using Rosetta. *Methods Enzymol.*, 383:66–93, 2004.

10. M. Collins. Parameter estimation for statistical parsing models: theory and practice of distribution-free methods. In *New Developments in Parsing Technology*, pp. 1–38. Kluwer, Norwell, MA, 2001.

11. N. Cristianini and J. Shawe-Taylor. *An Introduction to Support Vector Machines.* Cambridge University Press, Cambridge, 2000.

12. J. A. Cuff and G. J. Barton. Evaluation and improvement of multiple sequence methods for protein secondary structure prediction. *Proteins: Struct. Funct. Genet.*, 34:508–519, 1999.

13. A. Fiser, R. K. Do, and A. Sali. Modeling of loops in protein structures. *Protein Sci.*, 9:1753–1773, 2000.

14. D. Frishman and P. Argos. Seventy-five percent accuracy in protein secondary structure prediction. *Proteins: Struct. Funct. Genet.*, 27:329–335, 1997.

15. M. Gribskov, A. D. McLahlan, and D. Eisenberg. Profile analysis: detection of distantly related proteins. *Proc. Natl. Acad. Sci. USA*, 84:4335–4358, 1987.

16. S. Henikoff and J. G. Henikoff. Amino acid substitution matrices from protein blocks. *Proc. Natl. Acad. Sci. USA*, 89:10915–10919, 1992.

17. Y. Hou, W. Hsu, M. L. Lee, and C. Bystroff. Efficient remote homology detection using local structure. *Bioinformatics*, 19(17):2294–2301, 2003.

18. Y. Hou, W. Hsu, M. L. Lee, and C. Bystroff. Remote homolog detection using local sequence-structure correlations. *Proteins: Struct. Funct. Bioinf.*, 57:518–530, 2004.

19. T. Jaakola, M. Diekhans, and D. Haussler. A discriminative framework for detecting remote protein homologies. *J. Comput. Biol.*, 7(1):95–114, 2000.

20. D. T. Jones. Genthreader: an efficient and reliable protein fold recognition method for genomic sequences. *J. Mol. Biol.*, 287:797–815, 1999.

21. D. T. Jones, W. R. Taylor, and J. M. Thorton. A new approach to protein fold recognition. *Nature*, 358:86–89, 1992.

22. D. T. Jones. Protein secondary structure prediction based on positive-specific scoring matricies. *J. Mol. Biol.*, 292:195–202, 1999.

23. K. Joo, J. Lee, S. Kim, I. Kum, J. Lee, and S. Lee. Profile-based nearest neighbor method for pattern recognition. *J. Korean Phys. Soc.*, 54(3):599–604, 2004.

24. E. Huang, K. T. Simmons, C. Kooperberg, and D. Baker. Assembly of protein tertiary structures from fragments with similar local sequences using simulated annealing and Bayesian scoring functions. *J. Mol. Biol.*, 268:209–225, 1997.

25. W. Kabsch and C. Sander. Dictionary of protein secondary structure: pattern recognition of hydrogen-bonded and geometrical features. *Biopolymers*, 22:2577–2637, 1983.

26. K. Karplus, C. Barrett, and R. Hughey. Hidden Markov models for detecting remote protein homologies. *Bioinformatics*, 14:846–856, 1998.

27. K. Karplus, C. Barrett, and R. Hughey. Hidden Markov models for detecting remote protein homologies. *Bioinformatics*, 14:846–856, 1998.

28. K. Karplus, R. Karchin, J. Draper, J. Casper, Y. Mandel-Gutfreund, M. Diekhans, and R. Hughey. Combining local-structure, fold-recognition, and new fold methods for protein structure prediction. *Proteins: Struct. Funct. Genet.*, 53:491–496, 2003.

29. G. Karypis. Better kernels and coding schemes lead to improvements in SVM-based secondary structure prediction. Technical Report 05-028. Department of Computer Science, University of Minnesota, Minneapolis, MN, 2005.

30. A. Krogh, M. Brown, I. Mian, K. Sjolander, and D. Haussler. Hidden Markov models in computational biology: applications to protein modeling. *J. Mol. Biol.*, 235:1501–1531, 1994.

31. R. Kuang, E. Ie, K. Wang, K. Wang, M. Siddiqi, Y. Freund, and C. Leslie. Profile-based string kernels for remote homology detection and motif extraction. *Comput. Syst. Bioinf.*, pp. 152–160, 2004.

32. J. Lee, S. Kim, K. Joo, I. Kim, and J. Lee. Prediction of protein tertiary structure using Profesy, a novel method based on fragment assembly and conformational space annealing. *Proteins: Struct. Funct. Bioinf.*, 56:704–714, 2004.

33. C. Leslie, E. Eskin, and W. S. Noble. The Spectrum kernel: a string kernel for SVM protein classification. *Proc. Pacific Symposium on Biocomputing*, pp. 564–575, 2002.

34. C. Leslie, E. Eskin, W. S. Noble, and J. Weston. Mismatch string kernels for SVM protein classification. *Adv. Neutral Inf. Process. Syst.*, 20(4):467–476, 2003.

35. L. Liao and W. S. Noble. Combining pairwise sequence similarity and support vector machines for detecting remote protein evolutionary and structural relationships. *Proc. International Conference on Research in Computational Molecular Biology*, pp. 225–232, 2002.

36. M. Marti-Renom, M. Madhusudhan, and A. Sali. Alignment of protein sequences by their profiles. *Protein Sci.*, 13:1071–1087, 2004.

37. F. S. Matthews. The structure, function and evoluation of cytochromes. *Prog. Biophys. Mol. Biol.*, 45:1–56, 1975.

38. D. Mittelman, R. Sadreyev, and N. Grishin. Probabilistic scoring measures for profile–profile comparison yield more accurate short seed alignments. *Bioinformatics*, 19(12):1531–1539, 2003.

39. A. G. Murzin, S. E. Brenner, T. Hubbard, and C. Chothia. SCOP: a structural classification of proteins database for the investigation of sequences and structures. *J. Mol. Biol.*, 247:536–540, 1995.

40. C. H. Papadimitriou and K. Steiglitz. *Combinatorial Optimization: Algorithms and Complexity.* Prentice-Hall, Englewood Cliffs, NJ, 1982.

41. G. Pollastri and A. McLysaght. Porter: a new, accurate server for protein secondary structure prediction. *Bioinformatics*, 21:1719–1720, 2005.

42. G. Pollastri, D. Przybylski, B. Rost, and P. Baldi. Improving the prediction of protein secondary structure in three and eight classes using recurrent neural networks and profiles. *Proteins: Struct. Funct. Genet.*, 47:228–235, 2002.

43. A. Porollo, R. Adamczak, M. Wagner, and J. Meller. Maximum feasibility approach for consensus classifiers: applications to protein structure prediction. CIRAS, 2003.

44. D. Przybylski and B. Rost. Alignments grow, secondary structure prediction improves. *Proteins: Struct. Funct. Genet.*, 46:197–205, 2002.

45. H. Rangwala and G. Karypis. Profile based direct kernels for remote homology detection and fold recognition. *Bioinformatics*, 21(23):4239–4247, 2005.

46. V. Robles, P. Larranaga, J. M. Pena, E. Menasalvas, M. S. Perez, V. Herves, and A. Wasilewska. Bayesian network multi-classifiers for protein secondary structure prediction. *Art. Intell. Med.*, 31:117–136, 2004.

47. B. Rost. Unpublished.

48. B. Rost and V. A. Eyrich. EVA: large-scale analysis of secondary structure prediction. *Proteins: Struct. Funct. Genet.*, Suppl. 5:192–199, 2001.

49. B. Rost and C. Sander. Prediction of protein secondary structure at better than 70% accuracy. *J. Mol. Biol.*, 232:584–599, 1993.

50. B. Rost, C. Sander, and R. Schneider. Redefining the goals of protein secondary structure prediction. *J. Mol. Biol.*, 235:13–26, 1994.

51. B. Rost. Review: Protein secondary structure prediction continues to rise. *J. Struct. Biol.*, 134:204–218, 2001.

52. H. Saigo, J. P. Vert, N. Ueda, and T. Akutsu. Protein homology detection using string alignment kernels. *Bioinformatics*, 20(11):1682–1689, 2004.

53. A. A. Salamov and V. V. Solovyev. Protein secondary structure prediction using local alignments. *J. Mol. Biol.*, 268:31–36, 1997.

54. A. Semla, C. Venclovas, K. Fidelis, and B. Rost. A modified definition of SOV, a segment-based measure for protein secondary structure prediction assessment. *Proteins: Struct. Funct. Genet.*, 34:220–223, 1999.

55. T. F. Smith and M. S. Waterman. Identification of common molecular subsequences. *J. Mol. Biol.*, 147:195–197, 1981.

56. V. Vapnik. *Statistical Learning Theory*. Wiley, New York, 1998.

57. G. Wang and R. L. Dunbrack, Jr. Scoring profile-to-profile sequence alignments. *Protein Sci.*, 13:1612–1626, 2004.

58. Y. Wu and E. Y. Chang. Distance-function design and fusion for sequence data. *Proc. 13th ACM Conference on Information and Knowledge Management*, pp. 324–333, 2004.

9

PUBLIC GENOMIC DATABASES: DATA REPRESENTATION, STORAGE, AND ACCESS

ANDREW ROBINSON AND WENNY RAHAYU
Department of Computer Science and Computer Engineering, La Trobe University, Bundoora, Victoria, Australia

DAVID TANIAR
Clayton School of Information Technology, Monash University, Clayton, Victoria, Australia

Due to the introduction of mass genome sequencing machines and projects, the bioinformatics domain now has a large volume of sequence and annotation data to be stored and processed. This has opened up a new area which merges the advanced technology in databases and information management with the diverse data types covering DNA and protein sequences, gene expression, cellular role, taxonomic data, and so on, in a biology domain. This unique area is identified as that of genomic databases.

Initially, sequencing projects were slow and produced very small amounts of data. Thus, hand-edited files or spreadsheets were a feasible option for data storage. With the advent of mass genome sequencing projects, however, masses of data were being produced at an ever-increasing rate; this meant that these formats simply could not handle the data efficiently, and thus new methods needed to be found.

The initial storage formats grew out of the hand-edited files mentioned above. They were simple in nature but had a standard format that could be computer handled and thus remove or reduce the amount of hand editing needed before. The usefulness of such formats soon came under scrutiny and it was realized that more data needed to

Knowledge Discovery in Bioinformatics: Techniques, Methods, and Applications, Edited by Xiaohua Hu and Yi Pan

be stored. To solve these data storage shortages, more elaborate file formats were developed to store many attributes of annotation about each record.

The annotation-rich file formats quickly grew to unmanageable files that were near impossible to search or use effectively. Thus, there was a move to get the benefits of a database management system, such as the relational database systems.

In this chapter we present a review of public genomic databases, focusing on data representation, data storage, and data access. In Section 9.1 we discuss the data representation formats used in files for genomic data and in Section 9.2 describe the storage methods used by genetic databases. Section 9.3 covers the methods currently used to access the genetic data. Finally, in Sections 9.4 and 9.5 we provide a discussion and summary, respectively.

9.1 DATA REPRESENTATION

As public databases grew in popularity and complexity, so did the data representation and structure. Initially, the data stored in public databases was just the sequence data and a simple one-line description. After this it was realized that more annotation data were needed to expand the usefulness of the data. To overcome these issues, major public databases developed their own data representation format within a text file. This formatting was (and still is) well accepted by the bioinformatics community. However, with introduction of the XML (eXtensible Markup Language) standard [1] for data representation, many public databases have developed pseudostandards for XML biological data storage. In addition to these database-specific XML formats, several other organizations not associated with public databases have also put forth pseudostandards.

This section is devoted to describing the developments of evolving data representations. It begins with the flat file format of FASTA. After this we move on to the Genbank and Swiss-Prot formats. Following this is a discussion of many of the XML formats in the biological area.

9.1.1 FASTA Format

The FASTA file format was designed initially for the FASTA searching program, but it is used very widely for other tasks. The FASTA format is typically a sequence-centric format in that it provides the sequence with minimal annotations. This fact makes it ideal for processing that requires the sequence and its unique id. Each record in the FASTA file consists of two main compulsory sections: the annotation and sequence sections.

Annotation is identified by a line of text beginning with the greater-than symbol (>). The format specifies no other structure limitations on the annotation. However, there are many other pseudostandards for the annotations, such as comma (,)- or pipe (|)-separated fields. Usually, annotation up to the first space is the unique identifier, but it is not required. Even the entire annotation can be missing (e.g., only a greater-than symbol is provided to separate the records).

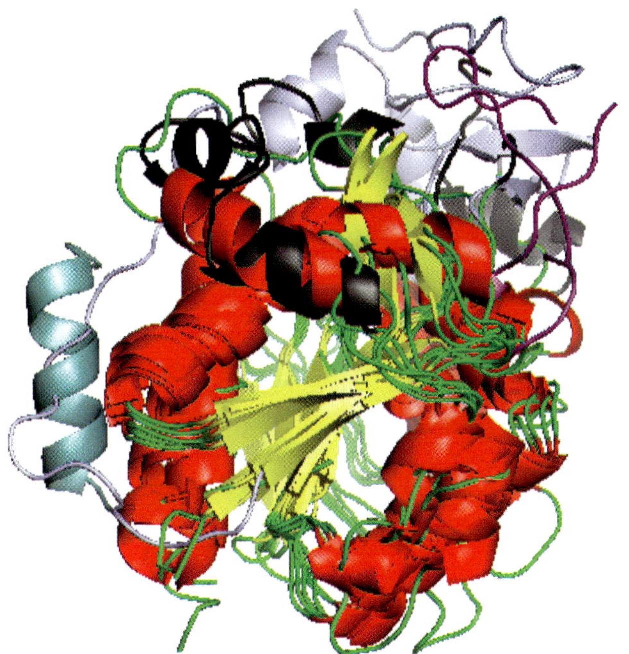

FIGURE 11.1 Superposition of tertiary structures of members of the DJ-1/PfpI family, depicting the well-conserved structural core of the members of the family as well as the variation at the surface. The conserved secondary-structural elements of all six PDB structures are colored similarly: helixes in red, sheets in blue, and loops in green. The insertions specific to each group are colored differently: 1pe0 in light teal, 1oy1 in purple, 1qvz in black, and 1izy in light blue. Generated using PyMOL [20].

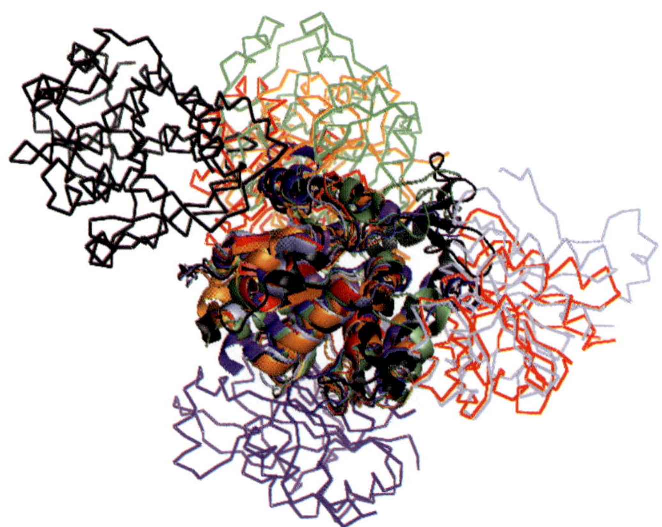

FIGURE 11.2 Superposition of quaternary structures of members of the DJ-1/PfpI family, illustrating the entire range of surfaces utilized for oligomerization by the members of the family, despite the well-conserved tertiary structure. The superimposed GAT domains are shown as cartoons, whereas the rest of the oligomeric structures are shown as ribbons. 1g2i is red, 1oi4 is light blue, 1pe0 is orange, 1oy1 is green, 1qvz is purple and 1izy is blue. Generated using PyMOL [20].

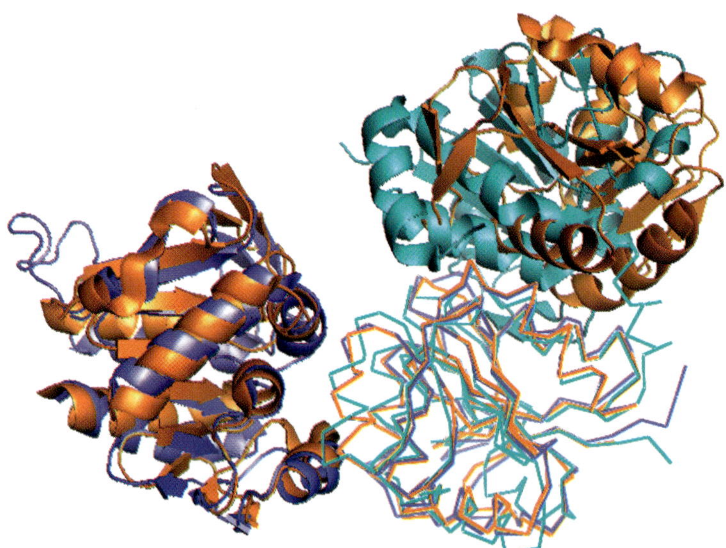

FIGURE 11.3 Superposition of the quaternary structures of 1g2i, 1oi4 and 1pe0. Although the three sequences are close homologs, they display slight variation in the orientation of their interacting surfaces. The superposed monomers are displayed as ribbons, whereas the rest of the oligomer is displayed as cartoons. 1g2i is orange, 1oi4 blue, and 1pe0 teal. Generated using PyMOL [20].

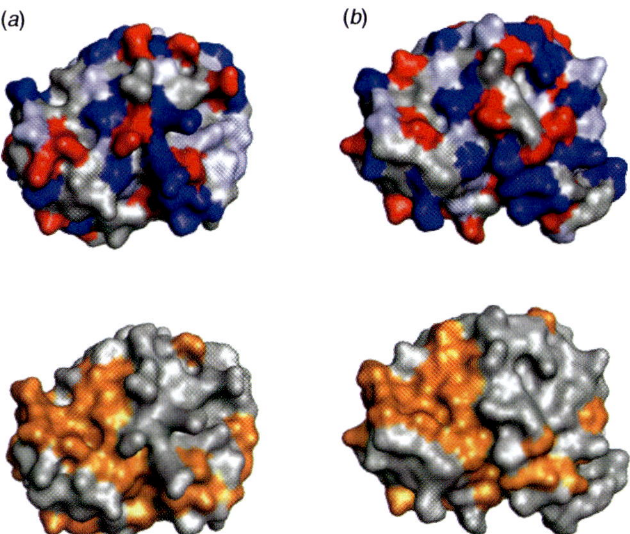

FIGURE 11.4 Conserved interface between (*a*) intracellular protease, 1g2i and (*b*) hypothetical protein YhbC, 1oi4. In the representations at the top of the figure, aliphatic and aromatic residues are represented in light gray, polar residues in light blue, positively charged residues in blue, and negatively charged residues in red. In the representations at the bottom, residues conserved between 1g2i and 1oi4 are displayed in orange, whereas variable residues are displayed in light gray. Generated using PyMOL [20].

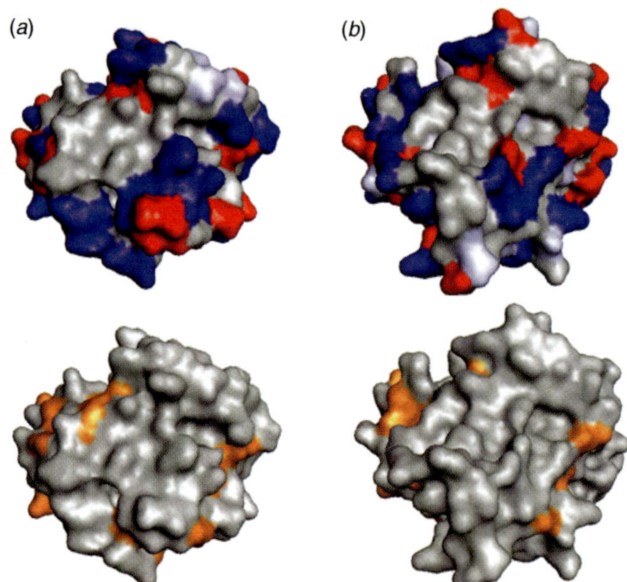

FIGURE 11.5 Topologically equivalent, orientationally variant interface between (*a*) intracellular protease, 1g2i and (*b*) DJ-1 protein, 1pe0. In the representations at the top of the figure, aliphatic and aromatic residues are represented in light gray, polar residues in light blue, positively charged residues in blue, and negatively charged residues in red. In the representations at the bottom, residues conserved between 1g2i and 1pe0 are displayed in orange, whereas variable residues are displayed in light gray. Generated using PyMOL [20].

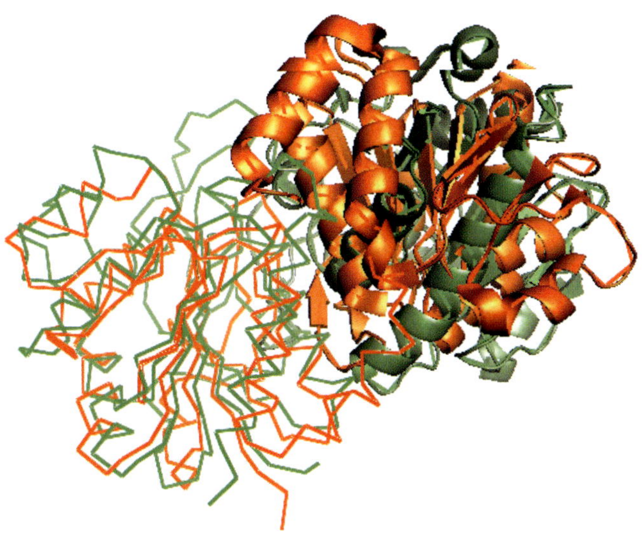

FIGURE 11.6 Superposition of the quaternary structures of divergent members of the DJ-1 family, putative sigma cross-reacting protein and DJ-1. The superposed monomers are displayed as ribbons, whereas the rest of the oligomer is displayed as cartoons. 1pe0 is orange and 1oy1 is green. Generated using PyMOL [20].

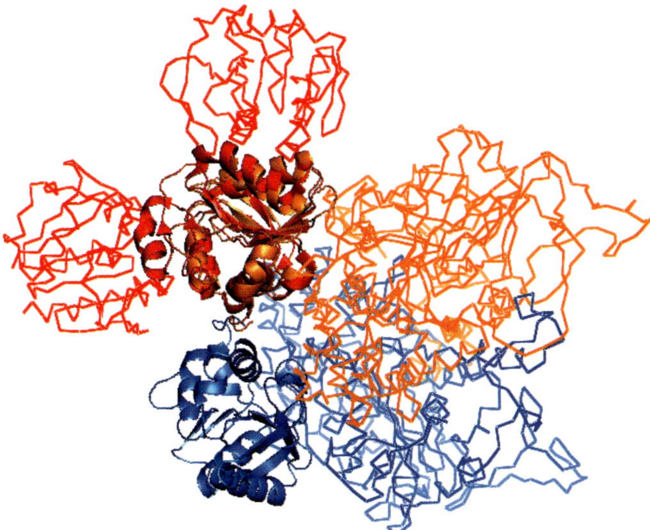

FIGURE 11.7 Quaternary structures of the GAT domains of 1g2i and 1cf9, proteins that belong to two different families within the same superfamily. The remote homologs do not even have topologically equivalent interacting surfaces, a feature observed in the protein–protein interfaces of many families related at the superfamily level. The three monomers of 1g2i are colored in red. Two molecules of catalase (1cf9) are shown, one orange and the other blue. One of the GAT domains in 1g2i and both GAT domains in 1cf9 are depicted as cartoons; the rest of the protein is depicted using ribbons. Generated using PyMOL [20].

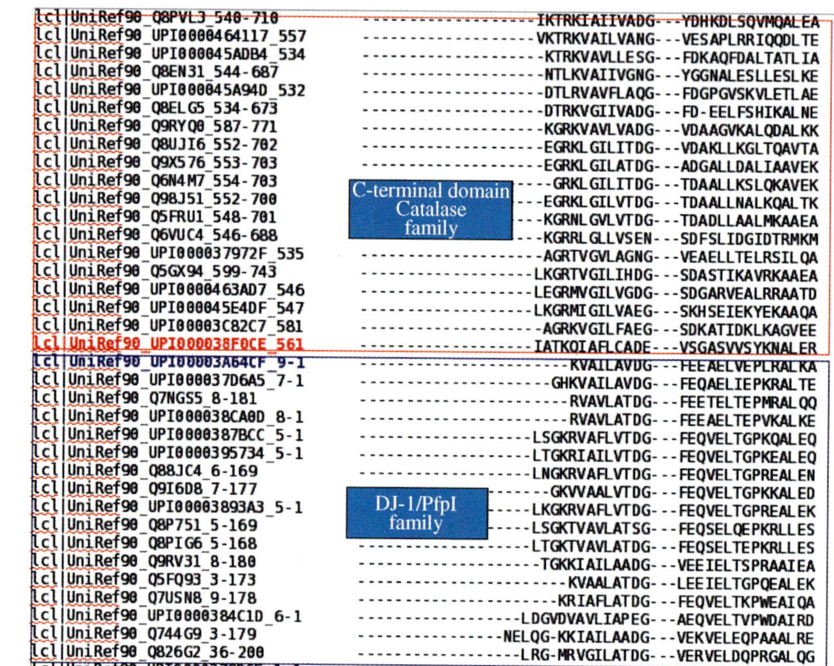

FIGURE 11.8 Drift in interfaces between members of different families in the same superfamily, depicting a clear-cut visual distinction between sequences from different families. The sequences forming the boundary of each of the families are highlighted using a color specific to the family.

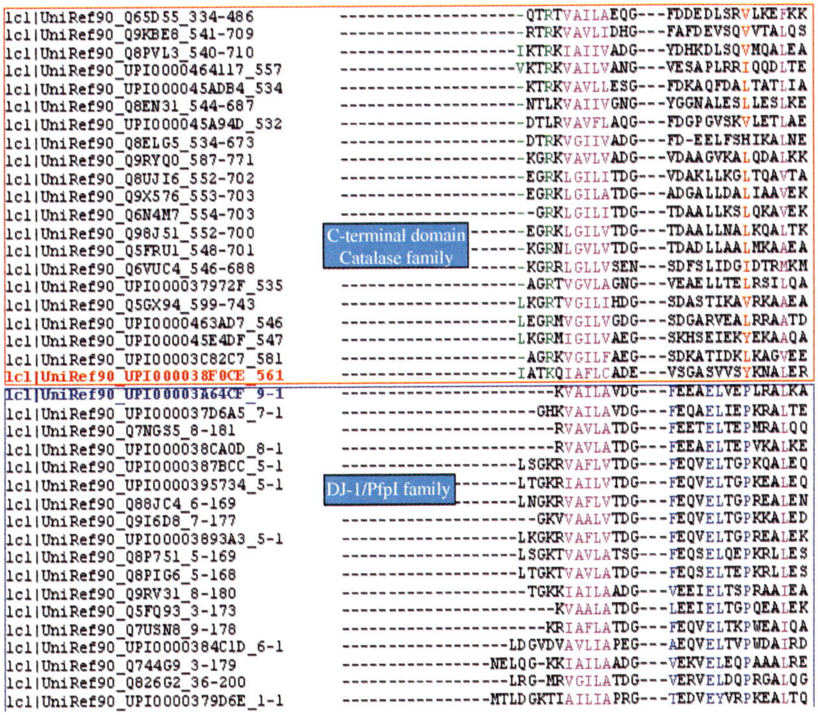

```
lcl|UniRef90_Q65D55_334-486          --------------------QTRTVAILAEQG---FDDEDLSRVLKEFKK
lcl|UniRef90_Q9KBE8_541-709          -------------------RTRKVAVLIDHG---FAFDEVSQVVTALQS
lcl|UniRef90_Q8PVL3_540-710          ------------------IKTRKIAIIVADG---YDHKDLSQVMQALEA
lcl|UniRef90_UPI0000464117_557       ------------------VKTRKVAILVANG---VESAPLRRIQQDLTE
lcl|UniRef90_UPI000045ADB4_534       -------------------KTRKVAVLLESG---FDKAQFDALTATLIA
lcl|UniRef90_Q8EN31_544-687          -------------------NTLKVAIIVGNG---YGGNALESLLESLKE
lcl|UniRef90_UPI000045A94D_532       -------------------DTLRVAVFLAQG---FDGPGVSKVLETLAE
lcl|UniRef90_Q8ELG5_534-673          -------------------DTRKVGIIVADG---FD-EELFSHIKALNE
lcl|UniRef90_Q9RYQ0_587-771          -------------------KGRKVAVLVADG---VDAAGVKALQDALKK
lcl|UniRef90_Q8UJI6_552-702          -------------------EGRKLGILITDG---VDAKLLKGLTQAVTA
lcl|UniRef90_Q9X576_553-703          -------------------EGRKLGILATDG---ADGALLDALIAAVEK
lcl|UniRef90_Q6N4M7_554-703          --------------------GRKLGILITDG---TDAALLKSLQKAVEK
lcl|UniRef90_Q98J51_552-700          ------EGRKLGILVTDG---TDAALLNALKQALTK
lcl|UniRef90_Q5FRU1_548-701          ------KGRNLGVLVTDG---TDADLLAALMKAAEA
lcl|UniRef90_Q6VUC4_546-688          ------KGRRLGLLVSEN---SDFSLIDGIDTRMKM
lcl|UniRef90_UPI000037972F_535       -------------------AGRTVGVLAGNG---VEAELLTELRSILQA
lcl|UniRef90_Q5GX94_599-743          ------------------LKGRTVGILIHDG---SDASTIKAVRKAAEA
lcl|UniRef90_UPI0000463AD7_546       ------------------LEGRMVGILVGDG---SDGARVEALRRAATD
lcl|UniRef90_UPI000045E4DF_547       ------------------LKGRMIGILVAEG---SKHSEIEKYEKAAQA
lcl|UniRef90_UPI00003C82C7_581       -------------------AGRKVGILFAEG---SDKATIDKLKAGVEE
lcl|UniRef90_UPI000038F0CE_561       ------------------IATKQIAFLCADE---VSGASVVSYKNALER

lcl|UniRef90_UPI00003A64CF_9-1       ---------------------KVAILAVDG---FEEAELVEPLRALKA
lcl|UniRef90_UPI000037D6A5_7-1       --------------------GHKVAILAVDG---FEQAELIEPKRALTE
lcl|UniRef90_Q7NGS5_8-181            ---------------------RVAVLATDG---FEETELTEPMRALQQ
lcl|UniRef90_UPI000038CA0D_8-1       ---------------------RVAVLATDG---FEEAELTEPVKALKE
lcl|UniRef90_UPI0000387BCC_5-1       -------------------LSGKRVAFLVTDG---FEQVELTGPKQALEQ
lcl|UniRef90_UPI0000395734_5-1       -------LTGKRIAILVTDG---FEQVELTGPKEALEQ
lcl|UniRef90_Q88JC4_6-169            -------LNGKRVAFLVTDG---FEQVELTGPREALEN
lcl|UniRef90_Q9I6D8_7-177            --------------------GKVVAALVTDG---FEQVELTGPKKALED
lcl|UniRef90_UPI00003893A3_5-1       -------------------LKGKRVAFLVTDG---FEQVELTGPREALEK
lcl|UniRef90_Q8P751_5-169            -------------------LSGKTVAVLATSG---FEQSELQEPKRLLES
lcl|UniRef90_Q8PIG6_5-168            -------------------LTGKTVAVLATDG---FEQSELTEPKRLLES
lcl|UniRef90_Q9RV31_8-180            --------------------TGKKIAILAADG---VEEIELTSPRAAIEA
lcl|UniRef90_Q5FQ93_3-173            ---------------------KVAALATDG---LEEIELTGPQEALEK
lcl|UniRef90_Q7USN8_9-178            --------------------KRIAFLATDG---FEQVELTKPWEAIQA
lcl|UniRef90_UPI0000384C1D_6-1       -----------------LDGVDVAVLIAPEG---AEQVELTVPWDAIRD
lcl|UniRef90_Q744G9_3-179            ----------------NELQG-KKIAILAADG---VEKVELEQPAAALRE
lcl|UniRef90_Q826G2_36-200           ----------------LRG-MRVGILATDG---VERVELDQPRGALQG
lcl|UniRef90_UPI0000379D6E_1-1       ----------------MTLDGKTIAILIAPRG---TEDVEYVRPKEALTQ
```


FIGURE 11.9 Conservation of buried residues in different families in the same superfamily. Even though there is a clear distinction between the sequences from the two families, the residues at the core of the structures are generally well conserved. The sequences forming the boundary of each of the families are highlighted using a color specific to the family. The residues colored in pink indicate conserved common buried residues in the two families. Residues in red indicate conserved buried residues specific to the C-terminal domain catalase family. Residues in blue indicate conserved buried residues specific to the DJ-1/PfpI family. Residues in green indicate conserved interfacial residues specific to the C-terminal domain catalase family.

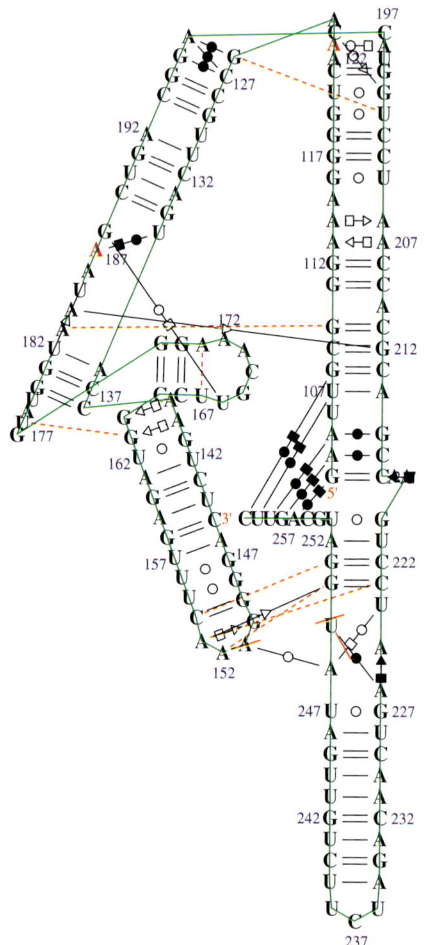

FIGURE 12.4 1GID chain A (P4-P6 RNA ribozyme domain).

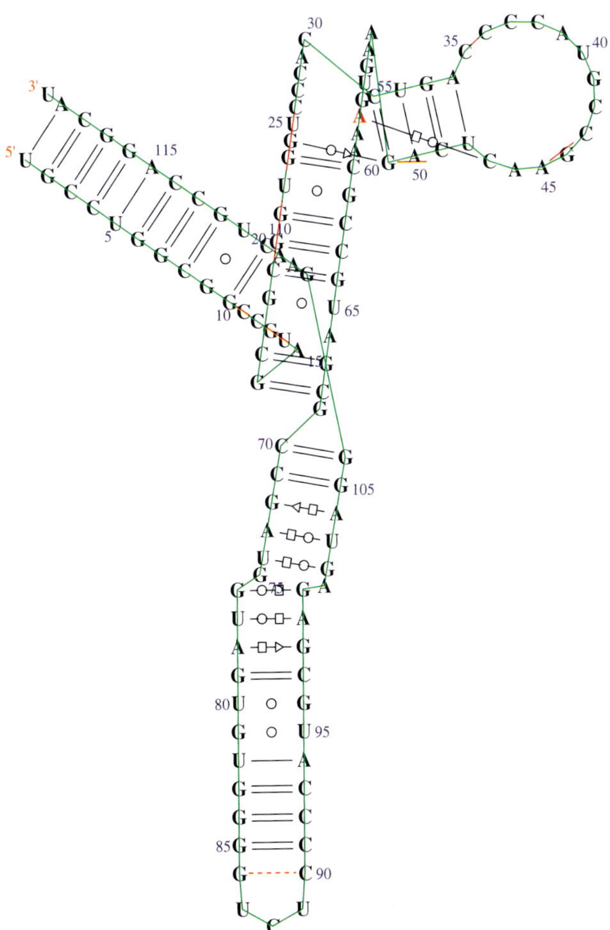

FIGURE 12.5   1C2X chain C (5S ribosomal RNA).

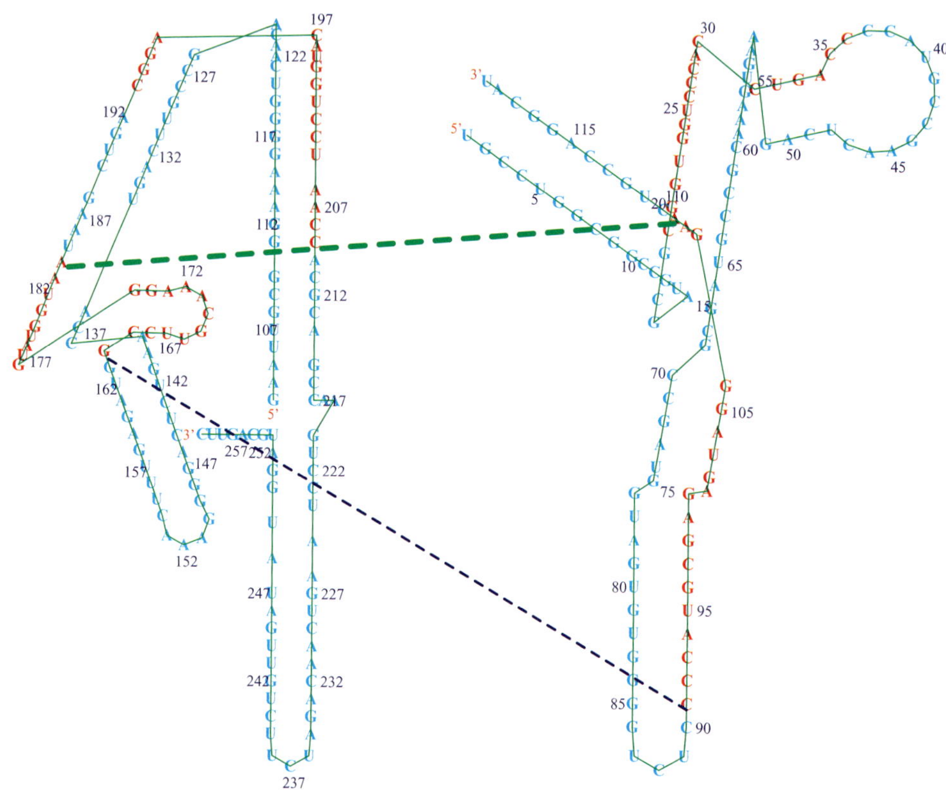

FIGURE 12.6   Output of RSview.

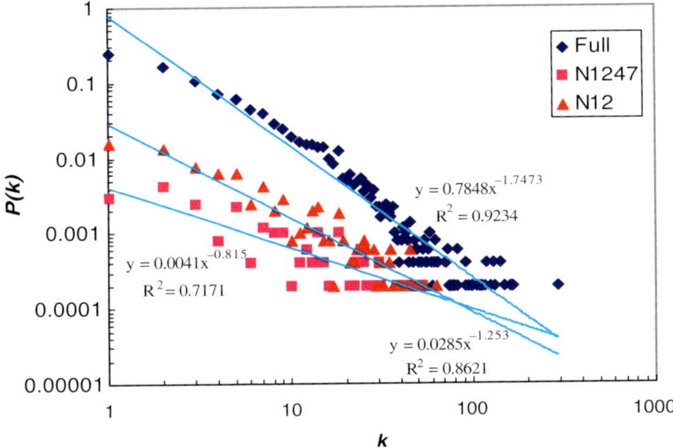

FIGURE 13.1 Degree distributions. The cyan lines show the power-law regression.

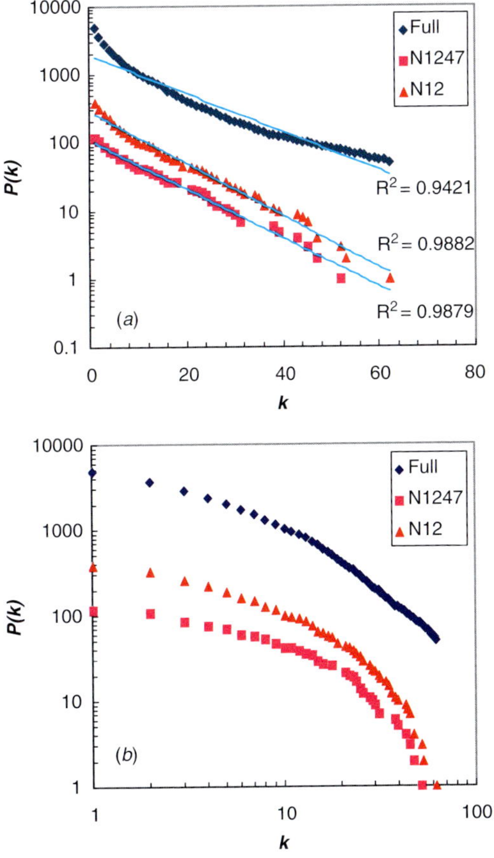

FIGURE 13.2 Cumulative degree distributions: (*a*) semilogarithmic plot with exponential regression; (*b*) log-log plot.

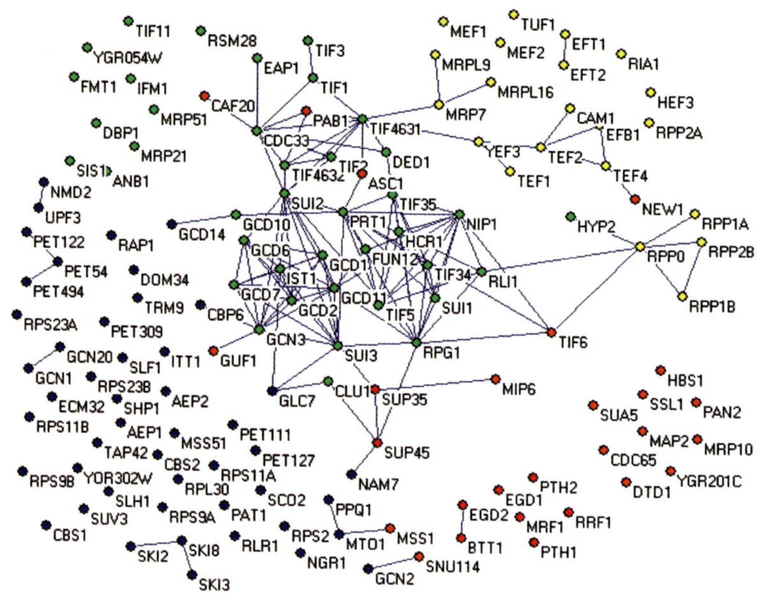

FIGURE 13.3 Synthetic protein translation network N1247S. All proteins are in MIPS functional categories of 12.04 (translation, orange), 12.04.01 (translation initiation, green), 12.04.02 (translation elongation, yellow), 12.04.03 (translation termination, red), and 12.07 (translational control, blue).

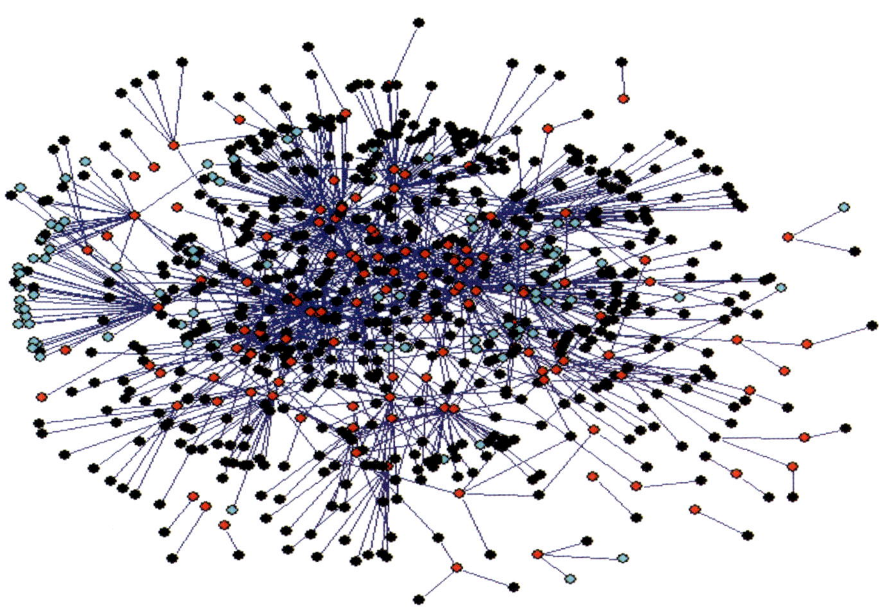

FIGURE 13.4 Synthetic protein translation network N1247SA. At least one of the interacting proteins is in N1247. Proteins in N1247 are indicated by red. Proteins in N12 but not in N1247 are indicated by cyan. All other proteins are shown in black.

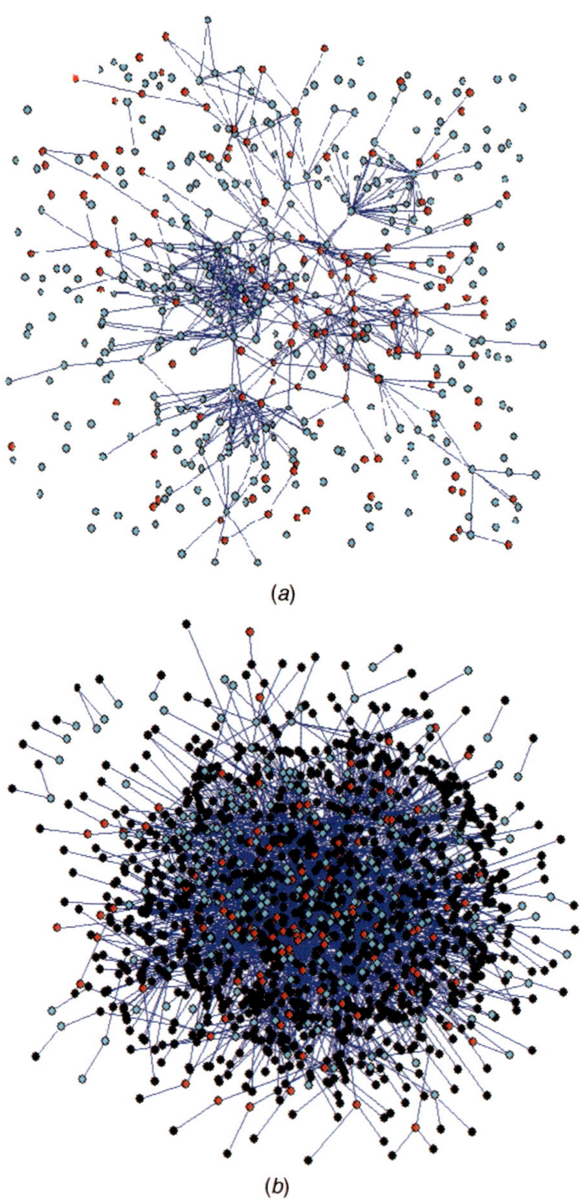

FIGURE 13.5 Synthetic protein translation networks: (*a*) network N12S represents proteins in N12; (*b*) network N12SA contains all proteins that are either in N12 or interacting with proteins in N12. For both networks, proteins in N1247 are in red, remaining N12 proteins are in cyan, all other proteins are in black.

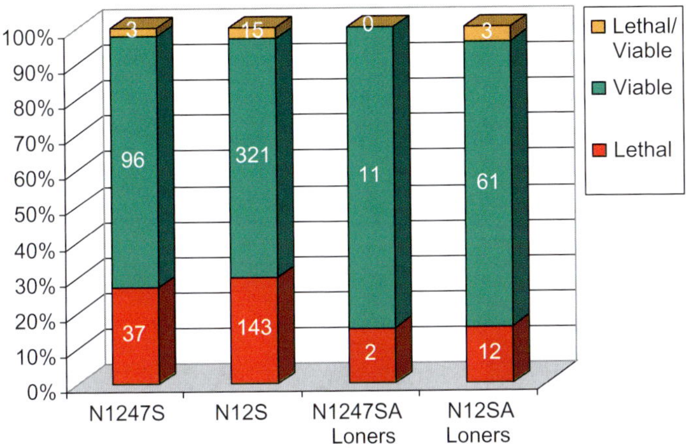

FIGURE 13.6 Essentiality of proteins in translation networks.

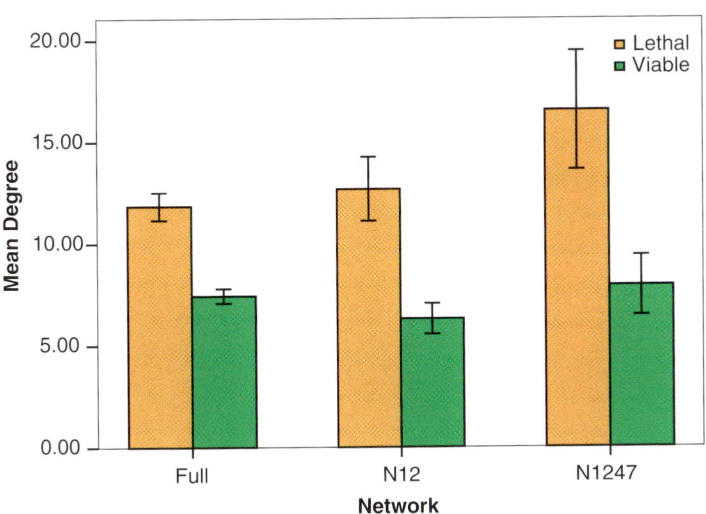

FIGURE 13.7 Essentiality of proteins in translation networks. Error bar at 95% confidence intervals. $p < 0.05$ between lethal and viable proteins in all networks (ANOVA test).

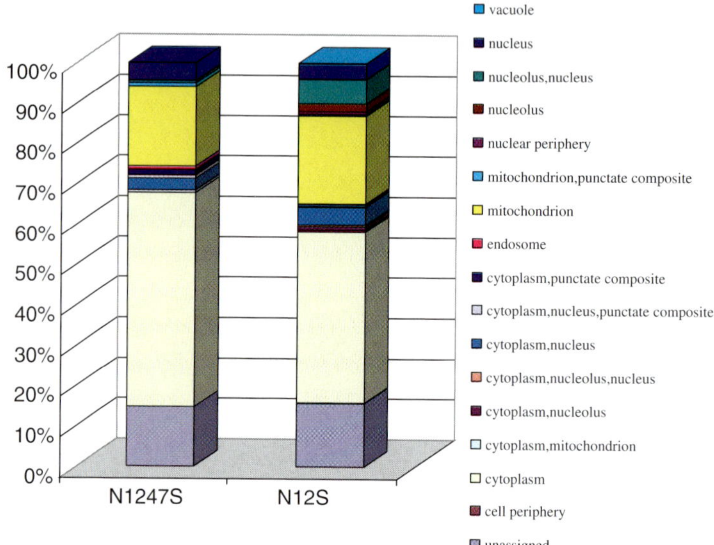

FIGURE 13.8 Cellular localization of proteins in a translation network.

```
>gi|84618082|emb|AM049264.1| Glomus intraradices mRNA for 14-3-3 protein
ATGGAACGCGAGAACCAGACCTCCAACAAAAAGGTTTATATGGCTAAGCTCGCTGAGCAAGCCGAACGTT
ATGATGAAATGGTATCTTATATGAAGGATGTTGCAAAGCTTGGAGTAGAACTTACTGTAGAAGAACGCAA
TCTCCTCTCAGTCGCTTATAAGAATGTAATTGGTGCCCGTCGAGCGTCTTGGAGAATTATTTCATCTATC
GAACAAAAGGAAGAAAATAAGGGTAATGAAACCCAAGTTCAGAAGATCAAAGCTTACCGTCAAAAAGTTG
AAACCGAACTTTCAGAAGTATGCTCGGATATCCTTGCCGTATTGGACGAGCATTTGATTCCATCTGCTGA
AGCCGGTGAATCTAAGGTCTTTTATTATAAGATGAAAGGTGATTATCACCGATATCTCGCTGAATTTGCT
TCGGGTGAAGGACGAAAAGAAGCCGCCACCCAGGCTCACGAAGCTTATAAACACGCAACAGATATCGCCC
AAACTGACCTTGCTCCAACTCATCCTATTCGTCTTGGACTTGCATTAAACTTTTCAGTTTTCTACTACGA
GATTCTCAATTCACCTGATCGCGCTTGTCATCTTGCAAAACAGGCATTTGATGATGCTATTGCTGAACTC
GATACTTTAAGTGAAGAATCGTATAAAGATAGCACATTAATTATGCAACTACTTAGAGATAACTTGACCC
TTTGGACTTCAGACTTGCAAGATGCAGAGGGTGAAAAACCGGAAGAAGCTAAACAAGAAGTTGAAGCTGA
AGAATCTAAGTAGCAATAGTTAGTTAATTTTATTAAAAAATATATATTTCGCACTTCATATATATAAAAA
AAAAAAAAAAAAAAAAAAAAAAAAAAAAAAAAAAAAAAAAAAAA
>gi|84617862|emb|AJ937963.1| Lactuca sativa partial mRNA for putative PIP2
   aquaporin
CACATCAACCCGGCGGTGACATTCGGTTTGTTTTTGGCGAGAAAGGTGTCGTTGATTAGGGCTGTAGGTT
ACATGATTGCACAGTGTTTGGGAGCAATCTGTGGTGTAGGGTTAGTGAAAGCTTTCCAGAGCTCATACTA
CGACCGGTACGGCGGCGGTGCCAACGAATTGGCCGACGGTTACAATAAGGGCACCGGACTCGGTGCTGAG
ATCATCGGCACCTTCGTCCTTGTTTACACTGTCTTCTCTGCCACCGATCCCAAGAGGAGTGCACGAGACT
CCCATGTTCCTGTATTGGCACCACTTCCCATCGGATTTGCTGTTTTCATGG
```

FIGURE 9.1 FASTA sample file showing two records.

The sequence section of a FASTA file record is identified by one or more lines that don't begin with a greater-than symbol. Unlike the annotation section, the sequence can be spread over multiple lines. The sequence is associated with the annotation directly preceding. Pseudostandards also exist with the sequence section; generally, it is accepted that sequences should be split by a new line every 70 to 80 characters to aid human readability [2].

Figure 9.1 shows an example of a FASTA file. It contains two records, each with an id and some comma-separated annotation. The sequence in this example is split at 70 characters. As can be seen in this figure, the FASTA file format is primarily for data (in this case the sequence)-intensive processing with the annotation provided so that human readers can understand the results of such processes.

9.1.2 Genbank Format

From the sequence-centric format of FASTA we now move to the Genbank file format. The Genbank format is primarily an annotation-centric format in that it concentrates on supplying detailed annotations for each sequence. The Genbank file format is used to store nucleotide sequences, but there is another flat file database called GenPept which is of the same format, which stores protein sequences. Since the formatting is the same, only the Genbank description is provided in this section.

Genbank files consist of three main sections: the headers, the features, and the sequence; and each section must be in that order. Within the sections there is a set order in which the annotation attributes exist, if they exist at all. The header section contains attributes such as keywords, sub-keywords, accession numbers, version numbers, source organism, taxonomic classification, comments, and references to journal articles and papers that reference this sequence.

The feature table section can contain zero or more subentries that have one of many types, such as source, protein, and CDS (coding sequence). The first line of the subentry contains the type of subentry followed by the start and end index to which this subentry refers in the sequence. The start and end index numbers are separated by a double period (..). The following lines of the subentry contain *name equals value pairs*. Names begin with a forward slash (/) and end at an equals symbol (=). Values begin at the equals symbol and are contained in double quotes ("). If the total length of the pair is greater than 80 characters, it is word-wrapped to the next line.

Finally, the sequence section begins with the word ORIGIN on a line by itself. As with the FASTA file, the sequence is presented in a format that makes it easier for human reading. Each line contains 60 base pairs which are split into groups of 10 characters. The beginning of each line contains a number that represents the index number of the first base pair on that line. The end of the sequence section is marked with a double forward slash (//), also on a line by itself. The double forward slash also marks the end of the record.

Figure 9.2 shows an example of a Genbank file. It contains two records, which are the same as those shown in the previous FASTA file. As can be seen, the Genbank format has much more expressiveness in terms of annotation. The data contained in the FASTA file are all contained within this Genbank file, with the addition of much more annotation. Given this, the Genbank file suits data processing based on the annotation more than just the sequence as for the FASTA file.

9.1.3 Swiss-Prot Format

An alternative annotation centric format to the Genbank format is the Swiss-Prot format. Swiss-Prot is operated and updated by the Swiss Institute of Bioinformatics (SIB) [3]. The Swiss-Prot data repository consists of only amino acid sequences that have been hand controlled for maximum quality and annotation. SIB believes that protein databases need to have three key aspects to measure them by: annotation, minimal redundancy, and integration with other databases [3]. To improve the annotation they try to include as many data about a sequence as possible. To improve the redundancy they combine as many sequences as possible and note any discrepancies. In the past they have combined sequences that are across different source organisms as a method to reduce redundancy. However, they have noticed that this caused confusion in the annotation of species-specific information. So recently, they have separated the sequences that are common over multiple organisms and noted the duplication in the organism species section of the annotation [4]. To improve the integration, they try to link as many other databases as possible [5]. Linking to specialized databases also adds to the level of annotation that the database can contain.

The task of annotation and verifying sequences is a rather time-consuming task and thus reduces the size of the database. So to overcome this, the scientists at the Swiss Institute of Bioinformatics (SIB) combined with EMBL to produce an automatic method of annotation. This is stored in another database called TrEMBL and is meant as a supplement to Swiss-Prot. As TrEMBL is a supplement to Swiss-Prot, it contains only sequences that are not already in Swiss-Prot. As the name suggests, the source of TrEMBL sequences, is the European Molecular Biology Laboratory (EMBL) database.

To conform to other databases standards and allow other databases (of the absorb locally type) to interact with the Swiss-Prot/TrEMBL databases, they produce regular updates. They produce four full yearly updates plus weekly difference updates which can be downloaded from their File Transfer Protocol (FTP) site.

```
LOCUS           AM049264                 884 bp    mRNA     linear    PLN 07-JAN-2006
DEFINITION  Glomus intraradices mRNA for 14-3-3 protein.
ACCESSION   AM049264
VERSION     AM049264.1  GI:84618082
KEYWORDS    14-3-3 gene; 14-3-3 protein.
SOURCE      Glomus intraradices
  ORGANISM  Glomus intraradices
            Eukaryota; Fungi; Glomeromycota; Glomeromycetes; Glomerales;
            Glomeraceae; Glomus.
REFERENCE   1
  AUTHORS   Porcel,R., Azcon,R., Cano,C., Bago,A. and Ruiz-Lozano,J.
  TITLE     Identification of a gene from the arbuscular mycorrhizal (AM)
            fungus Glomus intraradices encoding for a 14-3-3 protein, that is
            up-regulated by drought stress during the AM symbiosis
  JOURNAL   Unpublished
REFERENCE   2  (bases 1 to 884)
  AUTHORS   Ruiz-Lozano,J.
  TITLE     Direct Submission
  JOURNAL   Submitted (13-JUL-2005) Ruiz-Lozano J., Microbiology, Consejo
            Superior de Investigaciones Cien, Prof. Albareda, 1, Granada,
            Granada, SPAIN
FEATURES             Location/Qualifiers
     source          1..884
                     /organism="Glomus intraradices"
                     /mol_type="mRNA"
                     /db_xref="taxon:4876"
                     /tissue_type="extraradical mycelium"
     gene            1..884
                     /gene="14-3-3"
     CDS             1..783
                     /gene="14-3-3"
                     /function="regulatory protein"
                     /codon_start=1
                     /product="14-3-3 protein"
                     /protein_id="CAJ16742.1"
                     /db_xref="GI:84618083"
                     /translation="MERENQTSNKKVYMAKLAEQAERYDEMVSYMKDVAKLGVELTVE
                     ERNLLSVAYKNVIGARRASWRIISSIEQKEENKGNETQVQKIKAYRQKVETELSEVCS
                     DILAVLDEHLIPSAEAGESKVFYYKMKGDYHRYLAEFASGEGRKEAATQAHEAYKHAT
                     DIAQTDLAPTHPIRLGLALNFSVFYYEILNSPDRACHLAKQAFDDAIAELDTLSEESY
                     KDSTLIMQLLRDNLTLWTSDLQDAEGEKPEEAKQEVEAEESK"
     3'UTR           784..884
                     /gene="14-3-3"
ORIGIN
        1 atggaacgcg agaaccagac ctccaacaaa aaggtttata tggctaagct cgctgagcaa
       61 gccgaacgtt atgatgaaat ggtatcttat atgaaggatg ttgcaaagct tggagtagaa
      121 cttactgtag aagaacgcaa tctcctctca gtcgcttata agaatgtaat tggtgcccgt
      181 cgagcgtctt ggagaattat ttcatctatc gaacaaaagg aagaaaataa gggtaatgaa
      241 acccaagttc agaagatcaa agcttaccgt caaaaagttg aaaccgaact ttcagaagta
      301 tgctcggata tccttgccgt attggacgag catttgattc catctgctga agccggtgaa
      361 tctaaggtct tttattataa gatgaaaggt gattatcacc gatatctcgc tgaatttgct
      421 tcgggtgaag gacgaaaaga agccgccacc caggctcacg aagcttataa acacgcaaca
      481 gatatcgccc aaactgacct tgctccaact catcctattc gtcttggact tgcattaaac
      541 tttttcagttt tctactacga gattctcaat tcacctgatc gcgcttgtca tcttgcaaaa
      601 caggcatttg atgatgctat tgctgaactc gatactttaa gtgaagaatc gtataaagat
      661 agcacattaa ttatgcaact acttagagat aacttgaccc tttggacttc agacttgcaa
```

FIGURE 9.2   Genbank file.

```
       721 gatgcagagg gtgaaaaacc ggaagaagct aaacaagaag ttgaagctga agaatctaag
       781 tagcaatagt tagttaattt tattaaaaaa tatatatttc gcacttcata tatataaaaa
       841 aaaaaaaaaa aaaaaaaaaa aaaaaaaaaa aaaaaaaaaa aaaa
//
LOCUS       AJ937963                 331 bp    mRNA    linear    PLN 07-JAN-2006
DEFINITION  Lactuca sativa partial mRNA for putative PIP2 aquaporin (pip2
            gene).
ACCESSION   AJ937963
VERSION     AJ937963.1  GI:84617862
KEYWORDS    PIP2 aquaporin; pip2 gene.
SOURCE      Lactuca sativa
  ORGANISM  Lactuca sativa
            Eukaryota; Viridiplantae; Streptophyta; Embryophyta; Tracheophyta;
            Spermatophyta; Magnoliophyta; eudicotyledons; core eudicotyledons;
            asterids; campanulids; Asterales; Asteraceae; Cichorioideae;
            Cichorieae; Lactuca.
REFERENCE   1
  AUTHORS   Porcel,R., Aroca,R., Azcon,R. and Ruiz-Lozano,J.
  TITLE     Do PIP aquaporins play a role in the enhanced drought tolerance of
            arbuscular mycorrhizal plants?
  JOURNAL   Unpublished
REFERENCE   2  (bases 1 to 331)
  AUTHORS   Ruiz-Lozano,J.
  TITLE     Direct Submission
  JOURNAL   Submitted (11-APR-2005) Ruiz-Lozano J., Microbiology, Consejo
            Superior de Investigaciones Cien, Prof. Albareda, 1, Granada, SPAIN
FEATURES             Location/Qualifiers
     source          1..331
                     /organism="Lactuca sativa"
                     /mol_type="mRNA"
                     /cultivar="Romana"
                     /db_xref="taxon:4236"
                     /tissue_type="roots"
     gene            <1..>331
                     /gene="pip2"
     CDS             <1..>331
                     /gene="pip2"
                     /codon_start=1
                     /product="putative PIP2 aquaporin"
                     /protein_id="CAI79105.1"
                     /db_xref="GI:84617863"
                     /translation="HINPAVTFGLFLARKVSLIRAVGYMIAQCLGAICGVGLVKAFQS
                     SYYDRYGGGANELADGYNKGTGLGAEIIGTFVLVYTVFSATDPKRSARDSHVPVLAPL
                     PIGFAVFM"
ORIGIN
         1 cacatcaacc cggcggtgac attcggtttg tttttggcga gaaaggtgtc gttgattagg
        61 gctgtaggtt acatgattgc acagtgtttg ggagcaatct gtggtgtagg gttagtgaaa
       121 gctttccaga gctcatacta cgaccggtac ggcggcggtg ccaacgaatt ggccgacggt
       181 tacaataagg gcaccggact cggtgctgag atcatcggca ccttcgtcct tgtttacact
       241 gtcttctctg ccaccgatcc caagaggagt gcacgagact cccatgttcc tgtattggca
       301 ccacttccca tcggatttgc tgttttcatg g
//
```

FIGURE 9.2   (*Continued*).

The Swiss-Prot file format contains very similar amounts of annotation and has a similar structure to the Genbank format. Once again, this format is record based; each record has one sequence and some annotation about the sequence. It contains the same three main sections that the Genbank file contains: header, feature table, and sequence sections. As with Genbank, each part of the annotation has a specific order that it must follow.

The beginning of a record is marked with the ID line. The header section, which contains annotation about the sequence as a whole, comes first. The most common header fields are unique id's, accession numbers, dates/version numbers, organism/

```
ID   GRAA_HUMAN     STANDARD;      PRT;    262 AA.
AC   P12544;
DT   01-OCT-1989 (Rel. 12, Created)
DT   01-OCT-1989 (Rel. 12, Last sequence update)
DT   13-SEP-2005 (Rel. 48, Last annotation update)
DE   Granzyme A precursor (EC 3.4.21.78) (Cytotoxic T-lymphocyte proteinase
DE   1) (Hanukkah factor) (H factor) (HF) (Granzyme-1) (CTL tryptase)
DE   (Fragmentin-1).
GN   Name=GZMA; Synonyms=CTLA3, HFSP;
OS   Homo sapiens (Human).
OC   Eukaryota; Metazoa; Chordata; Craniata; Vertebrata; Euteleostomi;
OC   Mammalia; Eutheria; Euarchontoglires; Primates; Catarrhini; Hominidae;
OC   Homo.
OX   NCBI_TaxID=9606;
RN   [1]
RP   NUCLEOTIDE SEQUENCE [MRNA].
RC   TISSUE=T-cell;
RX   MEDLINE=88125000; PubMed=3257574;
RA   Gershenfeld H.K., Hershberger R.J., Shows T.B., Weissman I.L.;
RT   "Cloning and chromosomal assignment of a human cDNA encoding a T cell-
RT   and natural killer cell-specific trypsin-like serine protease.";
RL   Proc. Natl. Acad. Sci. U.S.A. 85:1184-1188(1988).
...
CC   -!- FUNCTION: This enzyme is necessary for target cell lysis in cell-
CC       mediated immune responses. It cleaves after Lys or Arg. May be
CC       involved in apoptosis.
...
DR   EMBL; M18737; AAA52647.1; -; mRNA.
DR   EMBL; BC015739; AAH15739.1; -; mRNA.
DR   EMBL; U40006; AAD00009.1; -; Genomic_DNA.
DR   PIR; A31372; A31372.
DR   PDB; 1HF1; Model; @=29-262.
DR   PDB; 1OP8; X-ray; A/B/C/D/E/F=29-262.
DR   PDB; 1ORF; X-ray; A=29-262.
DR   IntAct; P12544; -.
DR   MEROPS; S01.135; -.
DR   Ensembl; ENSG00000145649; Homo sapiens.
DR   HGNC; HGNC:4708; GZMA.
DR   H-InvDB; HIX0004862; -.
DR   MIM; 140050; -.
DR   GO; GO:0001772; C:immunological synapse; TAS.
DR   GO; GO:0005634; C:nucleus; TAS.
DR   GO; GO:0004277; F:granzyme A activity; IDA.
DR   GO; GO:0005515; F:protein binding; IPI.
DR   GO; GO:0042803; F:protein homodimerization activity; IDA.
DR   GO; GO:0006915; P:apoptosis; TAS.
DR   GO; GO:0006922; P:cleavage of lamin; IDA.
DR   GO; GO:0006955; P:immune response; TAS.
DR   InterPro; IPR001254; Peptidase_S1_S6.
DR   InterPro; IPR001314; Peptidase_S1A.
DR   Pfam; PF00089; Trypsin; 1.
DR   PRINTS; PR00722; CHYMOTRYPSIN.
DR   PROSITE; PS50240; TRYPSIN_DOM; 1.
DR   PROSITE; PS00134; TRYPSIN_HIS; 1.
DR   PROSITE; PS00135; TRYPSIN_SER; 1.


KW   3D-structure; Apoptosis; Cytolysis; Direct protein sequencing;
KW   Glycoprotein; Hydrolase; Protease; Serine protease; Signal; T-cell;
KW   Zymogen.
FT   SIGNAL        1     26
FT   PROPEP       27     28       Activation peptide.
FT                                /FTId=PRO_0000027393.
...
SQ   SEQUENCE   262 AA;  28969 MW;  DA87363A0D92BAF4 CRC64;
     MRNSYRFLAS SLSVVVSLLL IPEDVCEKII GGNEVTPHSR PYMVLLSLDR KTICAGALIA
     KDWVLTAAHC NLNKRSQVIL GAHSITREEP TKQIMLVKKE FPYPCYDPAT REGDLKLLQL
     TEKAKINKYV TILHLPKKGD DVKPGTMCQV AGWGRTHNSA SWSDTLREVN ITIIDRKVCN
     DRNHYNFNPV IGMNMVCAGS LRGGRDSCNG DSGSPLLCEG VFRGVTSFGL ENKCGDPRGP
     GVYILLSKKH LNWIIMTIKG AV
//
```

FIGURE 9.3   Swiss-Prot sample file showing one record.

species, references to journals/papers, comments, database references and keywords. Next is the feature table, which contains specific annotation to a specified part of the sequence. It is fairly similar to the Genbank format in that each feature contains a type, start–end index numbers, and qualifiers that may have name equals value pairs. As with Genbank, long lines are word-wrapped for human readability. The final section is the sequence marked with the line beginning with SQ. The SQ line contains a few summary attributes, such as the total length of the sequence. The actual sequence characters begin on the following line(s) with 60 characters per line. As with the Genbank file, the lines are split into 10 character pieces separated by a space; the Swiss-Prot file does not have an index at the beginning of the line. The end of the sequences is marked by a double forward slash (//), which also marks the end of the record.

Figure 9.3 shows a sample Swiss-Prot file that contains a single protein sequence. More sequences can be added to the Swiss-Prot file simply by appending each to the end of the file. This sample has been modified to remove many repeated attributes from the annotation. The ellipses denote where some repeated attributes have been removed.

9.1.4 XML Format

In previous sections we covered data formats that define their own structure to which the data are presented. To allow for easier machine, and more important, human readability, there has been a move to the XML standard of data structuring. Although there is agreement that XML is a good step forward, there has been much disagreement as to what structure of elements best expresses the types of data typically found in this domain. Thus, many formats have been put forward by varying groups or organisations. Due to the fact that there are so many formats out there, this section can cover only a few of the popular ones.

Since XML formats were developed after the flat file formats, many XML formats have adopted the main sections: headers, features, and sequence. In this section two of the more popular formats are discussed: first, the Biological Sequence Markup Language (BSML) by Lambook Inc. and then INSDseq by the National Center for Biotechnology Information (NCBI).

A cut-down example of the BSML format is shown in Figure 9.4. Repeated elements have been removed to conserve space since they provide no extra information as to the format of the data. BSML is a format provided by Labbook inc. It was developed initially as a standard XML format that Labbook's tools used for data exchange and manipulation. In the past the XEMBL project used the BSML and AGAVE formats to distribute the sequence data. However, the XEMBL database now has moved in favor of the INSDseq and EMBLxml XML formats.

The BSML format almost follows the grouping that flat files used except that the reference data are now included in the feature table. The bulk of the header information is included as attributes in the sequence element, with the exception of the references, as mentioned above. The feature table has a direct mapping from the flat file format to the feature-table element. The sequence data are mapped to the seq-data element.

```
<!DOCTYPE Bsml SYSTEM "BSML2_2.DTD">
<Bsml>
  <Definitions>
    <Sequences>
      <Sequence id="GBNAMEATAF001535" molecule="dna" representation="raw"
          topology="linear" strand="std-not-set" length="27502">
        <Feature-tables>
          <Feature-table>
            <Reference class="Submission">
              <RefAuthors>
                "Till","S." and "McCombie","W.R."
              </RefAuthors>
              <RefTitle>
                Direct Submission
              </RefTitle>
              <RefJournal>
                Submitted (29-APR-1997)
              </RefJournal>
            </Reference>
            ...
            <Feature display-auto="1" id="FTR1" class="GENE" value-type="gene">
              <Qualifier value-type="gene" value="AGAA.1"/>
              <Qualifier value-type="pseudo" value="FALSE"/>
              <Interval-loc startpos="0" endpos="865" complement="1"/>
            </Feature>
            ...
          </Feature-table>
          <Feature-table class="seq-set">
            <Feature id="FTR11" value-type="seq-set">
              <Qualifier value-type="class" value="nuc_prot"/>
            </Feature>
          </Feature-table>
          ...
        </Feature-tables>
        <Seq-data>
atcgaccaaactgtttggtgtctcgggtaagaaatacccgccaagcgtcataagcagagctggaaaagcagctaaacca
agagagagtctccatccccaaggcttaagctgttgagtaccgtaattaaccatgttcgctgtaaagatcccgatagttg
tagctagctgaaacatcatgtttaaaccacctcgtagatgagttggtgccacttctgataaatacaaaggaaccgcctg
aaagcaaaaaccaaagcaaggtaactttcattagagatgagattggtttattattggtgacaccaactaaacaattgtc
taacctgatttccaaatccaataccaacaccaagcatgatccgtccggcaagaagcatagctaagttcacagctccagc
...
ggggaaaaaccgttaatttcattcgatttcattagccattgttattataatgacaggtgttgcacaatataattatagt
taatttagtttattgcctactacgaacaaacctacaaatccaacgactgattgaaatgaaattgaagtagcatatacgt
aaattatagttatttggtaggagtatttgtttgatatagttaaatacaaacaattgaattttcaatttatattaaaca
attgacaaattgtataagatagagtatttatattaaacaattgaattttcaaaggtcgaaacttcagattagtatatt
cgaaa
        </Seq-data>
      </Sequence>
    </Sequences>
  </Definitions>
</Bsml>
```

FIGURE 9.4 BSML format with many repeated elements removed.

Since version 3.1 (their second release) was released, few changes have been noted on their Web site, and it is presumed that they no longer support this format [6]. Add to that the fact that the XEMBL project no longer supports the BSML format and it seems that the end is near, if it is not already here, for the BSML format. Another format, not in this condition, is the INSDSeq XML format. This format is a joint project developed by EMBL, DDBJ, and Genbank for nucleotide sequences. The INSDSeq format (Figure 9.5) follows the previous flat file formats much more closely in that it has header, feature, and sequence sections that contain the same information. The sections are grouped by an element with a simular name except for the header section. The header information is included as several elements within the sequence rather than being grouped within a header element like the others. The feature information is contained in many feature elements with in the feature-table element. The sequence information is contained in the sequence element.

```
<?xml version="1.0"?>
<!DOCTYPE INSDSeq PUBLIC "-//NCBI//INSD INSDSeq/EN"
     "http://www.ncbi.nlm.nih.gov/dtd/INSD_INSDSeq.dtd">
<INSDSeq_set>
<INSDSeq>
  <INSDSeq_locus>AM049264</INSDSeq_locus>
  <INSDSeq_length>884</INSDSeq_length>
  <INSDSeq_moltype>mRNA</INSDSeq_moltype>
  <INSDSeq_topology>linear</INSDSeq_topology>
  <INSDSeq_division>PLN</INSDSeq_division>
  <INSDSeq_update-date>07-JAN-2006</INSDSeq_update-date>
  <INSDSeq_create-date>07-JAN-2006</INSDSeq_create-date>
  <INSDSeq_definition>Glomus intraradices mRNA for 14-3-3
       protein</INSDSeq_definition>
  <INSDSeq_primary-accession>AM049264</INSDSeq_primary-accession>
  <INSDSeq_accession-version>AM049264.1</INSDSeq_accession-version>
  <INSDSeq_other-seqids>
    <INSDSeqid>emb|AM049264.1|</INSDSeqid>
    <INSDSeqid>gi|84618082</INSDSeqid>
  </INSDSeq_other-seqids>
  <INSDSeq_keywords>
    <INSDKeyword>14-3-3 gene</INSDKeyword>
    <INSDKeyword>14-3-3 protein</INSDKeyword>
  </INSDSeq_keywords>
  <INSDSeq_source>Glomus intraradices</INSDSeq_source>
  <INSDSeq_organism>Glomus intraradices</INSDSeq_organism>
  <INSDSeq_taxonomy>Eukaryota; Fungi; Glomeromycota; Glomeromycetes;
       Glomerales; Glomeraceae; Glomus</INSDSeq_taxonomy>
  <INSDSeq_references>
    <INSDReference>
      <INSDReference_reference>2 (bases 1 to 884)</INSDReference_reference>
      <INSDReference_authors>
        <INSDAuthor>Ruiz-Lozano,J.</INSDAuthor>
      </INSDReference_authors>
      <INSDReference_title>Direct Submission</INSDReference_title>
      <INSDReference_journal>Submitted (13-JUL-2005) Ruiz-Lozano J.,
          Microbiology, Consejo Superior de Investigaciones Cien, Prof.
          Albareda, 1, Granada, Granada, SPAIN</INSDReference_journal>
    </INSDReference>
    ...
  </INSDSeq_references>
  <INSDSeq_feature-table>
    <INSDFeature>
      <INSDFeature_key>gene</INSDFeature_key>
      <INSDFeature_location>1..884</INSDFeature_location>
      <INSDFeature_intervals>
        <INSDInterval>
          <INSDInterval_from>1</INSDInterval_from>
          <INSDInterval_to>884</INSDInterval_to>
          <INSDInterval_accession>AM049264.1</INSDInterval_accession>
        </INSDInterval>
      </INSDFeature_intervals>
      <INSDFeature_quals>
        <INSDQualifier>
```

FIGURE 9.5   INSDseq format with many repeated elements removed.

The two formats presented above both contain simular amounts of information; however, they are presented in quite varied ways. It is generally considered that element names and attributes are used for metadata and element values are used for the actual data [7]. Metadata are data that are used to describe data (e.g., "firstname" is metadata and "andrew" is actual data). In the first example, BSML goes against this "good practice" in that it provides many data as values of attributes. An example is

```
            <INSDQualifier_name>gene</INSDQualifier_name>
            <INSDQualifier_value>14-3-3</INSDQualifier_value>
          </INSDQualifier>
        </INSDFeature_quals>
    </INSDFeature>
    <INSDFeature>
      <INSDFeature_key>CDS</INSDFeature_key>
      <INSDFeature_location>1..783</INSDFeature_location>
      <INSDFeature_intervals>
        <INSDInterval>
          <INSDInterval_from>1</INSDInterval_from>
          <INSDInterval_to>783</INSDInterval_to>
          <INSDInterval_accession>AM049264.1</INSDInterval_accession>
        </INSDInterval>
      </INSDFeature_intervals>
      <INSDFeature_quals>
        <INSDQualifier>
          <INSDQualifier_name>gene</INSDQualifier_name>
          <INSDQualifier_value>14-3-3</INSDQualifier_value>
        </INSDQualifier>

        ...
        <INSDQualifier>
          <INSDQualifier_name>db_xref</INSDQualifier_name>
          <INSDQualifier_value>GI:84618083</INSDQualifier_value>
        </INSDQualifier>
        <INSDQualifier>
          <INSDQualifier_name>translation</INSDQualifier_name>
          <INSDQualifier_value>MERENQTSNKKVYMAKLAEQAERYDEMVSYMKDVAKLGVELTVE
ERNLLSVAYKNVIGARRASWRIISSIEQKEENKGNETQVQKIKAYRQKVETELSEVCSDILAVLDEHLIPSAEAG
ESKVFYYKMKGDYHRYLAEFASGEGRKEAATQAHEAYKHATDIAQTDLAPTHPIRLGLALNFSVFYYEILNSPDR
ACHLAKQAFDDAIAELDTLSEESYKDSTLIMQLLRDNLTLWTSDLQDAEGEKPEEAKQEVEAEESK
          </INSDQualifier_value>
        </INSDQualifier>
      </INSDFeature_quals>
    </INSDFeature>
    ...
  </INSDSeq_feature-table>
  <INSDSeq_sequence>ATGGAACGCGAGAACCAGACCTCCAACAAAAAGGTTTATATGGCTAAGCTCGCTG
AGCAAGCCGAACGTTATGATGAAATGGTATCTTATATGAAGGATGTTGCAAAGCTTGGAGTAGAACTTACTGTAG
AAGAACGCAATCTCCTCTCAGTCGCTTATAAGAATGTAATTGGTGCCCGTCGAGCGTCTTGGAGAATTATTTCAT
CTATCGAACAAAAGGAAGAAAATAAGGGTAATGAAACCCAAGTTCAGAAGATCAAAGCTTACCGTCAAAAAGTTG
AAACCGAACTTTCAGAAGTATGCTCGGATATCCTTGCCGTATTGGACGAGCATTTGATTCCATCTGCTGAAGCCG
GTGAATCTAAGGTCTTTTATTATAAGATGAAAGGTGATTATCACCGATATCTCGCTGAATTTGCTTCGGGTGAAG
GACGAAAAGAAGCCGCCACCCAGGCTCACGAAGCTTATAAACACGCAACAGATATCGCCCAAACTGACCTTGCTC
CAACTCATCCTATTCGTCTTGGACTTGCATTAAACTTTTCAGTTTTCTACTACGAGATTCTCAATTCACCTGATC
GCGCTTGTCATCTTGCAAAACAGGCATTTGATGATGCTATTGCTGAACTCGATACTTTAAGTGAAGAATCGTATA
AAGATAGCACATTAATTATGCAACTACTTAGAGATAACTTGACCCTTTGGACTTCAGACTTGCAAGATGCAGAGG
GTGAAAAACCGGAAGAAGCTAAACAAGAAGTTGAAGCTGAAGAATCTAAGTAGCAATAGTTAGTTAATTTTATTA
AAAAATATATATTTCGCACTTCATATATATAAAAAAAAAAAAAAAAAAAAAAAAAAAAAAAAAAAAAAAAAAAAAA
AAAA</INSDSeq_sequence>
</INSDSeq>
...
<INSDSeq_set>
```

FIGURE 9.5 (*Continued*).

the id and molecule attributes in the sequence element. It seems that Lambook Inc., the creators of BSML, have gone with the policy of using attributes unless the data are of multivalue or have subelements.

In contrast to BSML, INSDseq format is a "no attribute" XML format. This format follows good practices much more closely. However, they have made one exception, in that some metadata are included as element values. The element INSDQualifier_name has a metadata value. This appears to be intentional, so that the INSDSeq format is extendable; however, it does allow for mistakes that are not detectable by a DTD or XML Schema Document if the user misspells that element's value.

In the past the XEMBL project used the AGAVE XML format as well as the BSML format described above. As with BSML, the AGAVE format has been abandoned by XEMBL. The AGAVE format was created by Double Twist Inc., which has since gone out of business [8]. Around the time of changing to the INSDSeq format, the EMBL produced another more expressive XML format called EMBLxml [9,10]. AGAVE and EMBLxml as well as many others, such as GAME and BIOML, are not discussed here since most methods for using XML storage are shown here by examples.

9.2   DATA STORAGE

Initially, databases consisted of simple flat files in the formats described in Section 9.1. As for the size and detail of public databases, it was noted that a file database is simply too difficult to maintain and update, so many databases chose to store and maintain their records in a relational or other more advanced database system. We begin this section by discussing methods used by public databases and follow with a detailed look at methods used for multidatabases repositories.

Not much is known about the actual structure of the relational databases used for public databases except that until 2003 the Swiss-Prot database was maintained in flat file databases in the format shown previously. After realizing that this task is troublesome and inefficient, they transferred their data to a relational database structure [5]. The database format has not been published. However, distribution has been and still is via flat file. These data are accessible via the Swiss-Prot File Transfer Protocol (FTP) server.

9.2.1   Multidatabase Repositories

The data formats used to represent multidatabase repositories are generally ones contained in a database management system of some type. However, there are many ways to do this, three of the most common of which are presented below. The first way presented is one that defines a set relational database structure and absorbs all data into this format; the example used here is the PRINTS-S database. The second method used to represent the data, also in a relational database system, is to absorb all data as is from the external databases, adjusting the structure of the local database to fit. BioMolQuest is used to show this method. The final method is similar to the first in that it converts the external data into a single format that is stored locally except that instead of using a relational database system, it uses an object-oriented database.

The PRINTS-S database absorbs the OWL database, which itself absorbs six other databases [12] and the Swiss-Prot/TrEMBL databases. The original version of PRINTS was a single flat ASCII file. As the structure and size of the database grew, it became too complex to manipulate and search, so a simple relational database was implemented. Figure 9.6 shows an ER diagram for the PRINTS-S implementation. It is quite a simple structure that has three entities with relationships between each of them. The entities are fingerprint, sequence, and motif. Entities are represented by the square boxes and relationships by double-headed arrows. The attributes of each entity are represented by bubbles with a line running to the entity with which they are associated. The number next to each relationship is the cardinality. The 1 cardinality next to the fingerprint means "a motif has 1 fingerprint" and an M next to the motif means "a fingerprint can be in many motifs." The many-to-many (M-to-M) relationship between fingerprint and sequence is slightly different from the rest, as it has a name and some attributes. It is represented by the diamond.

Relational databases can be used in many ways and degrees of complexity; keeping it simple in this case allows for easy understanding but does decrease the types of data that can be stored within.

An alternative way to use a relational database is shown in the BioMolQuest database system. The method for incorporating public data for BioMolQuest [13] is to download all the current listings of the database sequences from all the databases that are to be included and to store them locally. They then run parsing scripts that have

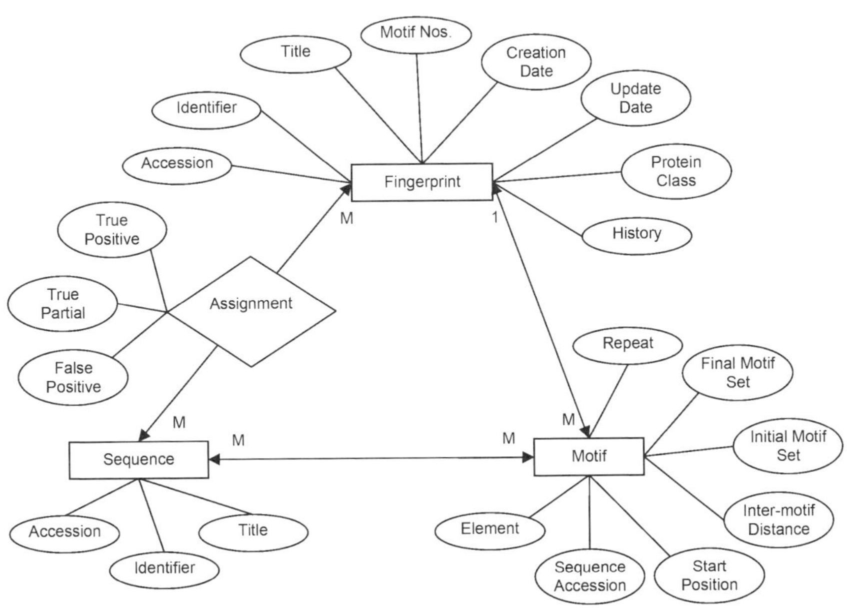

FIGURE 9.6 ER diagram of the PRINTS-S database. (From ref. 12.)

been custom built for each type of input format that has been downloaded to convert it to a compatible format for their database. The parsing scripts also collect and construct the data needed for the references between the data in all the databases.

The tables in this database fall into three main areas: legacy, indexes, and cross-reference tables. The legacy tables are used to store all the information that is absorbed from other databases: at least one for each source and some extras to store multivalue attributes. The parsing scripts also create some indexes; these are stored in the index tables and are used to speed up searches on the database. The final group of tables are the cross-reference tables; they provide links between the legacy data absorbed from the sources (i.e., provide the actual integration of the system).

The database part of this system is contained by a MySQL DBMS database server. The parsers are implemented in Perl, and the Web interface is done in CGI Perl.

This method provides a more adaptable format that allows for easier addition of extra attributes and types of data. However, it does have an increased level of complexity for the user to understand. It also increases the complexity of the database for each database it absorbs, since new tables are created for each external database used.

The Genome Information Management System (GIMS) database system uses the data warehouse model into which to absorb its data [14]. This system uses a similar method for data storage to PRINTS-S, except that the data are stored in an object-oriented database (OODB) with a rather complex set of objects for its structure, specified by UML notation. Figure 9.7 shows the system architecture for the GIMS system. In this system the user uses the interactive interface to get the results from the

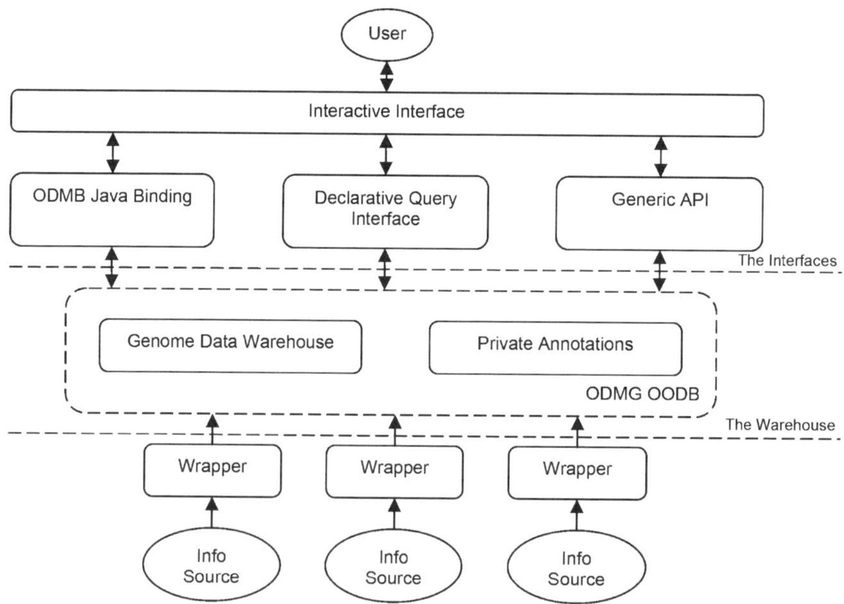

FIGURE 9.7   System architecture of the GIMS database system. (From ref. 14.)

system. The user interface uses the services of the three interface modules for the system: ODMB Java binding, declarative query interface, and generic API. The interface modules then interact with the OODB, which contains all the data collected from the public databases as well as the private annotations that the scientists have collated. Before the database can be used it must absorb the data from the external databases. It does this by using wrappers implemented for each database. The formatting of the data in the data warehouse is designed using UML notation and then converted in the OODB system. The actual UML design can be found in a paper by Paton et al. [14].

Provided here is just one example of the primary methods used to store data when integrating multiple data sources. There are many other implementations not mentioned here that simply use one of the methods explained above or a combination.

9.3 DATA ACCESS

Initially, public databases grew independently and no attempt was made to integrate data across many sources, other than absorbing the other relevant databases into themselves. In Section 9.3.1 we describe the nonintegrated interfaces provided by some public databases. From here it was then realized that integration is one of the key purposes of public databases in order to support research in a broad biological area. In Section 9.3.2 we describe the basic methods used by public databases to provide cross-reference integration. This cross-referencing works well for small data sets, but when the data sets become larger, a more automatable approach is needed. To solve these problems, multi-database access points were developed. In Section 9.3.3 we describe multidatabase access point methods and implementations. Finally, an alternative method to access and search data is presented in Section 9.3.4. This method involves using a tool to search the databases.

9.3.1 Single-Database Access Point

Initially, the public databases consisted of flat files available via Hyper Text Transfer Protocol (HTTP) or FTP. Since these are only really usable for mass comparison tools, searches were added to the public databases. The first versions just contained the text-based records from the flat files. This meant that many users found this difficult to understand and use, so the interfaces were developed further so that the results are nicer to use [15].

One example of this is the Genbank database system. The Genbank database is a collection of three different databases that are accessible from the NCBI Web site [16]. The three databases are called CoreNucleotide, dbGSS, and dbEST. The user can search these databases individually or all at once using the NCBI database search tool. To search all three sections the user selects Nucleotide. After making the database selection, the user types one or more keywords and clicks the Go button. See Figure 9.8 for a sample search for the keyword *brassica*.

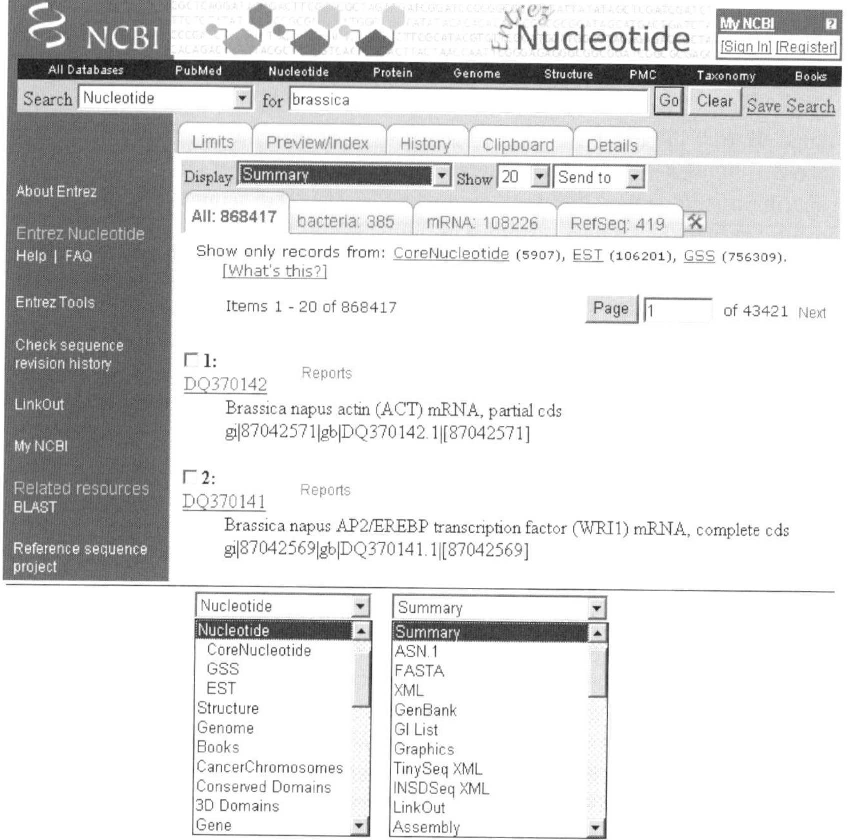

FIGURE 9.8   NCBI search form showing a sample search for the keyword *brassica*. The search and display combo boxes have been expanded at the bottom.

Once the user has done a search, he or she can select the display options. The user can choose to remove or keep specific results by checking or unchecking the check box at the beginning of each record. Optionally, the user can select the limits tab and exclude various results based on their type or modified date ranges, among others. They can select from many formats from publicly accepted, such as FASTA or ASN.1, or some in-house formats, such as Genbank or INSQSeq XML. After selecting the format, the user can choose whether to display them in HTML, text format on the screen, or as a file to download.

Another example of this is the Swiss-Prot (SRS interface) system. Figure 9.9 shows the three screens used to search the Swiss-Prot database system. The first section shows the initial screen, which allows the user to select which databases sections they wish to search. Once the user makes the section, he or she selects the continue button. The user is then presented with the search form depicted in the middle section of the figure. This screen allows the user to specify keywords and in which fields (or all) they should exist. The user is also given control over which fields

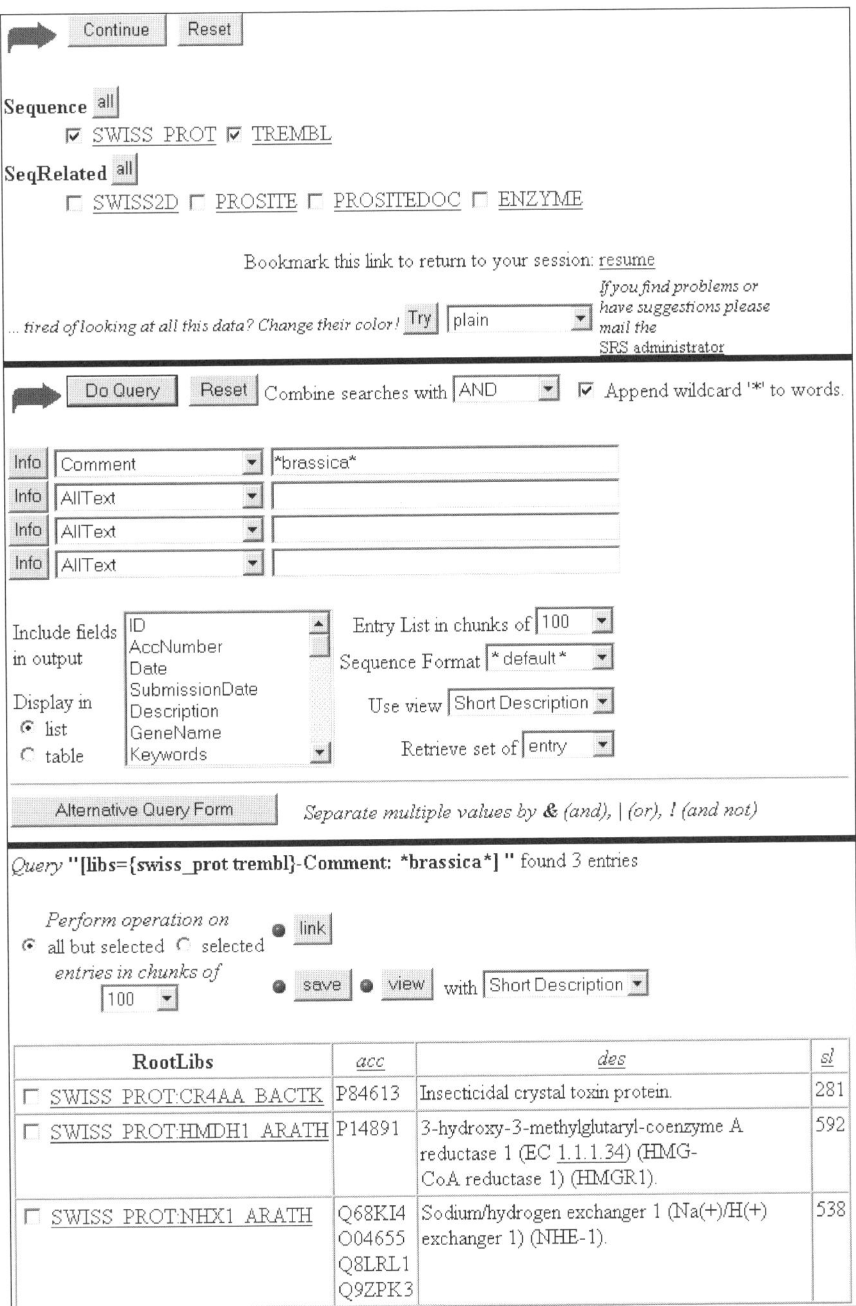

FIGURE 9.9 Swiss-Prot (SRS interface), showing the three steps involved in searching the database. Each screen is separated by a solid black line.

he or she is interested in, and thus which fields are displayed in the output. Other options include the output format (e.g., FASTA, Swiss-Prot) and how many results are displayed in one screen. Once the form is completed, the user clicks the Do Query button, which brings up the results screen shown in the third section of the figure. In this example the default (tabular) format is used.

9.3.2 Cross-Reference Databases

The primary integration between public databases is of the cross-reference type. Cross-references between public database records are the simplest type of data integration techniques. The majority of public database interfaces have adopted "nice" output and this form of integration.

The cross-references between public database records are implemented in one of two ways. The first is by HyperText links from one record to similar records in other public databases, and the second is to provide a record ID for the relevant record in the other database. This method is not limited to stand-alone databases; it is often used as a supplement to other database integration. Many of the other methods described in other sections of this chapter also use this method of database integration as a secondary method. Some go as far as exploiting these references to automatically expand the information provided to users by collecting the cross-referenced results on their behalf.

An example of cross-referencing is shown in Figure 9.10. This example contains its cross-references within the cross-references section in the second column of the figure. It contains links to the EMBL, PIR, and PRINTS databases, to just name a few. The example shown here provides a complete record from the Swiss-Prot database formatted in their nice-prot format. Swiss-Prot is one of the leaders in this type of integration, having a maximum of over 70 links to other databases for sequences contained within [3]. Swiss-Prot is operated and updated by the Swiss Institute of Bioinformatics (SIB).

Although this type of database integration is well suited to getting annotation for a sequence from many databases, it does not lend itself easily to multiple-sequence annotation retrieval. However, it does provide a useful supplement to help the user understand the results returned by other integration and searching systems.

9.3.3 Multiple-Database Access Points

The primary goal of multiple-database access points is to provide more information than a single database can provide by itself. There are two slightly different types of multidatabase access points; the differences are to do with where the data are stored. The first method absorbs the data from the external databases and stores them in the local database. This type of database access is summarized in Figure 9.11. The second type of multidatabase access is where all data are not stored locally. This method queries the external databases on behalf of the user, and only the necessary data are retrieved. Figure 9.12 summarizes this type of multidatabase access. Following the figures, a few examples of these methods are provided.

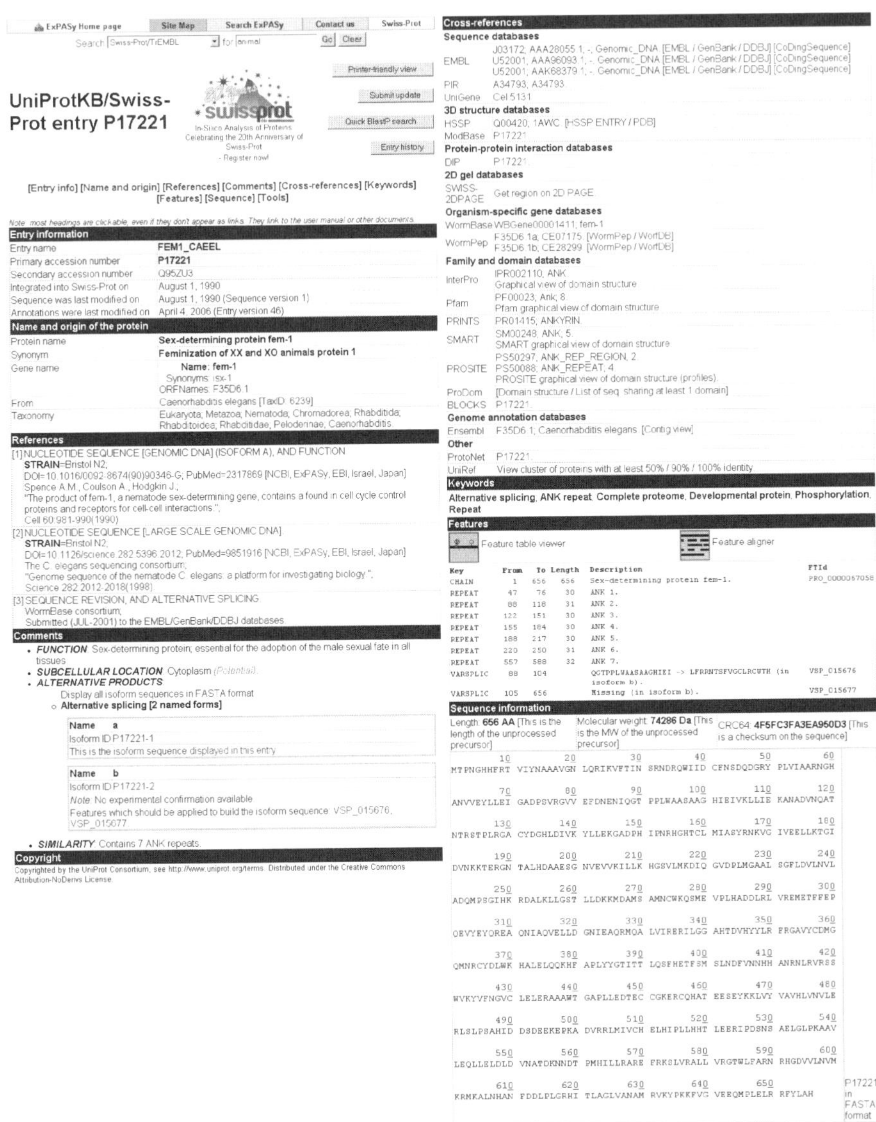

FIGURE 9.10 Example of hyperlinks to other databases. The cross-reference section at the top of the second column shows the links to results within other databases. This example is taken from the Australian home page for ExPASY, showing a result from the Swiss-Prot database.

Figure 9.11 shows the general database structure of the absorb-locally style of database access. The data from the public or external databases are absorbed and optionally formatted by the local database. The user then interacts with the local database.

For it to work, this structure needs to have an update scheme set out. The update scheme is the method that is used to update the records in the local database when

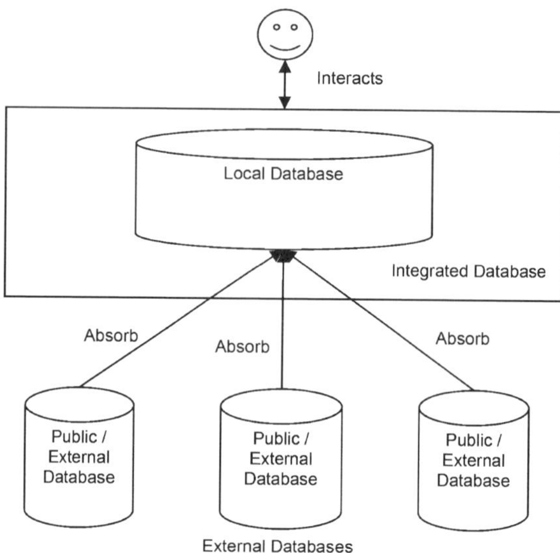

FIGURE 9.11   Basic database structure for the absorb-locally style of database access.

the external databases change. One method is to clear the local database and reabsorb the data from the external databases. This usually takes a while to complete when large amounts of data are being absorbed and requires a reasonable network or Internet connection at this time. As a result, some implementations of this style of integration use other methods that update the data in the local database (i.e., add new entries, update modified entries, and remove deleted entries). The method used is generally based on several things: the speed at which the external data can be obtained, the amount that gets changed, and how frequently the changes happen, just to name a few.

The key to this database design is that it downloads all of the public databases that the user wishes to integrate. This may be only a particular part of the public database for some databases. Parsing scripts are then run on these downloaded records to put them in the format required for their database to absorb.

Or simply, they recreate the public databases as one big database (e.g., if the external databases were relational databases, all the tables from these databases could be put into a single database locally). They then write their own supporting interface to do the required tasks with the data. These may be text searching, BLAST searching, or some other type of interaction. Some benefits of these are that doing searches is faster provided that a high-performance computer is available. It also reduces the Internet bandwidth used if a lot of searches are being completed.

Some examples of this type of database are a protein fingerprint database called PRINTS-S [12] (an extended version of the original PRINTS), BioMolQuest [13], and GIMS [14]. The majority of local laboratory information management system (LIMS) databases also use this method to search their in-house sequences against [17]. See Section 9.1 for a description of the internal data structures used for this type of multidatabase access.

The PRINTS-S database follows the typical search-result format for data access mentioned previously. It has two primary methods to access the data held within; both use this format. The first method is a BLAST search tool that allows users to match their sequences against those within the database to find annotations for their sequence. The search form contains three groups of parameters: search, output, and sequence. The search group contains options such as the cutoff score and which strand of DNA sequences to search, the output parameters set the number of results per page, the order, and the format used to display, and the sequence parameters specify what sequence type (or ID) the user is supplying.

The second way to access the PRINTS-S database is by means of the finger-PRINTScan access point. This is a very simple search that allows the user either to enter a database ID for the sequence or to provide a protein sequence and E-value threshold with which to search the database. Since the PRINTS-S database absorbs sequence data from the OWL database, the database ID must exist in the OWL database. When the search matches a record, it provides various forms of tabulated results. It provides simple and detailed data in a tabular format describing the fingerprints that were found and various scores of predicted accuracy as well as the actual protein sequence found.

Another example of this typical format is the BioMolQuest database. This database provides a World Wide Web (WWW) search engine–style interface to search protein data from many public protein databases. The interface allows users to select which database they wish to search, which fields they wish to search within, and the keywords they are searching for. Optionally, the user can specify simple comparisons using the simple operators greater than, equal to, less than, and combinations of these. Also, the user can specify how the keywords are used in the search with one keyword, all keywords, or any keyword or exact phrase if they provide more than one keyword.

The search method does keyword searches or simple comparisons on each of the integrated databases (local copy) and then expands the results using the cross-references that were provided with each record. See Section 9.3.2 for a discussion of cross-referenced database integration. The results are formatted in a tree structure and presented to the user in a tabular format.

Figure 9.12 shows the basic system architecture of the external query style of multidatabase access. The main distinguishing factor of this style is the fact that no data are actually stored long term on the local system (i.e., where the user interacts with the system). The user interacts with the system by inputting a query which is taken by the query constructor module, which then converts the query into one that is in the format that the external or public databases can use. It then sends these queries off to the required databases. The query executes on the external databases and returns the results to the system. This time the results formatter takes the result and converts them into the required format for the user.

This type of integration is distinguished by the fact that it stores few to no data locally and gets its results by querying other public databases. The main steps in this architecture are: get query from user, convert it so that it can be run on each of the external databases, run the query and retrieve the results, and format the results so that they are relevant and useful to the user. The differences between the major

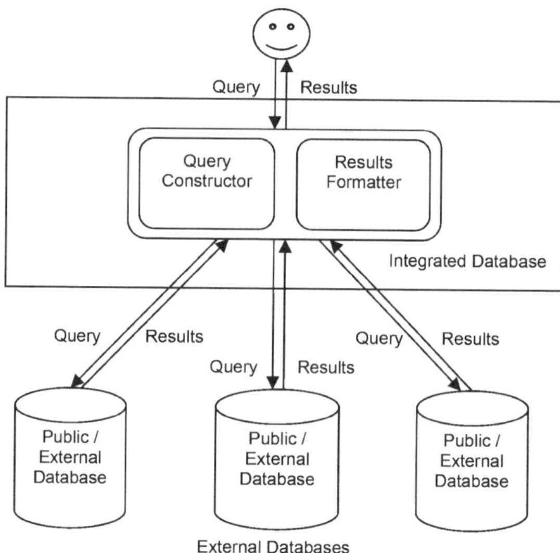

FIGURE 9.12   Database structure for the external query style of database access.

implementations of this type are mostly to do with how the user gives the query. This can vary from a SQL-like query constructed by the user to a very abstract system that allows the user to construct queries in terms of biological concepts.

The TAMBIS [18] system uses a graphical method to specify the query to attempt to provide a complex query language without introducing a huge learning curve. The TAMBIS database uses a single smart interface that gives the appearance that the data the user is searching are local. One of the aims of the project was to provide a seamless design for users, to avoid as much confusion as possible. The TAMBIS system integrates with Swiss-Prot, Enzyme, Cath, BLAST, and Prosite [19]. The authors produced two versions of this system, one with about 1800 biological concepts and the other with 250 concepts. The first wasn't actually usable but was just to show that you could implement all the concepts that they could think of in their internal concept representation system. The second was fully functioning, to prove that the rest of the system worked.

Figure 9.13 shows the system architecture of the TAMBIS system; the circle that looks like a face on the left is where the user interacts with the system, and the three cylinder-like objects on the right are the public databases. As can be seen, it is a rather complex and abstract system. The purpose of this is to make a clear distinction between the actual data source implementation and the biological concepts represented by the system. The main dataflow that the user is interested in is represented by the dark arrow, and the supporting flow is shown by the lighter arrow. (Note that the darkness doesn't represent the amount of data flow.)

In this system the user interacts with the conceptual query formulation module in terms of biological concepts known by the system. The user specifies the query in a graphical way by adding to the query concepts and limitations, represented as colored boxes with lines connecting. An example of this is a protein that has organism as a

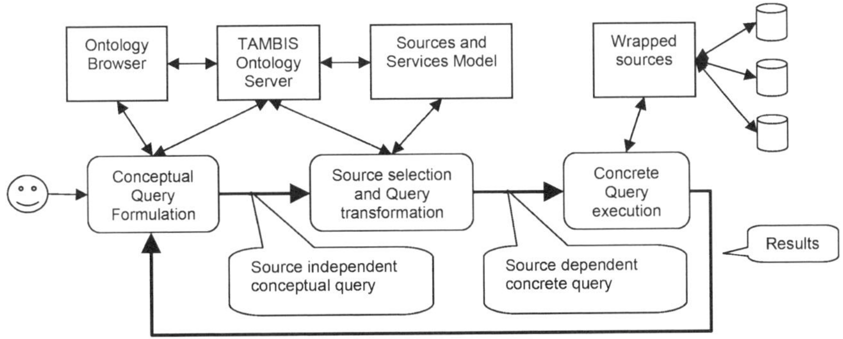

FIGURE 9.13 TAMBIS system architecture showing the flow between the processes. (From ref. 18.)

source. In this case, "protein" and "organism" are the concepts and "has organism source" is the connecting line between them. Additionally, concepts can have attributes such as "name" (e.g., "organism" might have the name "poecilia reticulata").

Once the query is specified, the conceptual query formulation module then constructs a source-independent query and pass it to the source selection and query transformation module. This module then works out which of the sources would produce reasonable results for the query. It also generates a query to get the required results from each relevant source and passes them onto the concrete query execution module. This module executes the query and passes the results back to the first module, conceptual query formation, to convert the results into biological concepts.

An alternative to the TAMBIS system is the Human Genome Project Chromosome 22's database system (HGPC22). The HGPC22 database [20] (Figure 9.14) implements a simplified and extended querying language using the Collection Programming Language (CPL) querying language on top of the extensible querying language called Kleishi. This implementation is constructed in two modules, the query constructor/optimizer and the database drivers. CPL provides the interface for this system to user interface programs, which can be written in languages such as Shell scripts, C (and other variants), Perl, Prolog, and HTML. Below the Kleishi query optimizer is an interface to the database drivers' module written in the ML programming language. In the database drivers' module there is one driver for every type of database, including the local data store. The public databases that are connected are Genbank, GDB, and Entrez.

Multipoint access points fit into one of two board categories: self-contained or central access points. The key difference is the way that the user interacts with the system; they can interact with the actual system themselves or interact via another program's user interface, respectively. The TAMBIS system is an example of the first group since it uses a self-contained interface for users to specify their query. The HGPC22 system uses the second methodology since the user interacts with an external system, which then uses the system's services to get the data required.

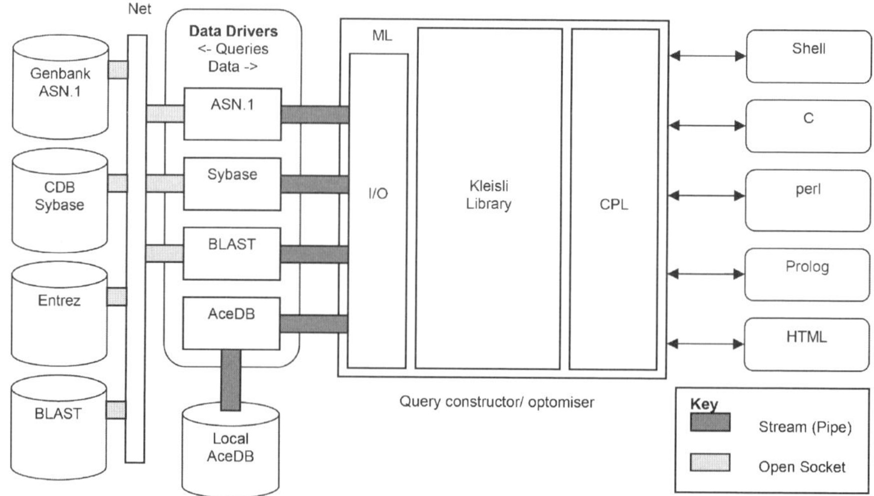

FIGURE 9.14   Architecture of the HGPC22 database. (From ref. 20.)

The benefits of using the self-contained system are that the interface is easier to build and can be coupled with the system directly and more efficiently. However, the benefit of using the second methodology is that reuse becomes much easier.

9.3.4   Tool-Based Interfaces

An alternative to searching the annotation for keywords is to use a specific tool to search the database. The most common tool used for public databases is the BLAST tool, which involves an exhaustive search to compare which sequences are similar in terms of sequence bases. Running a BLAST search compares each sequence on the query list with each sequence in the subject database, checking where they overlap. The overlaps do not have to overlap exactly, but there are penalties applied to every base that does not match and credits for those that do match. If after either of the sequences finishes and the current score is still above a threshold, they are considered as a overlap.

When BLAST is used as a tool to search public databases, the public database provides an interface (usually, Web based) to the BLAST tool. The subject database is part or all of the public database sequences. The search list sequences are usually provided by copying them into a text field on the form or uploading a file in a prespecified format. Often, the user is provided with many of the options that are available on the blast tool itself. Since there are many options, some interfaces do not allow some options to be changed to simplify the interface. Figure 9.15 shows an example input form for the BLAST search tool on the NCBI Web site. It gives an abundance of choices for input, options, and formatting.

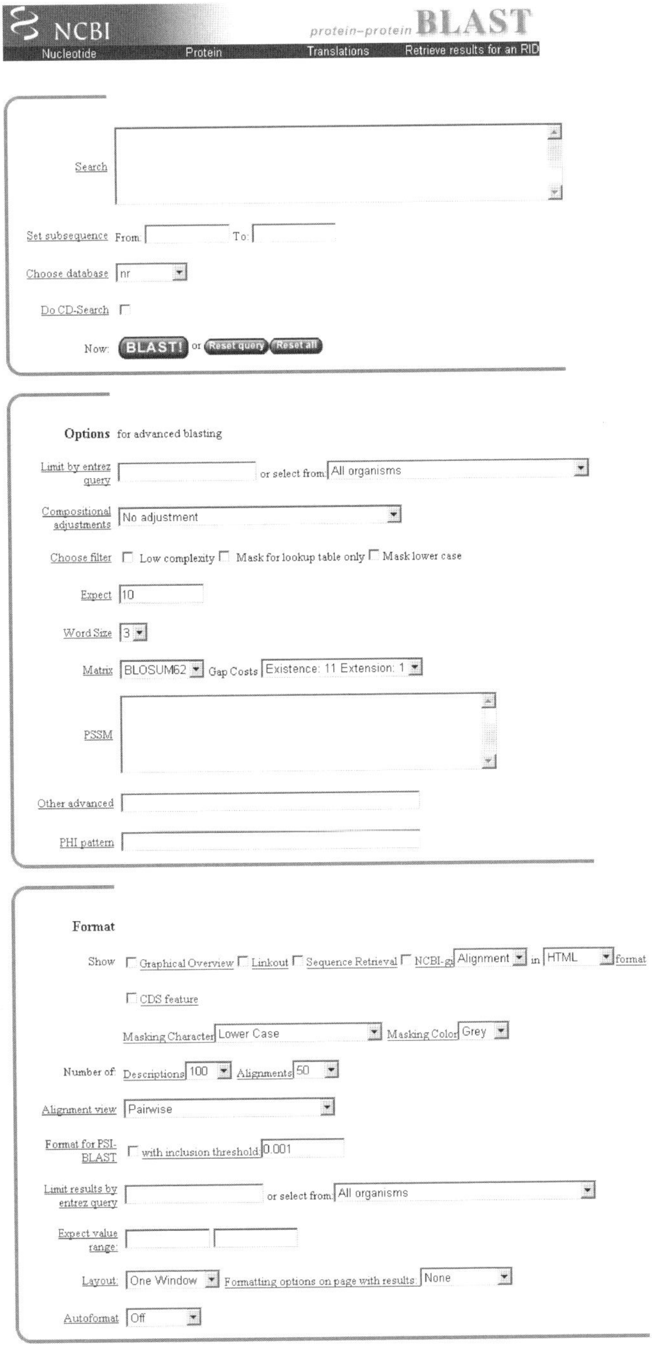

NCBI
protein–protein BLAST

Nucleotide   Protein   Translations   Retrieve results for an RID

Search

Set subsequence From: To:

Choose database nr

Do CD-Search

Now: BLAST! or Reset query Reset all

Options for advanced blasting

Limit by entrez query   or select from All organisms

Compositional adjustments No adjustment

Choose filter ☐ Low complexity ☐ Mask for lookup table only ☐ Mask lower case

Expect 10

Word Size 3

Matrix BLOSUM62 Gap Costs Existence: 11 Extension: 1

PSSM

Other advanced

PHI pattern

Format

Show ☐ Graphical Overview ☐ Linkout ☐ Sequence Retrieval ☐ NCBI-gi Alignment in HTML format

☐ CDS feature

Masking Character Lower Case   Masking Color Grey

Number of: Descriptions 100   Alignments 50

Alignment view Pairwise

Format for PSI-BLAST ☐ with inclusion threshold 0.001

Limit results by entrez query   or select from All organisms

Expect value range:

Layout: One Window   Formatting options on page with results: None

Autoformat Off

FIGURE 9.15 Simple BLAST search interface taken from the NCBI Web site. (From ref. 21.)

9.4 DISCUSSION

There are two main types of data formats: sequence centric and annotation centric. The type of data used generally dictates the methods that are used to store and maintain the data. As many public databases have found, maintaining annotation centric data in flat files is time consuming and not very efficient in terms of data use and unnecessary duplication. Thus, annotation-centric data formats are simply used as a backward-compatible transmission format for tools developed for the original data files. Sequence-centric data, on the other hand, generally do not need to be searched by hand and rarely need to be updated. This means that this type of data can be stored in flat files or in more advanced data management systems, such as a relational database management system. However, most annotation centric data are stored in the popular FASTA format within flat files simply because these data are used primarily for searching and matching tools that usually require this format as input.

The user interfaces style does not depend on the data format used; more commonly it is related to what users want to obtain from the system. There are many methods and procedures that a system can use to provide data access to users. Thus, every user interface seems to vary widely depending on which advantages a user needs and how he or she can handle the associated disadvantages to best deal with them.

9.5 CONCLUSIONS

Bioinformatics research has been developing, and will continue to develop, at an astounding rate. Together with this, the rate at which new data are being collected is also increasing, with no end in sight. With this thought, many possibilities in terms of data use come to mind.

There are three main aspects of genome databases discussed in this chapter: data representation, storage, and access. The representation types consist of data-centric, annotation-centric, and XML formats. The primary methods of storage are flat file storage and relational DBMS storage. The access methods consist of single-database accesses, cross-referencing, multidatabase access points, and tool-based interfaces.

The current trends are toward centralized access to multiple data sources: public databases or in-house (or confidential) data. The principal goal is to be as complete as possible in terms of annotation. This aim brings about the need to display masses of data in a useful manner. Satisfying this need brings about the second major trend: graphical representations and natural language interfaces.

REFERENCES

1. http://www.w3.org/XML/, XML eXtensible Markup Language, *W3C*.

2. R. A. Dwyer, *Genomic Perl: From Bioinformatics Basics to Working Code*, Chap. 6, Cambridge University, Cambridge, 2003.

3. http://www.expasy.ch/sprot/, ExPASy Swiss-Prot and TrEMBL.

4. http://us.expasy.org/sprot/relnotes/spwrnew.html, Swiss-Prot weekly release news headlines, release 48.7 of Dec. 20, 2005.

5. B. Boeckmann, A. Bairoach, R. Apweiler, M.-C. Blatter, A. Estreicher, E. Gasteiger, M. J. Martin, K. Michour, C. ODonovan, I. Phan, S. Pilbout, and M. Schneider, The Swiss-Prot protein knowledgebase and its supplement TrEMBL in 2003, *Nucleic Acids Res.*, pp. 365–370 (2003).

6. http://www.bsml.org/, Bioinformatic Sequence Markup Language, Lambook Inc.

7. http://www.w3schools.com/xml/xml_attributes.asp, XML attributes: XML tutorial, *w3schools.*

8. http://www.bio-itworld.com/archive/050702/survivor_sidebar_252.html, The demise of DoubleTwist :: Bio-IT World.

9. L. Wang, J.-J. Riethoven, and A. Robinson, XEMBL: distributing EMBL data in XML format, *Bioinformatics*, **18**(8), 1147–1148 (2002).

10. http://www.ebi.ac.uk/xembl/index.html, EMBL in XML.

11. B. Boeckmann, A. Bairoach, R. Apweiler, M.-C. Blatter, A. Estreicher, E. Gasteiger, M. J. Martin, K. Michour, C. ODonovan, I. Phan, S. Pilbout, and M. Schneider, The Swiss-Prot protein knowledgebase and its supplement TrEMBL in 2003, *Nucleic Acids Res.*, pp. 365–370 (2003).

12. T. K. Attwood, M. D. R. Croning, D. R. Flower, A. P. Lewis, J. E. Mabey, P. Scordis, J. N. Selly, and W. Wright, PRINTS-S: the database formerly known as PRINTS, *Nucleic Acids Res.*, **28**(1), 225–227 (2000).

13. Y. V. Bukhman and J. Skolnick, BioMolQuest: integrated database-based retrieval of protein structural and functional information, *Bioinformatics*, **12**(5), 468–478 (2001).

14. N. W. Paton, S. A. Hhan, A. Hayes, F. Moussouni, A. Brass, K. Eilbeck, C. A. Goble, S. J. Hubbard, and S. G. Oliver, Conceptual modelling of genomic information, *Bioinformatics*, **16**(6), 548–557 (2000).

15. E. Gasteiger, A. Gattiker, C. Hoogland, I. Ivanyi, R. D. Appel, and A. Bairoch, ExPASy: the proteomic server for in-depth protein knowledge and analysis, *Nucleic Acids Res.*, **31**(13), 3784–3788 (2003).

16. http://www.ncbi.nlm.nih.gov/Genbank/index.html, GenBank overview, *NCBI.*

17. J. Li, S.-K. Ng, and L. Wong, Bioinformatics adventure in database research, *Bioinformatics*, LNCS **2572**, Database Theory, pp. 31–46 (2003).

18. R. Stevens, P. Baker, S. Bechhofer, G. Ng, A. Jacoby, N. W. Paton, C. A. Goble, and A. Brass, TAMBIS: transparent access to multiple bioinformatics information sources, *Bioinformatics*, **16**(2), 184–185 (2000).

19. L. F. Bessa Seibel and S. Lifschitz, A genome databases framework, *Bioinformatics*, LNCS **2113**, Database and Expert Systems Applications, pp. 319–329 (2001).

20. P. Buneman, S. B. Davidson, K. Hart, C. Overton, and L. Wong, A data transformation system for biological data sources, *Proc. 21st VLDB Conference*, pp. 158–169 (1995).

21. http://www.ncbi.nlm.nih.gov/BLAST/Blast.cgi, NCBI Blast, *NCBI.*

10

AUTOMATIC QUERY EXPANSION WITH KEYPHRASES AND POS PHRASE CATEGORIZATION FOR EFFECTIVE BIOMEDICAL TEXT MINING

MIN SONG

Department of Information System, College of Computing Sciences, New Jersey Institute of Technology, Newark, New Jersey

IL-YEOL SONG

College of Information Science and Technology, Drexel University, Philadelphia, Pennsylvania

Handling a vast amount of unstructured data becomes a difficult challenge in text mining. To tackle this issue, we propose a novel text mining technique that integrates information retrieval (IR) techniques with text mining. In *relevance feedback*, a subfield of IR, relevance information is gathered from documents retrieved in a ranked list generated from an initial request. The relevance information is used to modify the search query and perform a further retrieval pass. The two main factors in relevance feedback are the source from which expansion terms are determined and the method of ranking expansion terms. These factors have a crucial impact on retrieval performance in *pseudo-relevance feedback*, an effective technique for the retrieval of more relevant documents without relevance feedback from users. In the pseudo-relevance feedback method, a small set of documents is retrieved using the original user query. These documents, whose relevance is assumed, are then used to construct an expanded query, which is used, in turn, to retrieve the set of documents that has actually been presented to the user.

Knowledge Discovery in Bioinformatics: Techniques, Methods, and Applications, Edited by Xiaohua Hu and Yi Pan

In this chapter we present a new and unique unsupervised query expansion technique that utilizes keyphrases and part-of-speech (POS) phrase categorization. We use keyphrases extracted from the retrieved documents to improve the term selection and query reranking for pseudo-relevance feedback. Keyphrase extraction is a process to extract important phrases in a document that an author or cataloger would assign as keyword metadata [18]. Keyphrases are extracted from the top N-ranked documents that have been retrieved and expansion terms are selected from the keyphrase list rather than from the entire document. The keyphrases selected are translated into disjunctive normal form (DNF) by the POS phrase categorization technique. We evaluate the keyphrases using the POS phrase categorization technique with Medline data. Retrieval results using Medline ad hoc tasks show that the use of keyphrases can improve pseudo-relevance feedback. We also explore a technique that combines synonymous terms from ontologies to keyphrases. However, there are mixed results when using ontologies such as WordNet and MeSH for the query expansion task.

In this chapter we make the following contributions. First, unlike most other query expansion techniques that use a single term selected with statistical-based term weighting, we use keyphrases as the basic unit for our query term. Phrase selection relies on the overall similarity between the query concept and phrases of the collection rather than on the similarity between a query and the collection's phrases [12]. We show that keyphrases extracted from retrieved documents better represent the core concepts of the documents. Second, we propose a new notion of POS phrase categories, which is used effectively to combine multiple keyphrases into the disjunctive and normal form (DNF) used for query expansion. Third, our techniques can make use of ontologies such as WordNet or MeSH to add more relevant phrases to the query. For WordNet, we employ a new word sense disambiguation technique. Our technique is novel in that it is based on the similarity between senses in WordNet and keyphrases extracted from retrieved documents. Fourth, we demonstrate that our proposed techniques are applicable to a variety of domains. We test our techniques on biomedical data collections. Finally, through extensive experiments, we validate the performance advantages of our techniques over those of other leading algorithms. The remainder of the chapter is organized as follows. In Section 10.1 we describe our keyphrase-based query expansion methods, in Section 10.2 describe query expansion with ontologies, in Section 10.3 outline the test data and report on experiments with Medline, and in Section 10.4 conclude the chapter.

10.1 KEYPHRASE EXTRACTION-BASED PSEUDO-RELEVANCE FEEDBACK

We test whether carefully selected keyphrases can be effective for pseudo-relevance feedback. In this section we discuss our techniques and procedures for query expansion. Then we give detailed descriptions of the techniques used in our approach for keyphrase extracting, query reweighting, and query translating. The architecture of the system is shown in Figure 10.1.

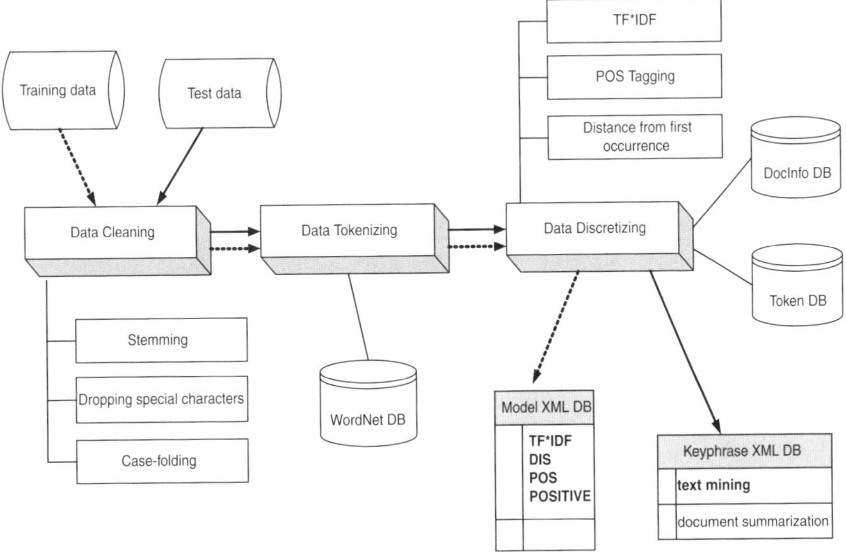

FIGURE 10.1 Data processing procedures for keyphrase extraction.

10.1.1 Keyphrase Extraction Procedures

Our keyphrase extraction consists of two stages: (1) building an extraction model and (2) extracting keyphrases. The input of the building extraction model stage consists of training data and that of the extracting keyphrases stage, test data or production data. These two stages are fully automated. Both training and test data are processed by three components: data cleaning, data tokenizing, and data discretizing. Detailed descriptions are provided below. These keyphrase-extraction procedures have proven effective in other information extraction studies (e.g., [5,21]).

Candidate Keyphrase Extraction Procedure   Input text is parsed into sentences. Candidate keyphrases are then selected within a sentence. Three rules were used to select candidate keyphrases: (1) A keyphrase is limited to a certain maximum length: three consecutive words in this research; (2) it cannot be a proper name (i.e., a single word that would appear with an initial capital); and (3) it cannot begin or end with a stop word (our stop word list from Okapi [26] consists of 256 unimportant terms). All continuous sequences of words in each document are evaluated as candidate phrases using these three rules.

Feature Selection   The following three feature sets were calculated for each candidate phrase: (1) term frequency $\times$ inverse document frequency (TF*IDF), (2) distance from first occurrence (DFO), and (3) part of speech (POS). TF*IDF is a well-established retrieval technique [19] for calculating the importance of a term in a document:

$$W_{ij} = tf_{ij} \log_2 \frac{N}{n}$$

where W_{ij} is the weight of term T_i in document D_j, tf_{ij} the frequency of term T_j in document D_j, N the number of documents in a collection, and n the number of documents in which term T_j occurs at least once.

The distance from first occurrence (DFO) is calculated as the number of phrases that precede the phrase's first appearance, divided by the number of phrases in the document:

$$\text{DFO} = \frac{\sum w_{i-1}}{\text{NP}}$$

where w_{i-1} is the number of phrases preceding the target phrase and NP is the total number of phrases in the document.

POS tagging assigns a POS such as noun, verb, pronoun, preposition, adverb, adjective or other lexical class marker to each word in a sentence. We combine four POS tagging techniques: NLParser, Link-Grammar, PCKimmo, and Brill's tagger to improve POS tagging accuracy. This combined approach to POS techniques enables us to assign the best tag to lexical tokens, constituting candidate phrases by utilizing optimal features of each POS technique [21].

Because the features selected in our approach are continuous, we need to convert them into nominal forms to apply our machine learning algorithm. From many possible discretization algorithms, we chose equal-depth (frequency) partitioning, which allows good data scaling [4]. Equal-depth discretization divides the range into N intervals, each containing approximately the same number of samples. The value of each feature, a candidate phrase, is replaced by the range to which the value belongs. Table 10.1 shows the results of discretization by equal-depth partitioning. The values shown in the table are derived from Medline data.

10.1.2 Keyphrase Ranking

Automatic query expansion requires a term-selection stage. The ranked order of terms is of primary importance in that the terms that are most likely to be useful are close to the top of the list. We reweight candidate keyphrases with information gain. Specifically, candidate keyphrases are ranked by GAIN(P), a measure of expected reduction in entropy based on the usefulness of attribute A. This is one of the most

TABLE 10.1 Discretization Table

Feature	\multicolumn{5}{c}{Discretization Range}				
	1	2	3	4	5
TF*IDF	<0.003	≥ 0.003 && $<$ 0.015	≥ 0.015 && $<$ 0.050	≥ 0.050 && $<$ 0.100	≥ 0.100
DFO	<0.150	≥ 0.150 && $<$ 0.350	≥ 0.350 && $<$ 0.500	≥ 0.500 && $<$ 0.700	≥ 0.700
POS	<0.001	≥ 0.001 && $<$ 0.200	≥ 0.200 && $<$ 0.700		

popular measures of association used in data mining. For instance, Quinlan [16] uses information gain for ID3 and its successor C4.5, which are widely used decision tree techniques. With ID3 and C4.5 we construct simple trees by choosing at each step the splitting feature that tells us the most about the training data. Mathematically, information gain is defined as

$$\text{GAIN}(P_i) = I(p,n) - E(P_i)$$

where P_i is the value of the candidate phrase that falls into a discretized range. $I(p,n)$ measures the information required to classify an arbitrary tuple:

$$I(s_1, s_2, \ldots, s_m) = -\sum_{i=1}^{m} \frac{s_i}{s} \log_2 \frac{s_i}{s}$$

S contains S_i tuples of class C_i for $i = (1, \ldots, m)$. $E(w)$ is the entropy of attribute W with values of $s_1, s_2, \ldots, s_v$, where s_1 is the probability that W_1 will occur:

$$E(w) = \sum_{j=1}^{v} \frac{s_{1j} + \cdots + s_{mj}}{s} I(s_{ij}, \ldots, s_{mj})$$

Each candidate phrase extracted from a document is ranked by the probability calculated with GAIN(P). In our approach, $I(p,n)$ is stated such that class p: candidate phrase = "keyphrase" and class n: candidate phrase = "non-keyphrase."

Many query reranking algorithms are reported in the literature [17,20]. These algorithms attempt to quantify the value of candidate query expansion terms. Formulas estimate the term value based on qualitative or quantitative criteria. The qualitative arguments are concerned with the value of the particular term in retrieval. On the other hand, the quantitative argument involves specific criteria such as a proof of performance. One example of the qualitative-based formula is the relevance weighting theory.

Although there are many promising alternatives to this weighting scheme in the IR literature [25], we chose the Robertson–Sparck Jones algorithm [18] as our base because it has been demonstrated to perform well, is naturally well suited to our task, and incorporating other term weighting schemes does not require changes to our model. The F4.5 formula was proposed by Robertson and Sparck Jones. It has been used widely in IR systems with some modifications (Okapi). Although a few more algorithms were derived from F4.5 formula by Robertson and Spark Jones, in this chapter, we modify the original for keyphrases as

$$P(w) = \log \frac{r + 0.5}{R - r + 0.5} \Big/ \frac{n - r + 0.5}{N - n - R + r + 0.5}$$

where $P(w)$ is keyphrase weight, N the total number of sentences, n the number of sentences in which that query terms cooccur, R the total number of relevant sentences, and r the number of relevant sentences in which the query terms co-occur. We combine information gain with the modified F4.5 formula to incorporate keyphrase properties gained:

$$KP(r) = \sqrt{\frac{GAIN(p)P(w)}{2}}$$

All candidate keyphrases are reweighted by $KP(r)$ and the N top-ranked keyphrases are added to the query for the next pass. N is determined by the size of the documents retrieved.

10.1.3 Query Translation into DNF

A major research issue in IR is how to ease the user's role of query formulation through automating the process of query formulation. There are two essential problems to address when searching with online systems: (1) initial query formulation, which expresses the user's information need; and (2) query reformulation, which constructs a new query from the results of a prior query [19]. The latter effort implements the notion of relevance feedback in IR systems and is the topic of this section.

An algorithm for automating Boolean query formulation was first proposed in 1970. This method employs a term weighting function first described by Frants and Shapiro [6] to determine the importance of terms that have been identified. The terms were then aggregated into sub-requests and combined in a Boolean expression in disjunctive normal form (DNF). Other algorithms that have been proposed to translate a query to DNF are based on classification [7], decision trees [2], and thesauri [22]. Hearst [9] proposed a technique for constructing Boolean constraints, which was revisited by Mitra et al. [14].

Our POS category-based translation technique differs from others in that ours is unsupervised and is easily integrated into other domains. In our technique, four phrase categories are defined: (1) the ontology phrase category, (2) the nonontology noun phrase category, (3) the nonontology proper noun phrase category, and (4) the verb phrase category. Phrases that have corresponding entities in ontologies such as WordNet and MeSH belong to the ontology phrase category. We include the verb phrase category as a major category because important verb phrases play a role in improving retrieval performance [8]. Keyphrases within a category are translated into DNF, and categories are then translated into conjunctive normal form.

The sample keyphrases for query 350 for Medline are shown in Figure 10.2. As explained earlier, within the same category the phrases are combined with the OR Boolean operator. Between categories, the terms are combined with the AND Boolean operator. Thus, the query shown in Figure 10.2 is translated as follows:

(occupational health OR workplace disorders OR physical injury OR computer terminal OR terminals activity) AND (computer screen OR workers computer) AND report.

```
<keyphrases id="350">
    <keyphrase weight="0.39282" category="2" >
        computer screen </keyphrase>
    <keyphrase weight="0.38114" category="1" >
        occupational health </keyphrase>
    <keyphrase weight="0.38566" category="1" >
        workplace disorders </keyphrase>
    <keyphrase weight="0.38432" category="1" >
        physical injury</keyphrase>
    <keyphrase weight="0.38427" category="1" >
        computer terminal </keyphrase>
    <keyphrase weight="0.38320" category="2" >
        workers computer </keyphrase>
    <keyphrase weight="0.38293" category="4">
        report </keyphrase>
    <keyphrase weight="0.38174" category="1" >
        terminals activity </keyphrase>
</keyphrases>
```

FIGURE 10.2 Sample keyphrases extracted for query expansion.

10.2 QUERY EXPANSION WITH WordNet

For the N top-ranked keyphrases, our technique can traverse ontologies such as
WordNet. If there is a corresponding phrase, the keyphrase is categorized as an
ontology phrase category. With WordNet, we encounter a complication with multiple
senses of a given phrase. To tackle this problem we introduce a straightforward word
sense disambiguation technique, which is based on similarities between WordNet
phrases and the keyphrases extracted by our technique. In WordNet, a group of
synonyms with the same meaning compose a *synset*. Synsets are linked to each other
through relationships such as hyponyms, hypernyms, and holonyms. If no synsets are
found for a given phrase, we traverse down in the synset list to find the sysnet. For
multiple synsets, all the non-stopwords are captured from synonyms and their
descriptions, hyponyms and their descriptions, and other relations for each synset.
These terms and phrases are then compared with the keyphrase list by the similarity
function Sim(S):

$$\text{Sim}(S) = \sum_{i=1}^{M} \max_{j \in \{1,\dots,n_i\}} w(p_{ij})$$

where $w(p_{ij})$ is the frequency of phrase p_{ij} if it occurs in a synset S, and is 0, otherwise.
The synset with the highest similarity value is chosen, and synonyms from the synset
are added for query expansion.

10.3 EXPERIMENTS ON MEDLINE DATA SETS

To explore the flexibility and generality of our algorithms, we explored query expansion
for Medline articles. The task we selected was to retrieve documents containing
protein–protein interaction pairs. The data sets are composed of abstracts collected
from the Medline database. Medline contains more than 12 million documents.

```
<init_query>
    <terms protein1="MAP4" protein2="Mapmodulin"/>
    <terms protein1="WIP" protein2="NCK"/>
    <terms protein1="GHR" protein2="SHB"/>
    <terms protein1="SHIP" protein2="DOK"/>
    <terms protein1="LNK" protein2="GRB2"/>
    <terms protein1="CRP" protein2="Zyxin"/>
</init_query>
```

FIGURE 10.3   Initial query used for protein–protein interaction tasks.

To measure the accuracy rate, we count the number of documents retrieved from Medline that contain protein–protein pairs. The protein names are collected from the database of interacting proteins and protein–protein interaction databases. For protein–protein interaction tasks, we use PubMed as the underlying IR engine. Initial queries consist of three to five protein–protein interaction pairs. Figure 10.3 shows the initial query used to retrieve the documents from PubMed.

The experimental results for Medline are shown in Figure 10.4. Our three algorithms improve the retrieval performance on the tasks of retrieval documents containing protein–protein interaction pairs. The KP+C algorithm gives the best average precision. As shown in Table 10.2, our keyphrase-based technique combined with the POS phrase category produces the highest average precision. Our two algorithms (KP and KP+C) improve the retrieval performance on tasks involving retrieval documents containing protein–protein interaction pairs. As with our TREC results, the KP+C algorithm gives the best average precision. The worst performance in both average precision and precision among the top 20 was produced by a rule-based algorithm (SLP).

We also explored the effect of a sequence of query expansion iterations. Table 10.3 shows the results for five query expansion iterations. The second column is the number of documents retrieved from Medline for each iteration. The third column shows the number of documents containing protein–protein pairs. The fourth column

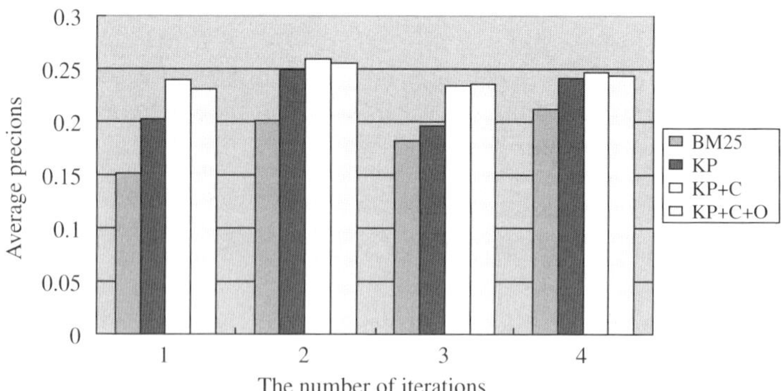

FIGURE 10.4   Experimental results for Medline with our four query expansion algorithms.

TABLE 10.2 Results for Medline with Four Query Expansion Algorithms

	Medline	
Algorithm	Average P	P@20
BM25	0.1282	0.2727
SLP	0.1051	0.2366
KP	0.1324	0.2844
KP+C	0.1522	0.2996

TABLE 10.3 Query Expansion Iterations for Medline

Iteration	No. of Retrieved Documents	No. of Documents Containing Protein–Protein Pairs	F-Measure (%)
1	30	18	47.76
2	609	289	51.65
3	832	352	51.27
4	1549	578	53.69
5	1312	545	53.21

is the F-measure [23], for which we use $b = 2$ since recall is more important than precision when dealing with the retrieval of documents containing protein–protein interaction pairs. Our results show that the F-measure generally increases as the number of iterations increases.

10.4 CONCLUSIONS

We have demonstrated an effective query expansion technique based on keyphrases and the POS phrase category. Encouraged by previous studies on pseudo-relevance feedback we apply keyphrase extraction techniques to query expansion. Along with keyphrase extraction–based expansion, we employ a new word-sense disambiguation technique in using an ontology (e.g., WordNet) to add terms to the query. We also employ a POS phrase category–based Boolean constraint technique.

We demonstrate that our techniques yield significant improvements over the well-established BM25 algorithm when dealing with Medline data on protein–protein interaction tasks. Among the four algorithms implemented (i.e., BM25, KP, KP+C, and KP+C+O), the KP+C algorithm seems to be the best. The reason that KP+C+O is not superior to KP+C, as hypothesized, might be because these ontologies are applied to already enriched keyphrases.

Applying ontologies to the original query or to both the original query and keyphrases as carried out by Liu et al. [12] may improve the performance of KP+C+O. In future work we will employ a more fine-tuned word sense disambiguation technique such as that described in ref. 12 to further improve the retrieval accuracy with WordNet. In the same vein, in ref. 10 we applied a medical ontology, the Unified Medical Language System (UMLS), to entity tagging with some promising results. As a follow-up study, we are investigating whether UMLS improves the retrieval accuracy of the documents containing protein–protein pairs.

We are also interested in whether keyphrases help users understand the content of a collection and provide sensible entry points into it. In addition, we investigate whether and how keyphrases can be used in information retrieval systems as descriptions of the documents returned by a query, as the basis for search indexes, as a way of browsing a collection, and as a document-clustering technique.

REFERENCES

1. B. Billerbeck and J. Zobel (2004). Questioning query expansion: an examination of behavior and parameters, *Proc. 15th Australasian Database Conference*, pp. 69–76.

2. K. C. Chang, H. Garcia-Molina, and A. Paepcke (1996). Boolean query mapping across heterogeneous information sources, *IEEE Trans. Knowledge Data Eng.*, 8(4): 515–521.

3. W. W. Cohen and Y. Singer (1999). Simple, fast, and effective rule learner, *Proc. 16th National Conference on Artificial Intelligence and 11th Conference on Innovative Applications of Artificial Intelligence*, July 18–22, pp. 335–342.

4. J. Dougherty, R. Kohavi, and M. Sahami (1995). Supervised and unsupervised discretization of continuous features, *Proc. 12th International Conference on Machine Learning ICML'95*, Lake Tahoe, NV, pp. 194–202.

5. E. Frank, G. W. Paynter, I. H. Witten, C. Gutwin, and C. G. Nevill-Manning (1999). Domain-specific keyphrase extraction, *Proc. 16th International Joint Conference on Artificial Intelligence*, Morgan Kaufmann, San Francisco, CA, pp. 668–673.

6. V. I. Frants and J. Shapiro (1991). Algorithm for automatic construction of query formulations in Boolean form, *J. Am. Soc. Inf. Sci.*, 42(1): 16–26.

7. J. C. French, D. E. Brown, and N. H. Kim (1997). A classification approach to Boolean query reformulation, *J. Am. Soc. Inf. Sci.*, 48(8): 694–706.

8. S. Gauch, J. Wang, and S. M. Rachakonda (1997). A corpus analysis approach for automatic query expansion and its expansion to multiple databases, *ACM Trans. Inf. Syst.*, 17: 250–269.

9. M. A. Hearst (1996). Improving full-text precision on short queries using simple constraints, *Proc. Symposium on Document Analysis and Information Retrieval*.

10. X. Hu, T. Y. Lin, I. Y. Song, X. Lin, I. Yoo, and M. Song (2004). An ontology-based scalable and portable information extraction system to extract biological knowledge from a huge collection of biomedical web documents, *Proc. 2004 IEEE/ACM Web Intelligence Conference*, pp. 77–83.

11. A. M. Lam-Adesina and G. J. F. Jones (2001). Applying summarization techniques for term selection in relevance feedback, *Proc. 24th Annual International ACM SIGIR Conference on Research and Development in Information Retrieval*, pp. 1–9.

12. S. Liu, F. Liu, C. Yu, and W. Meng (2004). An effective approach to document retrieval via utilizing WordNet and recognizing phrases, *Proc. 27th Annual International Conference on Research and Development in Information Retrieval*, pp. 266–272.

13. R. Mihalcea and D. Moldovan (2000). Semantic indexing using WordNet senses, *ACL Workshop on IR and NLP.*

14. C. U. Mitra, A. Singhal, and C. Buckely (1998). Improving automatic query expansion, *Proc. 21st Annual International ACM SIGIR Conference on Research and Development in Information Retrieval*, pp. 206–214.

15. Y. Qiu and H. Frei (1993). Concept-based query expansion, *Proc. 16th Annual International ACM SIGIR Conference on Research and Development in Information Retrieval*, pp. 160–169.

16. J. R. Quinlan (1993). *Programs for Machine Learning*, Morgan Kaufmann, San Francisco, CA.

17. J. S. Ro (1988). Evaluation of the applicability of ranking algorithms, pts. I and II, *J. Am. Soc. Inf. Sci.*, 39 (73–78): 147–160.

18. S. E. Robertson and K. Sparck Jones (1976). Relevance weighting of search terms, *J. Am. Soc. Inf. Sci.*, 27: 129–146.

19. G. Salton, C. Buckley, and E. A. Fox (1983). Automatic query formulations in information retrieval, *J. Am. Soc. Inf. Sci.*, 34(4): 262–280, July.

20. W. K. H. Sager and P. C. Lockemann (1976). Classification of ranking algorithms, *Int. Forum Inf. Doc.*, 1: 12–25.

21. M. Song, I. Y. Song, and X. Hu (2004). Designing and developing an automatic interactive keyphrase extraction system with UML, *ASIST Annual Meeting*, Providence, RI, pp. 367–372.

22. R. Van Der Pol (2003). Dipe-D: a tool for knowledge-based query formulation, *Inf. Retrieval*, 6: 21–47.

23. C. J. Van Rijsbergen (1979). *Information Retrieval*, Butterworth, London.

24. E. M. Voorhees (1998). Using WordNet for text retrieval, in *WordNet, an Electronic Lexical Database*, C. Fellbaum, Cambridge, MA, Ed., MIT Press, pp. 285–303.

25. J. Xu and W. B. Croft (2000). Improving the effectiveness of information retrieval with local context analysis, *ACM Trans. Inf. Syst.*, 18(1): 79–112.

26. Okapi, http://www.soi.city.ac.uk/~andym/OKAPI-PACK/.

11

EVOLUTIONARY DYNAMICS OF PROTEIN–PROTEIN INTERACTIONS

L. S. SWAPNA

Molecular Biophysics Unit, Indian Institute of Science, Bangalore, India

B. OFFMANN

Laboratoire de Biochimie et Génétique Moléculaire, Université de La Réunion, La Réunion, France

N. SRINIVASAN

Molecular Biophysics Unit, Indian Institute of Science, Bangalore, India

Biological processes in a cell involve formation of innumerable molecular assemblies of a wide variety of proteins and other types of molecules. For example, response of a cell to an extracellular stimulus involves a cascade of interactions between proteins inside the cell. The signal originated from outside the cell eventually reaches the nucleus, resulting in an effect on gene expression. There are a number of pathways by which a signal from outside a cell could reach the nucleus with possibilities of cross-talks between pathways. These pathways are characterized by a series of specific protein–protein interactions and use of second messengers. Protein–protein interactions are also very common in the metabolic pathways and in interactions between cells. Understanding the molecular basis of complex formation between proteins as well as between modules or domains in multimodular protein systems is central to the development of strategies for the human intervention of such biological processes. Preventing the assembly of protein complexes characteristic of a pathological condition provides a means of inhibition of a variety of undesired cellular and viral

Knowledge Discovery in Bioinformatics: Techniques, Methods, and Applications, Edited by Xiaohua Hu and Yi Pan
Copyright © 2007 John Wiley & Sons, Inc.

events. Interfering with the natural processes by peptides and small molecules that modify or disrupt the structure of a protein–protein complex or assembly of misfolded proteins or quaternary structure of an oligomeric enzyme is an attractive direction [1,2]. Peptides, peptidomimetic, and nonpeptide organic inhibitors as drug leads have been used successfully to search for antagonists of cell-surface receptors [3–6].

It is well understood that there are several groups of distantly related proteins with low sequence similarity that retain the functional residues during evolution and hence are classified under the same superfamily to indicate a possible divergent evolutionary relationship. Considering such distantly related proteins it has been shown that the similarity in the structure of protein–protein interfaces gets poorer with falling sequence identity between the proteins that are compared [7]. However, as shown for the first time by us [8], in distantly related homologous protein structures (superfamily related), the interfaces formed between protein modules (subunits or domains) are, in general, not even topologically equivalent. This unexpected observation is quite opposite to what is observed in closely related homologous proteins. The question of residue conservation at the interface does not even arise, as the protein–protein interaction regions are not topologically equivalent and are located in different parts on the surface of the highly similar tertiary structures of distantly related proteins. This nonequivalence of interfaces formed by protein modules of distantly related homologs is noted for both homo- and hetero- complexes [8].

This result is unexpected, as protein–protein complex structures or quaternary structures of oligomeric proteins are often critical for the proper functioning of protein. As protein–protein interaction sites has been compared by considering proteins within a superfamily, they are probably distantly related, and most of these proteins have highly similar gross functions. Despite similarity in gross functions, the protein–protein interfaces are not structurally equivalent. This observation has important implications. It has been observed that variation in protein–protein interactions in members within a superfamily could serve as diverging points in otherwise parallel metabolic or signaling pathways [8]. It has been found that if the homologous are highly divergent, as in superfamilies, the domains tethered to the distant homologous domains are often not homologous and completely different [9,10].

As mentioned earlier, analysis of protein–protein interacting surfaces on a large scale has revealed the interesting observation that the interacting surfaces of protein families related at superfamily level (terms as defined by SCOP [11]) are generally not conserved even at the topological level [7,8]. However, the interacting surfaces among proteins within the same SCOP family are largely conserved [12]. But in a few families in which protein members show good sequence identity, their oligomerization states vary [13].

Differences in the location of interacting surfaces on a set of related proteins can arise because of any of the following factors:

1. *Tethering to another domain partner, resulting in differences in partnerships.* The tethered domain would influence which surface of the domain is accessible for interaction.

2. *Insertions or deletions on the surface.* Insertions or deletions can abolish the formation of an interface or cause the formation of a new interface.

3. *Drastic changes in amino acid residues at the potential interface.* Even in cases where there are no insertions, deletions, or tethered domains, a part of an interface is lost or gained in certain cases. This could be due to drastic changes in amino acid residues at the interface.

Domains of different families can be tethered to a domain of a given family to give rise to proteins of different functions. Each tethered domain interacts in a specific manner with the common domain, depending on the complementarity of their surface patches. Such a feature can alter the interaction interfaces on the structure of the common domain.

The aim of this analysis is to trace the drift of protein–protein interaction surfaces between families related at the superfamily level or between subfamilies related at the family level when their interacting surfaces are not equivalent. We have investigated such evolutionary drifts in protein–protein interfaces by studying the class I glutamine amidotransferase–like superfamily of proteins and its constituent families.

11.1   CLASS I GLUTAMINE AMIDOTRANSFERASE–LIKE SUPERFAMILY

Domains from the class I glutamine amidotransferase–like superfamily from the SCOP database are characterized by a common flavodoxinlike fold with a layer of parallel β-strands flanked by α-helical regions. These structural domains are grouped into six families (Table 11.1): the large class I glutamine amidotransferases (GAT) family, the A4 β-galactosidase middle domain, the DJ-1/PfpI domains, the catalase C-terminal domain, the Aspartyl dipeptidase PepE family, and the ThuA-like domain. Although they possess a very similar fold, these families do not share similar active sites in terms of both active site residues and structure (Table 11.1). One additional important distinguishing feature of these families is that they contain different additional structures. This implies that the common structural core of these domains can exist in different contexts, in combination with different interacting partners. This differentiates them functionally in an important manner. Three of these domain families are hereafter focused, due to the fact that a simple PSI-BLAST was able to link them: the DJ-1/PfpI family, the catalase C-terminal domain, and the class I glutamine amidotransferases family.

Inspection of the families with sequences related using PSI-BLAST [14] (in bold in Table 11.1) reveals the presence of a common core in different contexts. They exist as oligomers (as in family DJ-1/PfpI) or as domains tethered to other domains in complex proteins, as in catalase and carbamoyl phosphate synthetase. Analysis of the family containing oligomers (DJ-1/PfpI) reveals that although the

TABLE 11.1 Description of the Families Present in the GAT Superfamily[a]

Family	Distinguishing Feature	Protein Domains	PDB Entries
Class I glutamine amidotransferase–like	Contains a catalytic Cys–His–Glu triad	9	32
A4 β-galactosidase middle domain	Probable noncatalytic branch of the class I GAT family; overall fold is very similar, but the active site is not conserved	1	2
DJ-1/PfpI	Contains a catalytic triad or dyad different from the class I GAT triad	6	22
Catalase, C-terminal domain		1	14
Aspartyl dipeptidase PepE	Probable circular permutation in the common core; contains a catalytic Ser–His–Glu triad	1	2
ThuA-like (Pfam 06283)	Overall fold is very similar to that of the class I GAT family, but the putative active site has a different structure	1	1

[a]The family names in bold are linked via a simple PSI-BLAST run using PDB entries in each of the families as a query.

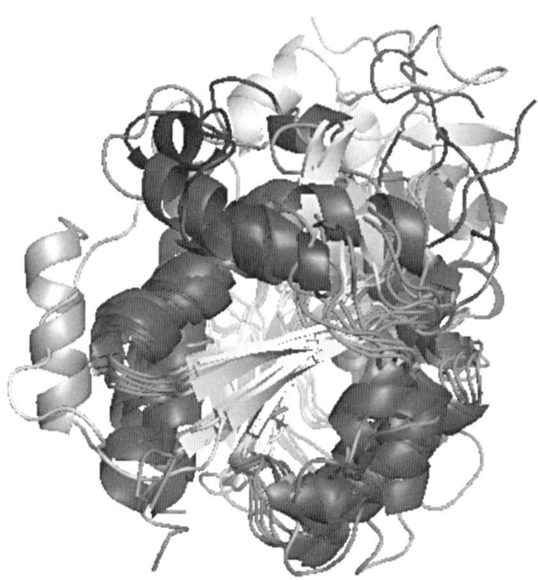

FIGURE 11.1 Superposition of tertiary structures of members of the DJ-1/PfpI family, depicting the well-conserved structural core of the members of the family as well as the variation at the surface. The conserved secondary-structural elements of all six PDB structures are colored similarly: helixes in red, sheets in blue, and loops in green. The insertions specific to each group are colored differently: 1pe0 in light teal, 1oy1 in purple, 1qvz in black, and 1izy in light blue. Generated using PyMOL [20]. (*See insert for color representation of figure.*)

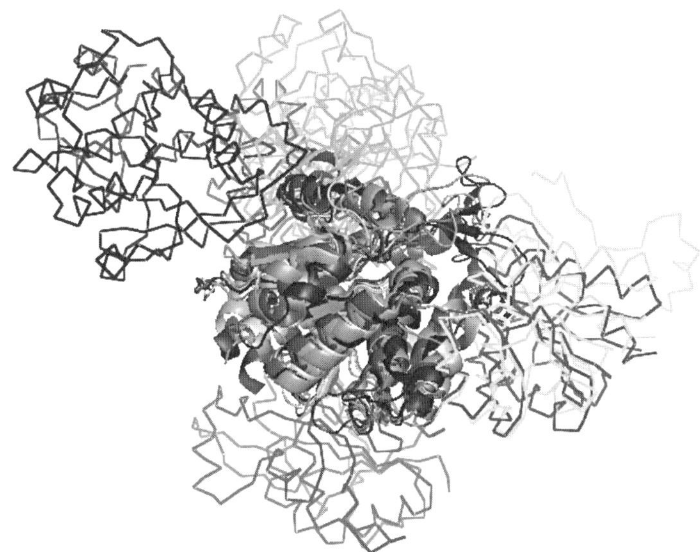

FIGURE 11.2 Superposition of quaternary structures of members of the DJ-1/PfpI family, illustrating the entire range of surfaces utilized for oligomerization by the members of the family, despite the well-conserved tertiary structure. The superimposed GAT domains are shown as cartoons, whereas the rest of the oligomeric structures are shown as ribbons. 1g2i is red, 1oi4 is light blue, 1pe0 is orange, 1oy1 is green, 1qvz is purple and 1izy is blue. Generated using PyMOL [20]. (*See insert for color representation of figure.*)

tertiary structures of the members are well conserved (Figure 11.1), the interacting surfaces of the different subfamilies within the family show variability in location and orientation (Figure 11.2), despite being related at the family level. This family is also fairly diverse, containing six different domains, and is well populated in the sequence as well as in structure space. Therefore, the DJ-1/PfpI family was chosen as the data set for in-depth study of drift in protein–protein interacting surfaces.

11.1.1 DJ-1/PfpI Family

The DJ-1/PfpI family is large and diverse, with various clades that contain examples of chaperones, intracellular proteases, RNA-binding proteins, AraC-type transcriptional regulators, and sigma cross-reacting proteins [15]. However, the functions of most of the proteins in this family are unknown. Most members exist as single domains, although a few are associated with other domains. The average sequence identity between the various members in this family is around 26%.

Crystal structures of different members of this family reveal the presence of a conserved core GAT domain fold that contains variable insertions of loops and additional secondary structures on the surface. The particular set of insertions in the core fold is specific to individual members of the family and may correlate with

TABLE 11.2   Subfamilies Within the DJ-1/PfpI SCOP Family

Protein Domain	PDB Code	Complete Gene	Oligomer	Length
Intracellular protease	1g2i	Yes	Dimer oftrimers	166
Hypothetical protein YhbO	1oi4	Yes	Dimer	172
DJ-1 (RNA binding protein regulatory subunit)	1pe0	Yes	Dimer	189
Putative sigma cross-reacting protein	1oy1	Yes	Dimer	220
HSP31	1izy/1n57	Yes	Dimer	283
Hypothetical protein ydr533Cp	1qvz	Yes	Dimer	237

functional differences among them. All characterized members of this family are oligomers, although the location and orientation of the monomers in the oligomers vary and appear to be related to the particular set of insertions in the core fold. Nearly all members of this family contain a reactive cysteine residue with an energetically unfavorable backbone conformation in a sharp turn between a β-strand and α-helix called a *nucleophilic elbow*. SCOP database classifies this family further into six protein domains, as summarized in Table 11.2. From the table it can be seen that all proteins other than intracellular proteases (1g2i) function in the dimeric state. Therefore, 1g2i, which forms a hexamer, is used as a reference oligomer in this study, as it contains two distinct and different interacting regions. The orientation of the interacting regions of all other structural members in this family is viewed and analyzed with respect to chain A of 1g2i.

11.1.2   Comparison of Quaternary Structures of DJ-1 Family Members

Superposition of the quaternary structures of the family members of DJ-1 shows that they possess varied interface surfaces, although their tertiary structures are remarkably similar. Since the members are sequentially less divergent (as they are within the same family) than in the case of families within a superfamily that differ in their oligomeric states, it might be possible to trace the drift in interfaces among these members. All the structural members of this family are oligomers; therefore, domain recruitment can be ruled out as a possible reason for variation of interacting surface in these cases. Insertions, deletions, and amino acid substitutions seem to be responsible for the different oligomerization states in the family.

11.2   DRIFTS IN INTERFACES OF CLOSE HOMOLOGS

The three PDB entries, intracellular protease (1g2i), hypothetical protein YhbO (1oi4), and DJ-1 (1pe0), are quite closely related to one another. When used as a query, each sequence picks up the other two members in the first round of a

PSI-BLAST search, within an E-value cutoff of 10^{-4}, indicating their close similarity.

The protein 1g2i is a member of a class of intracellular proteases, which have no clear sequence similarity to any other known protease family. These proteins are present in most bacteria and archaea. The active site is made up of the triad Cys-100, His-101, and Glu-74′ (residue from the other monomer). It contains two distinct interfaces on different surfaces of the protein, which repeat to form a cyclic functional ringlike hexameric structure. For this group of proteins, oligomerization is essential for function as the active site is formed by the interface of the monomers [16]. The active site resides in a cavity formed by the hexamer. The restricted access to the active site helps to regulate the entry of substrates into the chamber, thus providing a mechanism for preventing cytoplasmic proteins from undergoing unwanted proteolytic degradation. This class of intracellular proteases is speculated to specialize in the hydrolysis of small peptides that can freely permeate the small opening. This group of proteins is fairly well populated in the sequence space.

The protein 1oi4, the gene product of yhbO from *Escherichia coli*, has been annotated as a general stress protein despite the presence of the catalytic triad seen in intracellular proteases. *E. coli* YhbO is severalfold overexpressed in stationary phase, during hyperosmotic stress or acid stress. In addition, an *E. coli* strain deficient in yhbO is more sensitive than the parental strain to thermal, oxidative, hyperosmotic, pH, and UV stress-supporting the role of YhbO as a general stress-response protein. Furthermore, the authors could not detect any proteolytic or peptidolytic activity in the protein using classical biochemical substrates [17]. This protein shares very high sequence identity (47%) with 1g2i. However, it crystallizes as a dimer, whereas biochemical experiments determine the oligomeric state to be trimeric or hexameric.

The exact biological role of the DJ-1 protein 1pe0 is currently unknown. It has been implicated in various functions, such as oxidative stress response, fertilization in rat and mouse, oncogene in concert with Ras, and regulatory subunit of a 400-kDa RNA-binding protein complex, where its presence inhibits the binding of RNA by the complex. Mutations in this gene are also linked with autosomal recessive early onset familial Parkinson's disease, although the mechanism by which this is achieved is not known. Although the Cys residue in the active site of the PfpI proteases is conserved among all of the members of the DJ-1/PfpI family, the His and Glu/Asp residues are not conserved in the DJ-1 homologs. The DJ-1 protein crystallizes as a dimer. Its oligomerization state is further confirmed by gel filtration and light-scattering studies [18]. This group of proteins is also well populated in the sequence space.

The proteins 1g2i and 1oi4 are of almost equal length (170 residues); 1pe0 contains an extra C-terminal helix. The alignment of the sequences and structures shows good conservation. 1g2i and 1oi4 monomers share 47% sequence identity with each other, whereas 1g2i-1pe0 and 1oi4-1pe0 pairs share 24% and 28% sequence identity, respectively. However, despite their close similarity, their quaternary structures show differences in location and orientation (Figure 11.3).

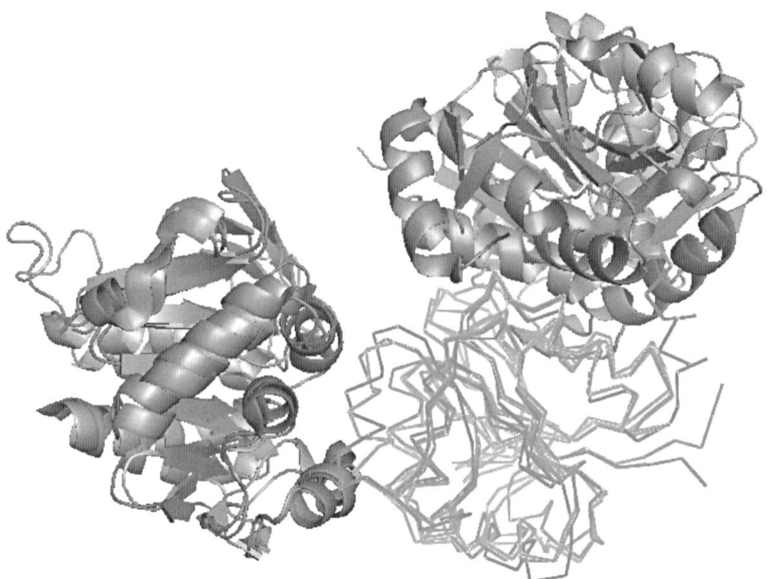

FIGURE 11.3   Superposition of the quaternary structures of 1g2i, 1oi4 and 1pe0. Although the three sequences are close homologs, they display slight variation in the orientation of their interacting surfaces. The superposed monomers are displayed as ribbons, whereas the rest of the oligomer is displayed as cartoons. 1g2i is orange, 1oi4 blue, and 1pe0 teal. Generated using PyMOL [20]. (*See insert for color representation of figure.*)

11.2.1   Comparison of Quaternary Structures of Intracellular Protease and Hypothetical Protein YhbO

The tertiary structures of the monomers of intracellular protease (1g2i) and hypothetical protein YhbO (1oi4) are almost identical, with an RMSD of 0.7 Å over 166 residues (almost the entire length of the protein). The intracellular protease 1g2i contains two distinct interacting surfaces. It exists as a ringlike structure in functional form, formed by a dimerization of trimers. Of the six intermolecular contacts, two types of interfacial contacts (formed by chains AB and chains AC, respectively) are biologically relevant, as they bury a large amount of surface area on oligomerization. In addition, many of the residues involved in interaction at these surfaces are also conserved among close homologs. The six monomers in the crystal are connected through the two major contacts to form a closed ring.

However, 1oi4, which shares 47% sequence identity with 1g2i, has been crystallized as a dimer. One of the interfaces (AC) is conserved completely, whereas the other interacting surface (AB) is absent (Figure 11.3). However, biochemical analysis (SDS-PAGE) of 1oi4 suggests that it exists as a trimer or a hexamer [17]. The reason that 1oi4 did not crystallize as a trimer or hexamer might have been because of an N-terminal histidine tag used with the protein, which could have sterically hindered the formation of the AB-like interface.

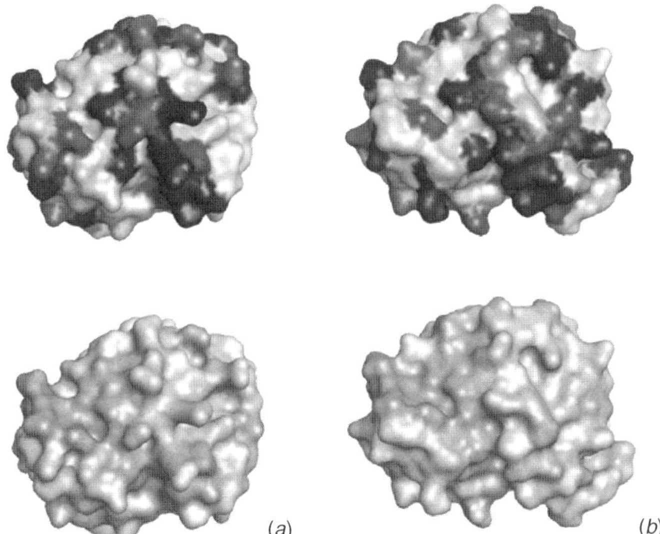

(a)   (b)

FIGURE 11.4 Conserved interface between (*a*) intracellular protease, 1g2i and (*b*) hypothetical protein YhbC, 1oi4. In the representations at the top of the figure, aliphatic and aromatic residues are represented in light gray, polar residues in light blue, positively charged residues in blue, and negatively charged residues in red. In the representations at the bottom, residues conserved between 1g2i and 1oi4 are displayed in orange, whereas variable residues are displayed in light gray. Generated using PyMOL [20]. (*See insert for color representation of figure.*)

To determine whether the nonconserved interface actually exists, the interacting residues were analyzed in the two monomers. An inspection of the alignment of 1g2i and 1oi4 monomers provides the following information. In the AC interface (Figure 11.4), the nature of 12 of 18 residues at this interface is conserved, whereas 3 of 18 residues show drastic substitutions. With respect to the AB interface, the nature of 7 of 16 residues at the interface is conserved, whereas another 7 residues at the interface show drastic substitutions. To further explore the probability of formation of the interface, a cross-model of the 1oi4 oligomeric complex was built. The 1oi4 dimer was placed in the coordinate space of the 1g2i A and B chains and the interface was analyzed for possible steric clashes or unfavorable interactions. From the analysis it was seen that there were some unfavorable interactions (data not shown), such as acidic residues in close proximity with each other. However, the data are insufficient to rule out the formation of an AB-like interface. Even if such an interface is formed, the nature of the interface could differ, due to the substitutions at the key residues involved in the interface.

The lesser representation of members belonging to the hypothetical protein YhbO subfamily and unclear annotation prevent clear-cut analysis of the available sequences to decide whether this protein contains both interfaces or only one.

11.2.2 Comparison of Quaternary Structures of Intracellular Protease and DJ-1

The tertiary structures of the monomers of intracellular protease (1g2i) and DJ-1(1pe0) are quite similar, with an RMSD of 1.7 Å over 166 residues (almost the entire length of the protein). Superposition of the quaternary structures of 1g2i and 1pe0 reveals that one of the interfaces of 1g2i (AC) is totally absent in 1pe0. The interacting surface of the other interface of 1g2i (AB) overlaps with the dimeric interface of 1pe0; however, the interacting region is slightly shifted and also oriented differently to accommodate the extra C-terminal hydrophobic helix in case of 1pe0 (Figures 11.3 and 11.5).

Analysis of the structural superposition of the 1g2i and 1pe0 monomers with respect to the nonconserved AC interface reveals that the C-terminal extra helix is too far away from the interacting surface to hinder its formation. However, there are drastic substitutions in 10 of 14 interacting positions, as compared to conservation of the nature of only 3 of 14 interacting positions. To determine whether the nonconserved AC-like interface can exist in 1pe0, a cross-model of 1pe0 in the structure space of the nonconserved interface of 1g2i was generated. Analysis of the cross-modeled interface revealed the presence of many unfavorable interactions (data not

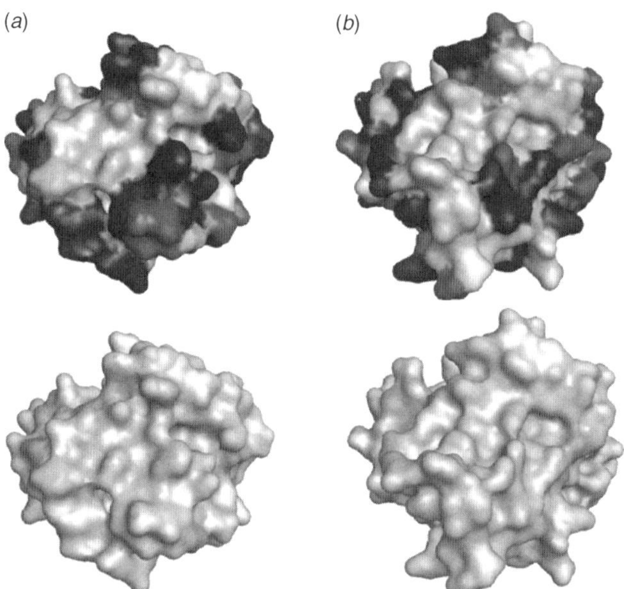

FIGURE 11.5 Topologically equivalent, orientationally variant interface between (*a*) intracellular protease, 1g2i and (*b*) DJ-1 protein, 1pe0. In the representations at the top of the figure, aliphatic and aromatic residues are represented in light gray, polar residues in light blue, positively charged residues in blue, and negatively charged residues in red. In the representations at the bottom, residues conserved between 1g2i and 1pe0 are displayed in orange, whereas variable residues are displayed in light gray. Generated using PyMOL [20]. (*See insert for color representation of figure.*)

shown), such as two lysines in close proximity to each other, two acidic residues in close proximity to each other, and some steric clashes, implying that the unfavorable amino acid substitutions are responsible for the loss of this interface.

The presence of the extra C-terminal helix in 1pe0 prevents it sterically from adopting the same oligomeric conformation as formed between the A and B chains of 1g2i. The C-terminal helix contains many hydrophobic residues, which will result in a destabilized structure if left exposed. Therefore, the 1pe0 dimer accommodates interactions between the C-terminal helices to bury the hydrophobic stretch. This interaction, along with other amino acid substitutions in 1pe0 with respect to 1g2i, leads to differences in the orientation and interacting surfaces of their corresponding interfaces.

The pair of 1g2i-1pe0 provides an excellent example to trace two types of variations (change in position and orientation of an interface and loss of an interface) at the interfacial level.

11.2.3 Tracing the Shift Between the Topologically Conserved Intracellular Protease and DJ-1 Interfaces

Intracellular proteases cluster together as a group in the multiple-sequence alignment and in the corresponding phylogenetic tree (not shown). They contain a characteristic conserved set of residues, cysteine and histidine, one after the other. The nonproteolytic group of this family, populated mostly by homologs of DJ-1, where the histidine of the conserved CH residues is mutated to a hydrophobic residue (alanine in RNA binding proteins), cluster into a separate group. The correctness of the alignment can be further ascertained by the clear demarcation of sequences of the two groups by the presence of the extra C-terminal region in sequences of the nonproteolytic group. As we parse the alignment, we notice that there is no sharp demarcation between the two groups; this provides a continuum of sequences to analyze the drift in interfaces of 1g2i and 1pe0. Tables 11.3 and 11.4 depict the drift in interfacial residues for 1g2i and 1pe0, respectively, from one cluster to another.

The most populated clusters are cluster 2 of the intracellular proteases subfamily (which contains the 1g2i protein) and cluster 6 of the DJ-1 subfamily (which contains the 1pe0 protein). The average sequence identity between sequences within cluster 2 is 49%, with the lowest value being 33%. The average sequence identity for cluster 6 is 38%, with the lowest being 13%. This reflects the larger sequence data set available in this subfamily and the extensive divergence. Clusters 1 and 3 form the extremes of the intracellular proteases subfamily; clusters 4 and 7 form the extreme ends of the DJ-1 subfamily. Other clusters in the alignment are not shown—for clarity in Tables 11.3, 11.4, and 11.5. Clusters 3 and 4 contain the closest sequences from the intracellular protease and DJ-1 subfamilies, respectively. The average sequence identity between sequences in these two clusters is 36%, with the lowest being 16%. In comparison, the average sequence identity between sequences of clusters 4 and 6, both of which belong to the DJ-1 subfamily, is 36%, and the lowest sequence identity between any proteins is 13%. These values indicate that a continuum exists which would enable us to trace the drift in protein–protein surfaces in this case.

TABLE 11.3   Drift of Interacting Residues Involved in the Topologically Equivalent Interface of 1g2i[a]

Annotation	Sequences	Important Interacting Pairs of Residues in 1g2i							
		R22-E25'	V14-I17'	H44-Y46'	E12-H44'	D13-H44'	Y18-E25'	L154-V14'	L154-I17'
Intracellular protease (Conserved "CH") Cluster 1	Q8CRV8_2-171	A–N	I–T	H–E	*E–H*	*D–H*	S–N	*L–I*	*L–T*
	1oi4a_	E–K	S–T	K–A	*E–K*	*D–K*	S–K	*L–S*	*L–T*
	Q5WER6_2-172	S–E	V–T	Q–A	E–Q	D–Q	D–E	*L–V*	*L–T*
	P80876_1-170	**A–E**	S–T	Q–A	E–Q	D–Q	E–E	I–S	I–T
	UPI0000390255_3-167	**A–E**	I–T	H–T	*E–H*	*D–H*	D–E	I–I	I–T
	Q7MQ54_70-248	Y–E	*L–L*	E–G	E–E	D–E	V–E	L–L	L–L
Intracellular protease (Conserved "CH") Cluster 2 *HEXAMER*	Q97KC8_6-150	*R–E*	*I–L*	*H–Y*	*E–H*	*D–H*	*Y–E*	#–I	#–L
	Q8ESW0_3-161	*R–E*	*L–W*	*Y–V*	E–Y	D–Y	*Y–E*	L–L	L–W
	Q72AQ9_1-171	*R–E*	*L–W*	*H–Y*	*E–H*	*D–H*	*Y–E*	L–L	L–W
	Q8YYF7_10-176	*R–E*	*A–I*	R–L	E–R	D–R	*Y–E*	L–A	L–I
	O66869_2-153	*R–E*	*V–I*	*K–M*	E–K	D–K	*Y–E*	#–V	#–I
	O28987_2-166	*R–E*	*L–F*	*K–Y*	E–K	D–K	*Y–E*	L–L	L–F
	1g2ia_	*R–E*	*V–I*	*H–Y*	*E–H*	*D–H*	*Y–E*	*L–V*	*L–I*
	Q9V1F8_2-166	*R–E*	*V–I*	*H–Y*	*E–H*	*D–H*	*Y–E*	*L–V*	*L–I*
	Q973K8_2-168	*R–E*	*I–L*	*H–Y*	*E–H*	*D–H*	*Y–E*	*L–I*	L–L
Intracellular protease (Conserved "CH") Cluster 3	Q5FQH7_6-198	M–L	Y–M	K–H	*E–K*	*D–K*	V–L	H–Y	H–M
	Q9M8R4_196-382	S–A	Y–K	K–H	*E–K*	*D–K*	V–A	H–Y	H–K
	Q869N4_1-188	M–L	Y–Y	L–H	E–L	D–L	A–L	H–Y	H–Y
DJ-1 group (Conserved "CA") Cluster 4	Q6F1K3_2-180	I–R	S–I	S–I	**E**–S	**E**–S	V–R	V–S	V–I
	Q6MTF0_2-181	V–R	I–V	A–I	**E**–A	**E**–A	T–R	S–I	S–V
	Q10356_3-188	I–R	I–S	S–V	**D**–S	**E**–S	A–R	A–I	A–S
	Q7P6I3_10-149	V–R	L–F	S–Q	**E**–S	T–S	A–R	#–L	#–F
	Q7WK03_63-243	I–R	T–L	F–L	**E**–F	T–F	V–R	S–T	S–L
	Q98CD4_38-218	I–R	T–L	H–L	*E–H*	T–H	M–R	S–T	S–L
DJ-1 group (Conserved "CA") Cluster 5	Q735E9_2-189	V–R	L–F	S–V	G–S	E–S	A–R	G–L	G–F
	Q81NK0_2-189	V–R	L–F	S–V	G–S	E–S	A–R	G–L	G–F
	Q9HMG4_3-187	V–T	L–I	R–L	D–R	E–R	G–T	G–L	G–I
DJ-1 group (Conserved "CA") Cluster 6 *DIMER*	Q8TY96_2-179	V–R	L–G	M–M	E–D	E–M	V–R	A–L	A–G
	Q8Z5U2_2-184	I–R	A–I	Y–I	E–Y	E–Y	V–R	V–A	V–I
	Q97IL1_1-168	I–R	V–I	A–I	E–A	E–A	T–R	A–V	A–I
	Q9PP32_1-174	V–R	A–I	A–I	E–A	E–A	G–R	A–A	A–I
	1pe0a_	V–R	M–V	R–V	E–R	E–R	I–R	S–M	S–V
	Q5W7N7_2-172	V–R	M–V	R–V	E–R	E–R	I–R	S–M	S–V
	UPI00004D6BE4_1-179	V–R	M–V	G–I	E–G	E–G	I–R	S–M	S–V
	Q61QH5_2-171	V–R	M–I	K–A	E–K	E–K	I–R	A–M	A–I
	Q7QSY3_4-181	V–R	I–V	S–V	E–S	E–S	T–R	A–I	A–V
	Q87M94_1-176	L–R	M–V	S–V	E–S	E–S	T–R	A–M	A–V
	Q7RFR5_3-172	V–R	V–I	S–N	E–S	E–S	T–R	A–V	A–I
	UPI00000A797B_189-345	V–R	M–I	#–#	E–#	E–#	T–R	A–M	A–I
	Q72RQ9_3-164	L–R	M–V	R–I	E–R	E–R	V–R	A–M	A–V
DJ-1 group (Conserved "CA") Cluster 7	Q738U4_2-180	V–W	V–S	T–F	E–T	A–T	V–W	A–V	A–S
	Q8RHW4_2-200	V–W	F–I	S–I	E–I	L–S	S–W	A–F	A–I

[a]Residues in bold denote the interfacial residue under consideration. Residue pairs in bold italic type indicate favorable interactions. Underlined residue pairs indicate unfavorable interactions. A prime indicates a residue from the other chain.

Analysis of the conservation of important interfacial residues for 1g2i and 1pe0 with respect to all homologs indicates that the drift in interfaces would have been a gradual process (Tables 11.3 and 11.4). Consider the following example: Arg-22 of 1g2i interacts with Glu-25′ of another monomer. This interaction is conserved

TABLE 11.4   Drift of Interacting Residues Involved in the Topologically Equivalent Interface of 1pe0[a]

Annotation	Sequences	Important Interacting Pairs of Residues in 1pe0						
		E15-R28'	V20-V50'	E16-R27'	E16-R28'	M17-F162'	V50-I52'	F162-F162'
Intracellular protease (Conserved "CH") Cluster 1	Q8CRV8_2-171	E-N	T-#	D-E	**D**-N	I-D	#-V	D-D
	1oi4a_	*E-K*	T-#	D-R	*D-K*	S-P	#-V	P-P
	Q5WER6_2-172	E-E	T-#	D-T	**D**-E	V-P	#-V	P-P
	P80876_1-170	E-E	T-#	D-K	**D**-E	S-P	#-V	P-P
	UPI0000390255_3-167	E-E	T-#	D-K	**D**-E	I-P	#-V	P-P
	Q7MQ54_70-248	E-E	L-#	D-R	**D**-E	L-P	#-L	P-P
Intracellular protease (Conserved "CH") Cluster 2	Q97KC8_6-150	E-E	L-#	D-R	**D**-E	I-#	#-V	#-#
	Q8ESW0_3-161	E-E	W-#	D-R	**D**-E	L-P	#-V	P-P
	Q72AQ9_1-171	E-E	W-#	D-E	**D**-E	L-P	#-C	P-P
	Q8YYF7_10-176	E-E	I-#	D-K	**D**-E	A-A	#-T	A-A
	O66869_2-153	E-E	I-#	D-K	**D**-E	V-#	#-**F**	#-#
	O28987_2-166	E-E	I-#	D-R	**D**-E	L-P	#-**V**	P-P
HEXAMER	1g2ia_	E-E	I-#	D-K	**D**-E	V-Y	#-**V**	Y-Y
	Q9V1F8_2-166	E-E	I-#	D-K	**D**-E	V-Y	#-**V**	Y-Y
	Q973K8_2-168	E-E	V-#	D-I	**D**-E	I-P	#-**V**	P-P
Intracellular protease (Conserved "CH") Cluster 3	Q5FQH7_6-198	E-L	M-H	D-L	**D**-L	Y-P	H-**F**	P-P
	Q9M8R4_196-382	E-A	K-H	D-L	**D**-A	Y-P	H-**F**	P-P
	Q869N4_1-188	E-L	Y-H	D-Q	**D**-L	Y-P	H-**I**	P-P
DJ-1 group (Conserved "CA") Cluster 4	Q6F1K3_2-180	*E-R*	*I-I*	E-R	*E-R*	S-**V**	I-N	*V-V*
	Q6MTF0_2-181	*E-R*	*V-V*	E-K	*E-R*	*I-M*	V-**I**	*M-M*
	Q10356_3-188	*E-R*	S-D	E-K	*E-R*	*I-M*	D-**M**	*M-M*
	Q7P6I3_10-149	*E-R*	**F**-N	T-K	T-R	**L**-#	N-**V**	#-#
	Q7WK03_63-243	*E-R*	*L-L*	T-R	*T-R*	*T-M*	*L-V*	*M-M*
	Q98CD4_38-218	*E-R*	*L-L*	T-R	*T-R*	*T-M*	*L-I*	*M-M*
DJ-1 group (Conserved "CA") Cluster 5	Q735E9_2-189	*G-R*	*F-V*	*E-K*	*E-R*	*L-L*	V-V	*L-L*
	Q81NK0_2-189	*G-R*	*F-V*	E-N	*E-R*	*L-V*	*V-V*	*V-V*
	Q9HMG4_3-187	D-T	*I-L*	*E-R*	*E-R*	*L-I*	*L-V*	*I-I*
DJ-1 group (Conserved "CA") Cluster 6	Q8TY96_2-179	*E-R*	*V-M*	*E-R*	*E-R*	*L-I*	*M-V*	*I-I*
	Q8Z5U2_2-184	*E-R*	*I-I*	*E-R*	*E-R*	*A-F*	*I-M*	*F-F*
	Q97IL1_1-168	*E-R*	*L-I*	*E-R*	*E-R*	*V-I*	*I-V*	*I-I*
	Q9PP32_1-174	*E-R*	*I-I*	*E-K*	*E-R*	*A-I*	*I-I*	*I-I*
DIMER	1pe0a_	*E-R*	*V-V*	*E-R*	*E-R*	*M-F*	*V-I*	*F-F*
	Q5W7N7_2-172	*E-R*	*V-V*	*E-R*	*E-R*	*M-F*	*V-I*	*F-F*
	UPI00004D6BE4_1-179	*E-R*	*V-I*	*E-R*	*E-R*	*M-F*	I-N	*F-F*
	Q61QH5_2-171	*E-R*	*I-A*	E-E	*E-R*	*M-F*	*A-I*	*F-F*
	Q7QSY3_4-181	*E-R*	*V-V*	*E-R*	*E-R*	*I-L*	*V-V*	*L-L*
	Q87M94_1-176	*E-R*	*V-V*	E-V	*E-R*	*M-L*	*V-L*	*L-L*
	Q7RFR5_3-172	*E-R*	*I-N*	*E-R*	*E-R*	*V-V*	N-V	*V-V*
	UPI00000A797B_189-345	*E-R*	*I-#*	*E-R*	*E-R*	*M-M*	#-#	*M-M*
	Q72RQ9_3-164	*E-R*	*L-I*	E-N	*E-R*	*M-F*	*I-I*	*F-F*
DJ-1 group (Conserved "CA") Cluster 7	Q738U4_2-180	E-W	S-**F**	A-G	A-W	*V-F*	*F-I*	*F-F*
	Q8RHW4_2-200	E-W	A-G	L-G	L-W	*F-L*	G-**L**	*L-L*

[a]Residues in bold denote the interfacial residue under consideration. Residue pairs in bold italic types indicate favorable interactions. Underlined residue pairs indicate unfavorable interactions. A gap in alignment is indicated by #. A prime indicates a residue from the other chain.

in all the cluster 2 proteins, which are annotated as intracellular proteases. However, in the case of 1pe0, the residue topologically equivalent to Glu-25 of 1g2i is Arg-27. This positively charged residue interacts with Glu-15' in the other chain. If the dimeric interface of 1g2i is maintained in case of 1pe0, Val-Arg would be present in place of Arg-Glu, which is clearly unfavorable. Similarly, if

TABLE 11.5 Drift of Interacting Residues Involved in the Nonconserved Interface of 1g2i[a]

Annotation	Sequences	Important Interacting Pairs of Residues in 1g2i		
		(E74–H101')	(R77–D126')	(E10–K43')
Intracellular protease (Conserved "CH") Cluster 1	Q8CRV8_2-171	*D–H*	*R–D*	*E–K*
	*1oi4a_*	*D–H*	*R–D*	*E–K*
	Q5WER6_2-172	*D–H*	*R–D*	L–K
	P80876_1-170	*D–H*	*R–D*	Y–K
	UPI0000390255_3-167	*D–H*	*R–D*	L–K
	Q7MQ54_70-248	*D–H*	*R–D*	G–M
Intracellular protease (Conserved "CH") Cluster 2	Q97KC8_6-150	Y–D	*R–D*	G–T
	Q8ESW0_3-161	*D–H*	*R–D*	*D–K*
	Q72AQ9_1-171	*D–H*	*R–D*	I–K
	Q8YYF7_10-176	*D–H*	*R–D*	A–K
	O66869_2-153	*D–H*	*R–D*	L–K
	O28987_2-166	*E–H*	*R–D*	*E–K*
HEXAMER	*1g2ia_*	*E–H*	*R–D*	*E–K*
	Q9V1F8_2-166	*E–H*	*R–D*	Q–K
	Q973K8_2-168	*E–H*	*R–D*	*E–K*
Intracellular protease (Conserved "CH") Cluster 3	Q5FQH7_6-198	*E–H*	S–**E**	F–E
	Q9M8R4_196-382	*E–H*	A–N	Y–E
	Q869N4_1-188	*E–H*	*R–E*	Y–E
DJ-1 group (Conserved "CA") Cluster 4	Q6F1K3_2-180	V–A	#–Y	N–G
	Q6MTF0_2-181	V–A	L–Y	G–G
	Q10356_3-188	A–A	L–Q	G–M
	Q7P6I3_10-149	Y–G	L–E	G–S
	Q7WK03_63-243	#–A	R–E	A–L
	Q98CD4_38-218	#–A	R–E	G–L
DJ-1 group (Conserved "CA")	Q735E9_2-189	V–T	Q–E	G–T
	Q81NK0_2-189	A–T	Q–K	G–T
	Q9HMG4_3-187	V–T	E–D	G–A
DJ-1 group (Conserved "CA") Cluster 6	Q8TY96_2-179	P–S	L–E	G–G
	Q8Z5U2_2-184	S–S	L–A	G–S
	Q97IL1_1-168	A–A	L–E	G–G
	Q9PP32_1-174	M–A	L–G	G–G
	*1pe0a_*	A–A	L–K	G–S
DIMER	Q5W7N7_2-172	A–A	L–K	G–S
	UPI00004D6BE4_1-179	A–A	L–L	G–K
	Q61QH5_2-171	S–A	L–Q	G–A
	Q7QSY3_4-181	W–A	M–G	G–G
	Q87M94_1-176	A–A	F–H	G–A
	Q7RFR5_3-172	S–A	I–K	G–Q
	UPI00000A797B_189-345	A–A	M–D	G–R
	Q72RQ9_3-164	T–A	L–S	G–S
DJ-1 group (Conserved	Q738U4_2-180	F–V	D–R	G–C
	Q8RHW4_2-200	F–S	D–Y	G–C

[a]Residues in bold denote the interfacial residue under consideration. Residue pairs in bold italic type indicate favorable interactions. Underlined residue pairs indicate unfavorable interactions. A gap in alignment is indicated by #. A prime indicates a residue from the other chain.

the dimeric interface of 1pe0 is maintained in 1g2i, Glu–Glu would be interacting in place of Glu–Arg, which is repulsive. Thus, we see that the interacting partners have shifted favorably from R22-E25' in 1g2i to E15-R27' in 1pe0. When we monitor the conservation of the nature of residues at these positions, we see the following trend. Glu-15 is conserved in almost all the clusters, indicating that it might have some other important role, perhaps structural. Arg-22 is specific only to cluster 2 of intracellular proteases. However, its interacting partner Glu-25′ is present in many proteins of cluster 1, thus indicating that acquiring a favorable interacting pair was not an abrupt event. The nature of both of these residues is not conserved in the other cluster (cluster 3) belonging to intracellular proteases. The ionic residues are substituted by hydrophobic residues in this cluster, thus probably maintaining the interaction, although it is now much more weaker than the electrostatic interaction seen in cluster 2. The clusters in the DJ-1 subfamily (characterized by the conserved *CA* in place of the conserved *CH* in intracellular proteases) contain a hydrophobic residue in the position occupied by R22 in 1g2i, a feature also seen in cluster 3 of the intracellular proteases subfamily, thus depicting a continuation between members of the DJ-1 subfamily and the intracellular proteases subfamily. Arg-27′ is present only in the first three clusters of the DJ-1 subfamily, indicating that its role is also changing (substituted to Trp in cluster 7 of the DJ-1 subfamily).

Such gradual changes are also seen in other important interacting pairs: Val14-Ile17′ and Glu12-His44′ in 1g2i (Table 11.3) and Val20-Val50′ in 1pe0 (Table 11.4). In each of these cases, it is seen that a topologically equivalent interacting pair in the other family would be unfavorable, due to substitutions at that location. This substitution is not abrupt and not concerted between the interacting pair; the residues of the pair mutate independently and randomly. When the mutation offers a functional advantage, it is retained.

11.2.4 Tracing the Shift Between the Nonconserved Intracellular Protease and DJ-1 Interfaces

The interface formed by chains A and C of 1g2i, which is also exactly retained in case of 1oi4, is completely absent in 1pe0. The loss of this interface is not due to any insertions or deletions but seems to be due to unfavorable amino acid substitutions. To trace the loss of this interface in 1pe0, the important interacting residues of the AC interface of 1g2i were mapped onto the existing alignment and analyzed for drift in their conservation (Table 11.5).

In this case, the drift in substitutions of interacting pairs of residues is more abrupt and concerted than in the preceding case. None of the clusters in the DJ-1 subfamily have any of the interacting residues of the intracellular proteases. Probably, loss of the interface or gain of the interface was mediated by abrupt substitutions on the interacting surface. When we analyze all the interacting residues that qualify based on the 4-Å distance cutoff (data not shown), we find that some residues show a gradual drift, whereas others show an abrupt change.

11.3 DRIFTS IN INTERFACES OF DIVERGENT MEMBERS

The putative sigma cross-reacting protein (1oy1) also belongs to the DJ-1/PfpI family of proteins. It is functional as a homodimer. All the homologs of this protein contain the conserved cysteine residue in the nucleophilic elbow. However, the adjacent histidine residue, conserved in all intracellular proteases, is substituted for by an isoleucine residue, which is conserved in all homologs of 1oy1. In this respect, sigma cross-reacting proteins are closer to DJ-1 homologs, which contain a conserved alanine at that position. The length of 1oy1 homologs is around 220 residues; the increase in length is due to an insertion in the middle region of the protein. The portion inserted is involved in formation of the dimeric interface and is instrumental in the variation of the orientation of the homodimeric interface compared to that of 1pe0 and 1oi4, its closest structural homologs. This protein also lacks the extra C-terminal helix seen in 1pe0, which also precludes the need for the protein to adopt the dimeric interface present in 1pe0.

A PSI-BLAST run using 1oy1 as the query sequence picks up three closely related structural members (i.e., 1g2i, 1oi4, and 1pe0) only in the fourth round, at E-values within 10^{-8}, indicating that sigma cross-reacting proteins have diverged much more than the other three subfamilies. The sequence identity between 1oy1 and 1oi4 is 19% and the sequence identity between 1oy1 and 1pe0 is 18%. The average sequence identity within the sigma cross-reacting proteins group is 53%, with the lowest pairwise sequence identity being 35%. In comparison, the average sequence identity is 42 and 30% within the intracellular proteases and DJ-1 groups, respectively. Sigma cross-reacting proteins seem to be a well-conserved group which is not as well populated as the divergent DJ-1 group in the sequence space.

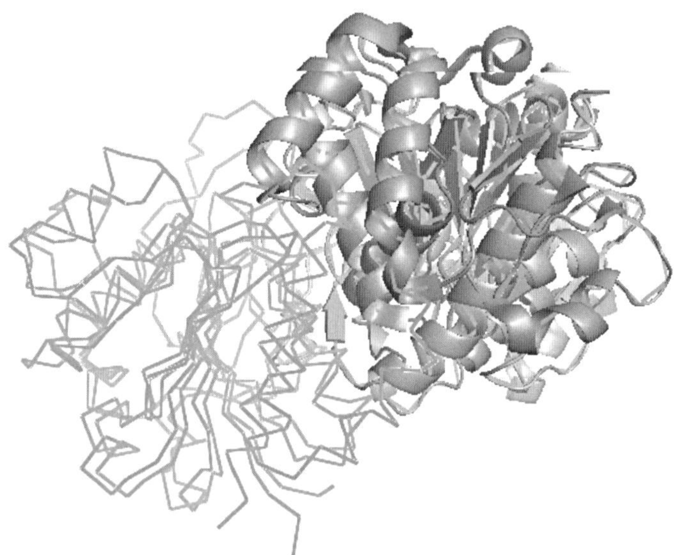

FIGURE 11.6 Superposition of the quaternary structures of divergent members of the DJ-1 family, putative sigma cross-reacting protein and DJ-1. The superposed monomers are displayed as ribbons, whereas the rest of the oligomer is displayed as cartoons. 1pe0 is orange and 1oy1 is green. Generated using PyMOL [20]. (*See insert for color representation of figure.*)

11.3.1 Comparison of the Quaternary Structures of Putative Sigma Cross-Reacting Protein and DJ-1

Superposition of the tertiary structures of 1oy1 and 1pe0 (Figure 11.6) gives an RMSD of 2.3 Å when 156 residues are aligned (about 70% of 1oy1), indicating that 1oy1 has diverged more than the other counterparts seen until now. Although 1oy1 and 1pe0 are functional homodimers and contain overlapping interacting surfaces, the orientation and interacting elements within the two structures are very different (Figure 11.6). This difference is due primarily to the insertion in the middle of the sequence characteristic of all sigma cross-reacting proteins and the C-terminal extra

TABLE 11.6   Drift of Interacting Residues Involved in Formation of the Topologically Equivalent Interface of 1oy1[a]

Annotation	Sequences	Important Interacting Pairs of Residues in 1oy1						
		D18-R72'	E21-R75'	I22-V26'	I22-I73'	L29-I73'	L29-I201'	L29-A205'
Intracellular protease (**Conserved "CH"**) Cluster 1	Q7NGS5_8-181	#-H	**E**-S	E-T	E-D	M-D	M-R	M-Q
	Q8PIG6_5-168	#-T	**E**-G	Q-T	Q-D	K-D	K-#	K-#
	Q8U8F0_4-167	#-K	**E**-G	R-R	R-N	Y-N	Y-#	Y-#
	Q826G2_36-200	#-L	**E**-A	R-D	R-E	R-E	R-#	R-#
	Q8FSF5_8-170	#-K	**E**-D	K-V	K-D	V-D	V-#	V-#
	Q5L013_3-183	#-#	**E**-I	A-Y	A-E	Y-E	Y-R	Y-L
HEXAMER	Q97BN0_1-184	#-#	**E**-V	S-M	S-Y	Y-Y	Y-R	Y-I
	Q7WAZ6_4-207	#-#	**E**-F	D-M	D-H	F-H	F-A	F-V
	Q6FDZ8_1-190	#-#	**E**-F	D-M	D-H	F-H	F-S	F-V
	Q7UM25_3-187	#-#	**E**-F	D-M	D-H	Y-H	Y-R	Y-L
	Q9M8R4_196-382	#-#	**E**-F	D-K	D-H	F-H	F-S	F-L
	Q5ZU31_7-196	#-#	**E**-F	D-Y	D-H	Y-H	Y-R	Y-I
	1g2ia_	#-#	**E**-V	D-I	D-Y	Y-Y	Y-R	Y-L
	Q8PSJ5_26-144	#-#	**E**-V	#-#	#-#	#-#	#-R	#-L
	Q65FN6_3-161	#-#	**E**-A	D-W	D-V	**V**-V	V-#	V-#
	Q8RB38_2-168	#-#	**E**-M	E-L	E-Y	Y-Y	Y-K	Y-K
	Q97KC8_6-150	#-#	**E**-V	D-L	D-Y	**L**-Y	L-#	L-#
	1oi4a_	#-A	**E**-T	D-T	D-S	**A**-S	A-R	**A-L**
	Q6GFI2_3-171	#-#	**E**-T	D-S	D-K	K-K	K-R	K-Q
	Q6FA92_1-181	#-T	S-D	D-N	D-S	**A**-S	A-N	**A-V**
Sigma cross-reacting protein (**Conserved "CI"**) Cluster 2	Q88CG1_5-220	**D-R**	**E-R**	**I-V**	**I-I**	**L-I**	**L-I**	**L-A**
	Q87IN9_2-216	**D-R**	**E-R**	**I-V**	**I-I**	**L-I**	**L-I**	**L-A**
	1oy1a_	**D-R**	**E-R**	**I-V**	**I-I**	**L-I**	**L-I**	**L-A**
	UPI0000384817_6-220	**D-R**	**E-R**	**I-V**	**I-I**	**L-I**	**L-I**	**L-I**
	Q6APY5_5-218	**D-R**	**E-R**	**I-T**	**I-I**	**L-I**	L-R	**L-I**
	P30042_20-238	**D-R**	**E-R**	**I-V**	**I-I**	**L-I**	L-#	L-#
DIMERIC	Q5ZM94_16-229	**D-R**	**E-R**	**I-V**	**I-I**	**L-I**	L-#	L-#
	UPI00004D3100_4-199	**D-R**	**E-R**	**I-V**	**I-I**	**L-I**	L-#	L-#
	Q90257_38-268	**D-R**	**E-R**	**I-V**	**I-I**	**M-I**	**M-Y**	**M-I**
DJ-1 group (**Conserved "CA"**) Cluster 3 *DIMERIC*	O51566_1-169	#-D	**E**-I	D-**I**	D-D	**I**-D	**I-F**	**I-M**
	Q7MWH8_3-152	#-D	**E**-M	E-**V**	E-R	**L**-R	L-#	L-#
	Q8VY09_62-252	#-D	**E**-I	E-**V**	E-V	**I**-V	I-L	**I-Q**
	Q8D159_11-178	#-D	**E**-L	E-**V**	E-T	**I**-T	I-L	**I-L**
	1pe0a_	#-D	**E**-L	E-**V**	E-T	**V**-T	V-L	**V-A**
	Q5W7N7_2-172	#-D	**E**-L	E-**V**	E-S	**V**-S	V-L	V-E
	Q61QH5_4-156	#-D	**E**-F	E-**I**	E-A	G-A	G-#	G-#
	Q7MBB7_3-149	#-D	**E**-L	E-**V**	E-T	**I**-T	I-#	**I-#**
	Q97IL1_1-168	#-D	**E**-I	E-**L**	E-K	**V**-K	V-L	**V-I**
	Q5FM46_3-182	#-D	**E**-V	E-**L**	E-K	**V**-K	V-Y	**V-A**
	Q6YR79_2-178	#-D	**E**-L	D-**L**	D-A	R-A	R-F	R-**V**

[a]Residue pairs in bold italic type indicate favorable interactions. Underlined residue pairs indicate unfavorable interactions. A gap in alignment is indicated by #. A prime indicates a residue from the other chain.

helix characteristic of DJ-1 group members. Both insertions are specific to their respective groups and are involved in their respective dimer interface.

11.3.2 Tracing the Shift Between the Interfaces of Putative Sigma Cross-Reacting Protein and DJ-1

The drastic substitutions of amino acids at the interface indicates why the interface of 1oy1 cannot be adopted by intracellular proteases and DJ-1 members, as the topologically equivalent interacting pairs would render the interaction unfavorable. From Table 11.6 we see that the transition of interacting pairs from one cluster to the other is quite abrupt in most of the cases studied here (D18-R72′, I22-V26′, I22-I73′). This may not necessarily indicate an abrupt change in the interacting pair per se. In contrast to the situation witnessed in the case of close homologs in Section 11.3.1, the sequences obtained from the BLAST search clustered into three specific groups, representing the three functional groups. No intermediate clusters are seen, which could probably have shed more light on the nature of the residues involved at the interface. The absence of any intermediate clusters could be due to the incompleteness of the protein sequence data available or may be an indicator that the intermediates were opted out of evolution.

11.4 DRIFTS IN INTERFACES AT EXTREME DIVERGENCE

Although sequence identity is very low at the superfamily level, PSI-BLAST was able to link three PDB members from three different families within the superfamily at a significant E-value. The proteins that were reversibly linked by PSI-BLAST are:

1. 1g2i: intracellular protease (DJ-1/PfpI family)
2. 1a9x: C-terminal domain of the small subunit of carbamoyl phosphate synthetase (class I glutamine amidotransferase–like family)
3. 1cf9: C-terminal domain of catalase (catalase, C-terminal domain family)

The intracellular protease 1g2i picks up 1cf9 as a hit in the second round of PSI-BLAST. However, only 60% of the query is aligned to the hit, with a sequence identity of 22%, indicating that these are remote homologs. The GAT domain of carbamoyl phosphate synthetase (1a9x) is an even more divergent member of this family, as 1g2i picks up this protein as a hit only in the fourth round of PSI-BLAST. The query coverage is 70% at a sequence identity of 15%. In both 1cf9 and 1a9x, the GAT domain occurs as a tethered domain.

We see that domain recruitment influences the interacting surface in two cases (1a9x, 1cf9). The tethered domains in both cases are from different SCOP folds. This phenomenon can virtually determine which surface of the protein surface is exposed for other proteins to interact. As the interacting partners are very different, their interfaces would be influenced by totally different factors. Therefore, it would not be possible to trace the variation of interfaces between these two families.

11.4.1 Comparison of Quaternary Structures of Intracellular Protease and Catalase

We attempted to trace the variation of the protein–protein interacting surface between the oligomeric family (DJ-1/PfpI family) and the closer of the other two families, the C-terminal domain of catalases. The GAT domain of catalase also has an oligomeric interface apart from the interface with its tethered domain.

The quaternary structure of catalase (1cf9) was superposed on intracellular protease (1g2i). Although the GAT domain of 1cf9 forms a symmetric oligomeric interface with the GAT domain of the adjacent catalase subunit, the interacting surfaces of the oligomers are not even topologically equivalent (Figure 11.7). In addition, the interacting surface of the GAT domain with its tethered domain does not overlap with the regions involved in formation of the two oligomeric interfaces in 1g2i. However, the specific region of GAT domain–GAT domain interface is probably determined by the larger interface formed between the larger N-terminal subunits of the catalase protein. The important and well-conserved interface between the nearby N-terminal catalase domains probably causes the smaller C-terminal GAT domains to interact as they do.

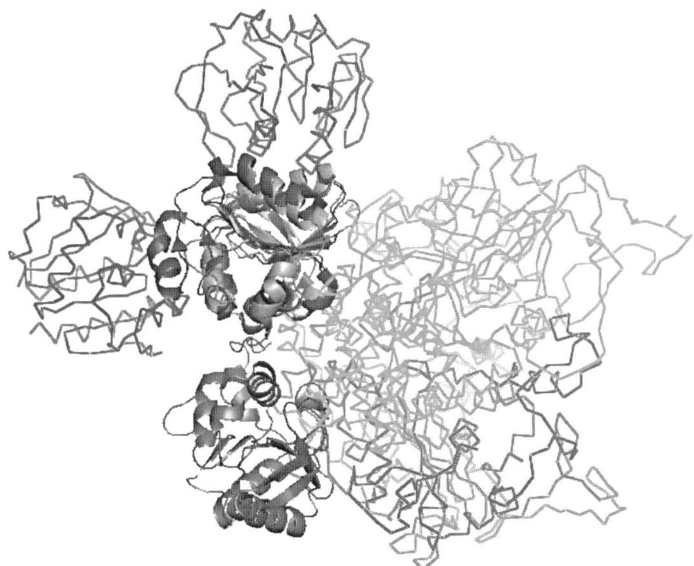

FIGURE 11.7 Quaternary structures of the GAT domains of 1g2i and 1cf9, proteins that belong to two different families within the same superfamily. The remote homologs do not even have topologically equivalent interacting surfaces, a feature observed in the protein–protein interfaces of many families related at the superfamily level. The three monomers of 1g2i are colored in red. Two molecules of catalase (1cf9) are shown, one orange and the other blue. One of the GAT domains in 1g2i and both GAT domains in 1cf9 are depicted as cartoons; the rest of the protein is depicted using ribbons. Generated using PyMOL [20]. (*See insert for color representation of figure.*)

11.4.2 Tracing the Drift in Protein–Protein Interaction Surface from One Family to Another: Intracellular Protease and Catalase

To trace the drift in interface from the catalase C-terminal GAT domain to the oligomeric GAT domain, a PSI-BLAST search was performed against the PALI+ database [19] with 1cf9a1 as the query, using an E-value of 0.01 and an H-value of 0.0001 for 20 rounds. The round at which the first member (sequence or structure) of the other family is picked up is analyzed. In this case, the first member of the DJ-1 family was picked up in round 2. All the sequences picked as hits in this round are used to obtain a multiple-sequence alignment. The sequences that form the boundary for the respective family and are closest to the next family are analyzed to follow whether the drift in amino acid residues is abrupt or gradual.

From Figure 11.8 we can see a distinct difference at the amino acid level between the closest sequences from the two families (from the sequences available in the UniRef90 database). The change in amino acids involved in interface formation is abrupt, although PSI-BLAST is able to establish an evolutionary connection between the two families. This could indicate one of two possibilities: the intermediate sequences have been lost in the course of evolution or that there is an abrupt change in residues involved at the protein–protein interface in the sequence space.

However, when we consider the conservation of buried residues (whose surface accessibility is <1%) between the two families, the results are different. Of the

```
                                                         C-terminal domain Catalase family
lcl|UniRef90_Q8PVL3_540-710         --------------------IKTRKIAIIVADG---YDHKDLSQVMQALEA
lcl|UniRef90_UPI0000464117_557      --------------------VKTRKVAILVANG---VESAPLRRIQQDLTE
lcl|UniRef90_UPI000045ADB4_534      ---------------------KTRKVAVLLESG---FDKAQFDALTATLIA
lcl|UniRef90_Q8EN31_544-687         --------------------NTLKVAIIVGNG---YGGNALESLLESLKE
lcl|UniRef90_UPI000045A94D_532      --------------------DTLRVAVFLAQG---FDGPGVSKVLETLAE
lcl|UniRef90_Q8ELG5_534-673         --------------------DTRKVGIIVADG---FD-EELFSHIKALNE
lcl|UniRef90_Q9RYQ0_587-771         --------------------KGRKVAVLVADG---VDAAGVKALQDALKK
lcl|UniRef90_Q8UJI6_552-702         --------------------EGRKLGILITDG---VDAKLLKGLTQAVTA
lcl|UniRef90_Q9X576_553-703         --------------------EGRKLGILATDG---ADGALLDALIAAVEK
lcl|UniRef90_Q6N4M7_554-703         -----------------GRKLGILITDG---TDAALLKSLQKAVEK
lcl|UniRef90_Q98J51_552-700         -----------------EGRKLGILVTDG---TDAALLNALKQALTK
lcl|UniRef90_Q5FRU1_548-701         -----------------KGRNLGVLVTDG---TDADLLAALMKAAEA
lcl|UniRef90_Q6VUC4_546-688         -----------------KGRRLGLLVSEN---SDFSLIDGIDTRMKM
lcl|UniRef90_UPI00037972F_535       --------------------AGRTVGVLAGNG---VEAELLTELRSILQA
lcl|UniRef90_Q5GX94_599-743         -------------------LKGRTVGILIHDG---SDASTIKAVRKAAEA
lcl|UniRef90_UPI0000463AD7_546      -------------------LEGRMVGILVGDG---SDGARVEALRRAATD
lcl|UniRef90_UPI000045E4DF_547      -------------------LKGRMIGILVAEG---SKHSEIEKYEKAAQA
lcl|UniRef90_UPI00003C82C7_581      --------------------AGRKVGILFAEG---SDKATIDKLKAGVEE
lcl|UniRef90_UPI000038F0CE_561      --------------------IATKQIAFLCADE---VSGASVVSYKNALER

                                                         DJ-1/PfpI family
lcl|UniRef90_UPI00003A64CF_9-1      --------------------KVAILAVDG---FEEAELVEPLRALKA
lcl|UniRef90_UPI00037D6A5_7-1       -------------------GHKVAILAVDG---FEQAELIEPKRALTE
lcl|UniRef90_Q7NGS5_8-181           ---------------------RVAVLATDG---FEETELTEPMRALQQ
lcl|UniRef90_UPI000038CA0D_8-1      ---------------------RVAVLATDG---FEEAELTEPVKALKE
lcl|UniRef90_UPI0000387BCC_5-1      ------------------LSGKRVAFLVTDG---FEQVELTGPKQALEQ
lcl|UniRef90_UPI0000395734_5-1      ------------------LTGKRIAILVTDG---FEQVELTGPKEALEQ
lcl|UniRef90_Q88JC4_6-169           ------------------LNGKRVAFLVTDG---FEQVELTGPREALEN
lcl|UniRef90_Q9I6D8_7-177           -----GKVVAALVTDG---FEQVELTGPKKALED
lcl|UniRef90_UPI00003893A3_5-1      -----LKGKRVAFLVTDG---FEQVELTGPREALEK
lcl|UniRef90_Q8P751_5-169           -----LSGKTVAVLATSG---FEQSELQEPKRLLES
lcl|UniRef90_Q8PIG6_5-168           -----LTGKTVAVLATDG---FEQSELTEPKRLLES
lcl|UniRef90_Q9RV31_8-180           ---------------------TGKKIAILAADG---VEEIELTSPRAAIEA
lcl|UniRef90_Q5FQ93_3-173           ---------------------KVAALATDG---LEEIELTGPQEALEK
lcl|UniRef90_Q7USN8_9-178           ---------------------KRIAFLATDG---FEQVELTKPWEAIQA
lcl|UniRef90_UPI0000384C1D_6-1      ---------------------LDGVDVAVLIAPEG---AEQVELTVPWDAIRD
lcl|UniRef90_Q744G9_3-179           ---------------------NELQG-KKIAILAADG---VEKVELEQPAAALRE
lcl|UniRef90_Q826G2_36-200          ---------------------LRG-MRVGILATDG---VERVELDQPRGALQG
```

FIGURE 11.8 Drift in interfaces between members of different families in the same superfamily, depicting a clear-cut visual distinction between sequences from different families. The sequences forming the boundary of each of the families are highlighted using a color specific to the family. (*See insert for color representation of figure.*)

```
lcl|UniRef90_Q65D55_334-486      -------------------QTRTVAILAEQG---FDDEDLSRVLKEFKK
lcl|UniRef90_Q9KBE8_541-709      -------------------RTRKVAVLIDHG---FAFDEVSQVVTALQS
lcl|UniRef90_Q8PVL3_540-710      ------------------IKTRKIAIIVADG---YDHKDLSQVMQALEA
lcl|UniRef90_UPI0000464117_557   -----------------VKTRKVAILVANG---VESAPLRRIQQDLTE
lcl|UniRef90_UPI000045ADB4_534   ------------------KTRKVAVLLESG---FDKAQFDALTATLIA
lcl|UniRef90_Q8EN31_544-687      -----------------NTLKVAIIVGNG---YGGNALESLLESLKE
lcl|UniRef90_UPI000045A94D_532   ------------------DTLRVAVFLAQG---FDGPGVSKVLETLAE
lcl|UniRef90_Q8ELG5_534-673      ------------------DTRKVGIIVADG---FD-EELFSHIKALNE
lcl|UniRef90_Q9RYQ0_587-771      ------------------KGRKVAVLVADG---VDAAGVKALQDALKK
lcl|UniRef90_Q8UJI6_552-702      ------------------EGRKLGILITDG---VDAKLLKGLTQAVTA
lcl|UniRef90_Q9X576_553-703      ------------------EGRKLGILATDG---ADGALLDALIAAVEK
lcl|UniRef90_Q6N4M7_554-703      ------------------GRKLGILITDG---TDAALLKSLQKAVEK
lcl|UniRef90_Q98J51_552-700      [ C-terminal domain  ]---EGRKLGILVTDG---TDAALLNALKQALTK
lcl|UniRef90_Q5FRU1_548-701      [ Catalase family    ]---KGRNLGVLVTDG---TDADLLAALMKAAEA
lcl|UniRef90_Q6VUC4_546-688                           ---KGRRLGLLVSEN---SDFSLIDGIDTRMKM
lcl|UniRef90_UPI00037972F_535    ------------------AGRTVGVLAGNG---VEAELLTELRSILQA
lcl|UniRef90_Q5GX94_599-743      ------------------LKGRTVGILIHDG---SDASTIKAVRKAAEA
lcl|UniRef90_UPI0000463AD7_546   ------------------LEGRMVGILVGDG---SDGARVEALRRAATD
lcl|UniRef90_UPI000045E4DF_547   ------------------LKGRMIGILVAEG---SKHSEIEKYEKAAQA
lcl|UniRef90_UPI00003C82C7_581   ------------------AGRKVGILFAEG---SDKATIDKLKAGVEE
lcl|UniRef90_UPI000038F0CE_561   ------------------IATKQIAFLCADE---VSGASVVSYKNALER
lcl|UniRef90_UPI00003A64CF_9-1   -------------------KVAILAVDG---FEEAELVEPLRALKA
lcl|UniRef90_UPI000037D6A5_7-1   ------------------GHKVAILAVDG---FEQAELIEPKRALTE
lcl|UniRef90_Q7NGS5_8-181        --------------------RVAVLATDG---FEETELTEPMRALQQ
lcl|UniRef90_UPI000038CAOD_8-1   --------------------RVAVLATDG---FEEAELTEPVKALKE
lcl|UniRef90_UPI0000387BCC_5-1   -----------------LSGKRVAFLVTDG---FEQVELTGPKQALEQ
lcl|UniRef90_UPI0000395734_5-1   [ DJ-1/PfpI family ]----LTGKRIAILVTDG---FEQVELTGPKEALEQ
lcl|UniRef90_Q88JC4_6-169                          ----LNGKRVAFLVTDG---FEQVELTGPREALEN
lcl|UniRef90_Q9I6D8_7-177        -------------------GKVVAALVTDG---FEQVELTGPKKALED
lcl|UniRef90_UPI00003893A3_5-1   -----------------LKGKRVAFLVTDG---FEQVELTGPREALEK
lcl|UniRef90_Q8P751_5-169        -----------------LSGKTVAVLATSG---FEQSELQEPKRLLES
lcl|UniRef90_Q8PIG6_5-168        -----------------LTGKTVAVLATDG---FEQSELTEPKRLLES
lcl|UniRef90_Q9RV31_8-180        -----------------TGKKIAILAADG---VEEIELTSPRAAIEA
lcl|UniRef90_Q5FQ93_3-173        -------------------KVAALATDG---LEEIELTGPQEALEK
lcl|UniRef90_Q7USN8_9-178        ------------------KRIAFLATDG---FEQVELTKPWEAIQA
lcl|UniRef90_UPI0000384C1D_6-1   ---------------LDGVDVAVLIAPEG---AEQVELTVPWDAIRD
lcl|UniRef90_Q744G9_3-179        ---------------NELQG-KKIAILAADG---VEKVELEQPAAALRE
lcl|UniRef90_Q826G2_36-200       ---------------LRG-MRVGILATDG---VERVELDQPRGALQG
lcl|UniRef90_UPI0000379D6E_1-1   ---------------MTLDGKTIAILIAPRG---TEDVEYVRPKEALTQ
```

FIGURE 11.9 Conservation of buried residues in different families in the same superfamily. Even though there is a clear distinction between the sequences from the two families, the residues at the core of the structures are generally well conserved. The sequences forming the boundary of each of the families are highlighted using a color specific to the family. The residues colored in pink indicate conserved common buried residues in the two families. Residues in red indicate conserved buried residues specific to the C-terminal domain catalase family. Residues in blue indicate conserved buried residues specific to the DJ-1/PfpI family. Residues in green indicate conserved interfacial residues specific to the C-terminal domain catalase family. (*See insert for color representation of figure.*)

22 aligned buried residues between the two families, 13 are common to the two families and well conserved. A subset of these conserved residues is depicted in Figure 11.9. Of the remaining residues, only in two cases is there family-specific conservation. This indicates that the core residues for the tertiary structure are well conserved, which probably explains why PSI-BLAST is able to establish the connection between members. The scenario is entirely different with respect to interacting residues, where the divergence is too high even to see a continuum at the sequence level.

11.5 CONCLUSIONS

Variation in protein–protein interacting surfaces can occur within members belonging to different families in the same superfamily or sometimes even within members in the same family. The variation can be in terms of slight orientational and spatial differences even though the gross topology of the interface is conserved (as generally seen in the case of closely related members), in the use of an entirely different surface for interaction (as seen in the case of more divergent members), or in complete loss of the interface. The above analysis tries to look at the drift in protein–protein interaction surfaces in each of the foregoing cases.

Analysis of variation in protein–protein interaction surfaces when they show only slight differences between homologous members indicates that the drift is gradual, as seen in case of the conserved interface between intracellular protease (1g2i) and DJ-1 (1pe0). There exist sequences containing many different intermediate combinations of the interacting residues involved in both sets of proteins.

Comparisons of homologs where an entire interface is lost seemingly show a different trend. The most prominent interacting residues show an abrupt shift between the two different subfamilies. However, inspection of the other interacting residues reveals that a gradual change is occurring generally, although a drastic change in the important (although quantitatively lesser) residues would have led to loss of interface.

Analysis of drift of protein–protein interaction surfaces in distant homologs is distinctly abrupt. However, this is probably because of two factors: either gaps in the current knowledge of sequence space or because the intermediate sequences have been opted out of evolution.

Acknowledgments

This research is supported by the Department of Biotechnology, India in the form of a computational genomics project. L.S.S. is supported by a fellowship from the Council of Scientific and Industrial Research, India. N.S. is a visiting professor at Université de la Réunion.

REFERENCES

1. V. Bellotti, P. Mangione, and M. Stoppini. Biological activity and pathological implications of misfolded proteins. *Cell Mol. Life Sci.*, **55**: 977–991 (1999).

2. R. Zutshi, M. Brickner, and J. Chmielewski. Inhibiting the assembly of protein–protein interfaces. *Curr. Opin. Chem. Biol.*, **2**: 62–66 (1998).

3. A. I. Archakov, V. M. Govorun, A. V. Dubanov, Y. D. Ivanov, A. V. Veselovsky, P. Lewi, and P. Janssen. Protein–protein interactions as a target for drugs in proteomics. *Proteomics*, **3**: 380–391 (2003).

4. R. C. Liddington. Structural basis of protein–protein interactions. *Methods Mol. Biol.*, **261**: 3–14 (2004).

5. L. Pagliaro, J. Felding, K. Audouze, S. J. Nielsen, R. B. Terry, C. Krog-Jensen, and S. Butcher. Emerging classes of protein–protein interaction inhibitors and new tools for their development. *Curr. Opin. Chem. Biol.*, **8**: 442–449 (2004).

6. L. O. Sillerud and R. S. Larson. Design and structure of peptide and peptidomimetic antagonists of protein–protein interaction. *Curr. Protein Pept. Sci.*, **6**: 151–169 (2005).

7. P. Aloy, H. Ceulemans, A. Stark, and R. B. Russell. The relationship between sequence and interaction divergence in proteins. *J. Mol. Biol.*, **332**: 989–998 (2003).

8. N. Rekha, S. M.. Machado, C. Narayanan, A. Krupa, and N. Srinivasan. Interaction interfaces of protein domains are not topologically equivalent across families within superfamilies: implications for metabolic and signaling pathways. *Proteins*, **58**: 339–353 (2005).

9. A. Krupa and N. Srinivasan. Repertoire of protein kinases encoded in the draft version of the human genome: atypical variations and uncommon domain combinations. *Genome Biol.*, **3**: 66.1–66.14 (2002).

10. C. Vogel, S. A. Teichmann, and J. Pereira-Leal. The relationship between domain duplication and recombination. *J. Mol. Biol.*, **346**: 355–365 (2005).

11. A. G. Murzin, S. E. Brenner, T. Hubbard, and C. Chothia. SCOP: a structural classification of proteins database for the investigation of sequences and structures. *J. Mol. Biol.*, **247**: 536–540 (1995).

12. D. Korkin, F. P. Davis, and A. Sali. Localization of protein-binding sites within families of proteins. *Protein Sci.*, **14**: 2350–2360 (2005).

13. M. M. Prabu, K. Suguna, and M. Vijayan. Variability in quaternary association of proteins with the same tertiary fold: a case study and rationalization involving legume lectins. *Proteins*, **35**: 58–69 (1999).

14. S. F. Altschul, T. L. Madden, A. A. Schaffer, J. Zhang, Z. Zhang, W. Miller, and D. J. Lipman. Gapped BLAST and PSI-BLAST: a new generation of protein database search programs. *Nucleic Acids Res.*, **25**: 3389–3402 (1997).

15. S. Bandyopadhyay and M. R. Cookson. Evolutionary and functional relationships within the DJ1 superfamily. *BMC Evol. Biol.*, **4**: 6 (2004).

16. X. Du, I. G. Choi, R. Kim, W. Wang, J. Jancarik, H. Yokota, and S. H. Kim. Crystal structure of an intracellular protease from *Pyrococcus horikoshii* at 2-Å resolution. *Proc. Natl. Acad. Sci. USA*, **97**: 14079–14084 (2000).

17. J. Abdallah, R. Kern, A. Malki, V. Eckey, and G. Richarme. Cloning, expression, and purification of the general stress protein YhbO from *Escherichia coli. Protein Expr. Purif.* (Epub ahead of print) (2005)

18. X. Tao and L. Tong. Crystal structure of human DJ-1, a protein associated with early onset Parkinson's disease. *J. Biol. Chem.*, **278**: 31372–31379 (2003).

19. V. S. Gowri, S. B. Pandit, P. S. Karthik, N. Srinivasan, and S. Balaji. Integration of related sequences with protein three-dimensional structural families in an updated version of PALI database. *Nucleic Acids Res.*, **31**: 486–488 (2003).

20. W. L. DeLano. *The PyMOL Molecular Graphics System*. DeLano Scientific, San Carlos, CA (2002).

12

ON COMPARING AND VISUALIZING RNA SECONDARY STRUCTURES

JASON T. L. WANG, DONGRONG WEN, AND JIANGHUI LIU

Department of Computer Science (J.T.L.W., D.M.) and Bioinformatics and Life Science Informatics Laboratory (J.L.), New Jersey Institute of Technology, Newark, New Jersey

Comparing and aligning RNA secondary structures is fundamental to knowledge discovery in biomolecular informatics [1–3]. It assists scientists in performing many important RNA mining operations, including the understanding of functions of RNA sequences, the detection of structural RNA motifs, and the clustering of RNA molecules, among others. In recent years, much progress has been made in RNA structure alignment and comparison. However, existing tools either require a large number of prealigned structures or suffer from high time complexities. This makes it difficult for the tools to process RNAs whose prealigned structures are unavailable or to process very large RNA structure databases. We present here an efficient method, called RSmatch, for comparing and aligning RNA secondary structures. Motivated by widely used algorithms for RNA folding, we decompose an RNA secondary structure into a set of atomic structure components that are further organized by a tree model to capture the structural particularities of RNA. RSmatch can find the optimal global or local alignment between two RNA secondary structures. The time complexity of RSmatch is $O(mn)$, where m is the size of the query structure and n that of the subject structure. We also present a visualization tool, called RSview, which is capable of displaying the output of RSmatch in a colorful and graphical manner. This tool will be useful to researchers interested in comparing RNA structures obtained from wet lab experiments or RNA folding programs for the purpose of discovering knowledge from RNA molecules. Related software can be downloaded from http://datalab. njit.edu/biodata/ssi/software.shtml.

12.1 BACKGROUND

Ribonucleic acid (RNA) plays various roles in the cell. Many functions of RNA are attributable to their structural particularities (herein called RNA motifs). RNA motifs have been studied extensively for noncoding RNAs (ncRNAs), such as transfer RNA (tRNA), ribosomal RNA (rRNA), small nuclear RNA (snRNA), and small nucleolar RNA (snoRNA) [4]. More recently, small interfering RNA (siRNA) and microRNA (miRNA) have been under intensive study [5]. Less well characterized are the structures in the untranslated regions (UTRs) of messenger RNAs (mRNAs) [6]. However, biochemical and genetic studies have demonstrated a myriad of functions associated with the UTRs in mRNA metabolism, including RNA translocation, translation, and RNA stability [7–9].

RNA structure determination via biochemical experiments is laborious and costly. Predictive approaches are valuable in providing guide information for wet lab experiments. RNA structure prediction is usually based on the thermodynamics of RNA folding or phylogenetic conservation of base-paired regions. The former uses thermodynamic properties of various RNA local structures, such as base-pair stacking, hairpin loop, and bulge, to derive thermodynamically favorable secondary structures. A dynamic programming algorithm is used to find optimal or suboptimal structures. The best known tools belonging to this group are MFOLD [10] and RNAFold in the Vienna RNA package [11,12]. Similar tools have been developed in recent years to predict higher-order structures, such as pseudoknots [13]. On the other hand, RNA structure prediction using phylogenetic information implies RNA structures based on covariation of base-paired nucleotides [14–17]. It is generally believed that methods using phylogenetic information are more accurate. However, their performance depends critically on the high-quality alignment of a large number of structurally related sequences.

Tools that align biosequences (DNA, RNA, protein), such as FASTA and BLAST, are valuable in identifying homologous regions, which can lead to the discovery of functional units, such as protein domains and DNA *cis* elements [18,19]. However, their success is more evident in the study of DNAs and proteins than of RNAs. This is mainly because the sequence similarity among DNAs and proteins can usually faithfully reflect their functional relationship, whereas additional structure information is needed to study functional conservation among RNAs. Therefore, in comparing RNA sequences, it is necessary to take into account both structural and sequential information.

Several tools are available that carry out RNA alignment and folding at the same time (Tables 12.1 and 12.2). The pioneer work by Sankoff [20], which involves simultaneous folding and aligning of two RNA sequences, has huge time and space complexities. FOLDALIGN [21] improves Sankoff's method by (1) scoring

TABLE 12.1 Time Complexities of RNA Secondary-Structure Analysis Tools

Tool Name	Running Time
Sankoff[a]	$O(N^6)$
FOLDALIGN[b]	$O(N^4)$
RAGA[c]	$O(M^2 N^3)$
rna_align[d]	$\min\{O(MN^3), O(M^3 N)\}$
Dynalign[e]	$O(M^3 N^3)$
stemloc[f]	$O(LM)$
Rsearch[g]	$O(M^3 N)$
RNAforester[h]	$O(|F_1||F_2|\deg(F_1)\deg(F_2)[\deg(F_1) + \deg(F_2)])$
CARNAC[i]	$O(N^6)$, $O(N^2)$
comRNA[j]	$O(M^N)$

[a]N is the average sequence length [20].
[b]N is the average length of a given set of RNAs [21].
[c]M and N are the lengths of the two given sequences [25].
[d]M and N are the lengths of the two given sequences [29].
[e]M is the maximum distance allowed to match two nucleotides, and N is the length of the shorter sequence [22].
[f]L and M are the two RNA sequence lengths (valid only in extreme cases) [35].
[g]M is the query length, and N is the subject sequence length [34].
[h]$|F_i|$ is the number of nodes in forest F_i, and $\deg(F_i)$ is the degree of F_i [30].
[i]N is the sequence length; the theoretical time complexity of $O(N^6)$ could be reduced significantly to $O(N^2)$ by preprocessing the sequences, as noted in [23].
[j]M is the maximum number of stems examined, and N is the total number of sequences under analysis. The comRNA's average running time can be improved significantly by carefully choosing parameters, as noted in [24].

structures based solely on the number of base pairs instead of the stacking energies in them, and (2) disallowing branch structures (junctions). Dynalign [22] reduces the time complexity by restricting the maximum distance allowed between aligned nucleotides in two structures. By taking into account local similarity, stem energy,

TABLE 12.2 Space Requirement of RNA Secondary-Structure Analysis Tools

Tool Name	Space Requirement
Sankoff	$O(N^4)$
FOLDALIGN	$O(N^4)$
RAGA	$O(M^2 N^2)$
rna_align	$O(MN^2)$
Dynalign	$O(M^2 N^2)$
stemloc	N/A
Rsearch	$O(M^3)$
RNAforester	$O(|F_1||F_2|\deg(F_1)\deg(F_2))$
CARNAC	$O(N^4)$, $O(N^2)$
comRNA	N/A

and covariations, Perriquet et al. [23] proposed CARNAC for pairwise folding of RNA sequences. Ji et al. [24] applied a graph-theoretical approach, called comRNA, to detect common RNA secondary structure motifs from a group of functionally or evolutionally related RNA sequences. One noticeable advantage of comRNA is its capability to detect pseudoknot structures. In addition, algorithms using derivative-free optimization techniques, such as genetic algorithms and simulated annealing, have been proposed to increase the accuracy in structure-based RNA alignment [25–27]. For example, Notredame et al. [25] presented RAGA to conduct alignment of two homologous RNA sequences when the secondary structure of one of them was known. As shown in Table 12.1, most of these methods suffer from high time complexities, making the structure-based RNA alignment tools much less efficient than sequence-based alignment tools.

Tools that search for optimal alignment for given structures include RNAdistance [28], rna_align [29], and RNAforester [30]. RNAdistance uses a tree-based model to coarsely represent RNA secondary structures, and compares RNA structures based on edit distance. In a similar vein, rna_align [29] models RNA secondary structures by nested and/or crossing arcs that connect bonded nucleotides. With the crossing arcs, rna_align is able to align two RNA secondary structures, one of which could contain pseudoknots. RNAforester extends the tree model to a forest model, which improves both time and space complexities significantly (Tables 12.1 and 12.2).

In addition, methods using stochastic context-free grammars (SCFGs) have been developed to compare two RNA structures. Original SCFG models [31,32] require a prior multiple-sequence alignment (with structure annotation) for training purposes. Thus, their applicability is limited to RNA types for which structures of a large number of sequences are available, such as snoRNA and tRNA [31,33]. However, Rsearch [34] and stemloc [35], both based on SCFG, are capable of conducting pairwise structure comparisons with no requirement for prealignment of sequences. Rsearch uses RIBOSUM substitution matrices derived from ribosomal RNAs to score the matches in single-stranded (ss) and double-stranded (ds) regions. The stemloc tool uses a "fold envelope" to improve efficiency by confining the search space involved in calculations. The time and space complexities of these two tools are also listed in Tables 12.1 and 12.2. Furthermore, pattern-based techniques such as RNAmotif, RNAmot, and PatSearch [6,36,37] have been used in database searches to detect similar RNA substructures. These tools represent RNA structures by a consensus pattern containing both sequence and structure information. One important advantage of these pattern-based tools is the ability of dealing with pseudoknots. See [41] for a more detailed comparison of these tools.

12.2 RSmatch

We present here a computationally efficient tool, called RSmatch, capable of aligning two RNA secondary structures both globally and locally. RSmatch does not require prior knowledge of structures of interest. It can uncover structural similarities by means of direct aligning at the structure level. Details concerning RSmatch can be found in ref. 41, where we demonstrate that RSmatch is faster and achieves comparable or higher accuracy than existing tools when applied to a number of known RNA structures, including simple stem-loops and complex structures containing junctions.

12.2.1 Secondary-Structure Decomposition

RSmatch models RNAs by a structure decomposition scheme similar to the loop model commonly used in algorithms for RNA structure prediction [38,39]. With this model, pseudoknots are not allowed. Our method differs from loop decomposition methods in that it completely decomposes an RNA secondary structure into units called *circles* (see Figure 12.1*a*). When the secondary structure is depicted on a plane, a circle is defined as a set of nucleotides that are reachable from one another without crossing a

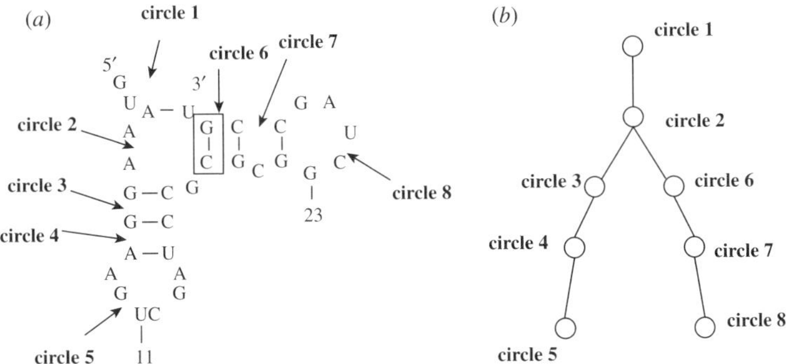

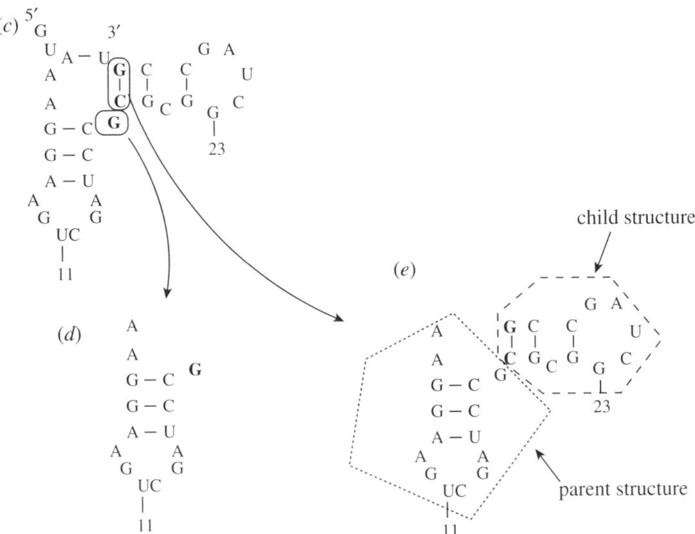

FIGURE 12.1 RNA structure decomposition (*a*, *b*) and partial structure determination (*c*, *d*, *e*).

base pair. As shown in Figure 12.1*a*, all circles are closed or ended by a base pair except the first circle (circle 1 in Figure 12.1*a*), which always contains the 5′- and 3′-most bases. Various types of RNA substructures, such as bulges, loops, and junctions, can be represented by circles, as shown in Figure 12.1*a*.

Circles of an RNA structure can be organized as a hierarchical tree according to their relative positions in the secondary structure, where each tree node corresponds to a circle (see Figure 12.1*b*). This tree organization is useful in deducing the structural relationship among circles and reflects the structural particularities of the particular RNA secondary structure. If two circles reside on the same lineage (path) in the tree, the circle appearing higher in the tree is called an *ancestor* of the other, and the latter is a *descendent* of the former. As a result, in the context of the hierarchical tree, two distinct circles fall into one of the following two categories, in order of decreasing closeness: (1) the two circles maintain an ancestor–descendent relationship, or (2) they share a common ancestor in the tree. For example, in Figure 12.1*b*, circle 2 is an ancestor of circle 5, whereas circle 6 does not have an ancestor–descendent relationship with circle 5 since they are not on the same lineage. The double-stranded region or stem of a structure is decomposed into a set of "degenerated" circles, each containing only two base pairs. As such, a stem of n bases in length will result in $n - 1$ consecutive degenerated circles. A base pair may have two associated circles; we name one circle the *parent circle* and the other the *child circle*, according to their positions in the hierarchical tree. For example, for the boxed C-G base pair in Figure 12.1*a*, circle 2 is its parent circle and circle 6 is its child circle.

12.2.2   Structure Alignment Formalization

Given an RNA secondary structure, we consider two types of *structure components*, single bases and base pairs, in the secondary structure. To integrate sequence and structure information, we introduce two constraints among the structure components: the precedence constraint and the hierarchy constraint. The *precedence constraint* is defined based on the precedence order among structure components, and the *hierarchy constraint* specifies the intercomponent relationship in the context of the hierarchical tree described above. The precedence order is determined by the 3′ bases of individual structure components; the one with its 3′ base closer to the RNA sequence's 5′-end precedes the other. For example, in Figure 12.1*a* the single base component U (marked as the 11th nucleotide) in circle 5 precedes the base-pair component C-G (boxed) in circle 6.

To capture the intercomponent relationship within the hierarchical tree context, we need to map each structure component to a circle in the tree. It is obvious that each single base can be mapped to a unique circle. However, a base pair could be mapped to two alternate circles: one parent circle and one child circle. To resolve this ambiguity, we always require mapping to the parent circle. The intercomponent relationship is then reduced to the intercircle relationship of three types: (1) ancestor–descendent, (2) common ancestor, and (3) identical circle.

Given two RNA secondary structures A and B, where A, referred to as the *query structure*, has m structure components $\{A^1, A^2, \ldots, A^m\}$ and B, referred to as the *subject structure*, has n structure components $\{B^1, B^2, \ldots, B^n\}$, the structure alignment between A and B is formalized as a conditioned optimization problem based on the two constraints above. Given a scoring scheme consisting of two matrices, one for matching two single bases and the other for matching two base pairs, find an optimal alignment between the two sets of structure components such that the aforementioned precedence and hierarchy constraints are preserved for any two matched component pairs (A^i, B^i) and (A^j, B^j). In other words, the two structure constraints between A^i and A^j must be equivalent to that between B^i and B^j, respectively. This formalization has an implicit biological significance in that a single-stranded region in one structure, if not aligned to a gap as a whole, will always align with a single-stranded region in the other structure. This alignment requirement is important because single-stranded regions are usually treated as functional units in binding to specific proteins.

12.2.3 Algorithmic Framework

A dynamic programming algorithm is employed in RSmatch. As with sequence alignment, the structure alignment could be either global or local. The difference lies only in the setup of initialization conditions; the algorithmic framework is the same since both global and local alignments must preserve the two constraints described above.

A scoring table is established with its rows and columns corresponding to the structure components of the two given RNA secondary structures. We organize the rows and columns in such a way that the precedence and hierarchy constraints are combined and easy to follow in the course of alignment computation. Specifically, we sort the structure components of each structure according to the precedence order defined above. It is straightforward that this arrangement of rows and columns preserves the precedence constraint automatically. However, preservation of the hierarchy constraint is much more complicated and can only be accomplished in the derivative analysis for each cell (entry) in the scoring table. We discuss the derivation when filling in the scoring table.

Each cell of the scoring table represents an intermediate comparison between two partial structures corresponding to the cell's row and column components (either single base or base pair), respectively. The partial structure with respect to a structure component c (single base or base pair) is a set of structure components S_c such that for any component $a \in S_c$, the following three structure constraints between c and a must be satisfied: (1) a precedes c; (2) by the hierarchy constraint, a is not an ancestor of c; and (3) c itself is included in S_c.

Furthermore, since a base pair could appear in two circles, its corresponding partial structure could be divided into two smaller substructures: parent structure and child structure. Formally, given a base pair component c, the *parent structure* of c is the set of structure components $P_c \subseteq S_c$ (excluding c itself) such that for any component $a \in P_c$, a's 3'-base is always 5' upstream of c's 5'-base; the *child structure* of c is the set of structure components $L_c \subseteq S_c$ (including c) such that for any component $a \in L_c$, a's 5'-base is always 3' downstream of c's 5'-base. It can be shown that $P_c \cup L_c = S_c$ and $P_c \cap L_c = \emptyset$. Examples of partial structures are given in

Figure 12.1c–e. As shown in parts (c) and (e), for a base pair, its child and parent structures together constitute the entire partial structure for the base pair.

As we will see in the following discussions, the concept of a partial structure and its by-products (parent structure and child structure) form the core of our algorithmic framework. We can solve the RNA structure alignment problem progressively by aligning small structures and expanding each of them one structure component at a time until all structure components are covered.

12.2.4 Basic Concepts and Definitions

Cells in the scoring table are processed row by row from top to bottom and from left to right within each row. By considering the row and column components, we have three types of cells: (1) a cell corresponding to two single bases, (2) a cell corresponding to one single base and one base pair, and (3) a cell corresponding to two base pairs. For (1), each cell stores the score of aligning the partial structures corresponding to the cell's row and column components respectively. For (2) and (3) we need to consider alignments involving the partial and child structures induced by the base-pair components. Notice that the parent structures of the base-pair components are excluded. It can be shown that each parent structure P_c of component c can always be considered as the partial structure S_x of some other component x, which means that we need to consider child and partial structures only in the alignment computation. Consequently, the three types of cells above have one, two, and four alignment scores, respectively.

A scoring scheme is required to score the match of two structure components. We define the scoring scheme as a function $g(a, b)$, where a and b represent two structure components that are matched with each other. Another important aspect of the alignment algorithm is to penalize the match involving gap(s). In the course of computation, one structure component (single base or base pair) could match with a gap, or an entire small structure (parent or child structure) could match with a large gap. Intuitively, the larger the gap, the heavier the penalty will be. In our implementation, we set an atomic penalty value, denoted as u, for the smallest gap equivalent to a single base. The penalty value for a large gap is proportional to its size in terms of the number of bases matched with the gap.

Let A^* be a small structure in the query RNA structure A and B^* a small structure in the subject RNA structure B. The score obtained by aligning the two structures A^* and B^*, denoted $f(A^*, B^*)$, is

$$f(A^*, B^*) = \sum_{a \in A^*, b \in B^*} g(a, b) + uG$$

where G represents the total number of gaps in aligning A^* and B^*.

12.2.5 Initialization

We assume that the row components (a's) are from the query RNA structure A and the column components (b's) from the subject RNA structure B. We focus on global alignment here; initializations for local alignment can be derived similarly. The initialization conditions deal with the cases where at least one of the structures under alignment is an empty structure $\emptyset$. This is equivalent to setting up the zeroth row or column in the scoring table. As discussed above, each base-pair component has two small structures to be considered: a child structure and a partial structure. Thus, the aforementioned three types of cells have one, two, and four initialization scores, respectively.

For a given structure component x (single base or base pair), let S_x represent its partial structure. If x is a base pair, we also use L_x to represent its child structure. We have $f(\emptyset, \emptyset) = 0$. Furthermore, for any structure components a and b, $f(S_a, \emptyset) = |S_a| \cdot u$, $f(\emptyset, S_b) = |S_b| \cdot u$, if a and b are base pairs, $f(L_a, \emptyset) = |L_a| \cdot u$ and $f(\emptyset, L_b) = |L_b| \cdot u$, where $|\cdot|$ represents the cardinality of the respective set.

12.2.6 Filling in the Scoring Table

The simplest cell type is the one whose row (column, respectively) component is a single base a (single base b, respectively). Let a^p denote the structure component that precedes a by precedence order established before. Formally, in matching the partial structure S_a with the partial structure S_b, there are only three possibilities: (1) a is aligned with b, (2) a is aligned with a gap, and (3) b is aligned with a gap. Thus, the score of matching S_a with S_b can be calculated as

$$f(S_a, S_b) = \max \begin{cases} f(S_{a^p}, S_{b^p}) + g(a,b) \\ f(S_{a^p}, S_b) + u \\ f(S_a, S_{b^p}) + u \end{cases} \tag{12.1}$$

The second cell type is the one formed by one single base and one base pair. There are actually two symmetric subtypes where either a or b is a base pair. Since the analysis is identical, we focus on the former case, where a is a base pair. As discussed before, besides the partial structure S_a we have to consider the child structure L_a for the base pair a. Thus, for this type of cell, we have to compute two alignment scores.

By the principle of dynamic programming, the smaller problem needs to be solved before the larger problem. Thus we first find the structure alignment between the child structure L_a and the partial structure S_b. There are only two possibilities: (1) the single base component b is aligned with a gap; and (2) the base pair a is aligned with a gap (see Figure 12.2a). Therefore, we have

$$f(L_a, S_b) = \max \begin{cases} f(L_a, S_{b^p}) + u \\ f(S_{a^p}, S_b) + 2u \end{cases} \tag{12.2}$$

In aligning the partial structure S_a with the partial structure S_b, to preserve precedence and hierarchy constraints simultaneously, there are only three possibilities:

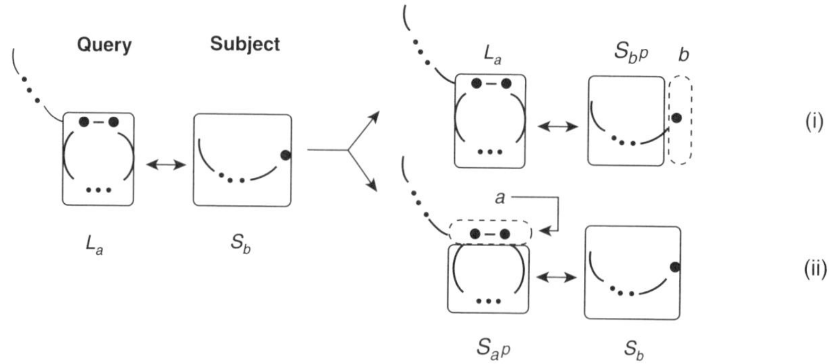

(a)

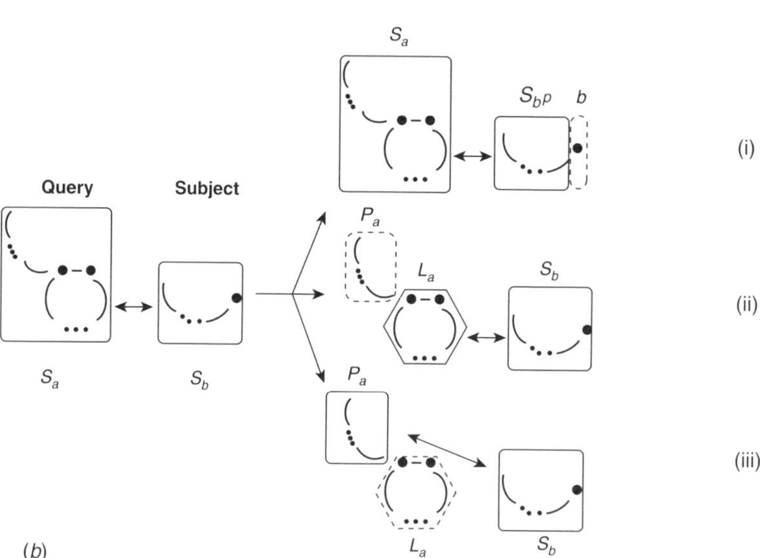

(b)

FIGURE 12.2   Optimal alignment of two RNA secondary structures.

(1) the single base b matches with a gap; (2) the partial structure S_b matches with the child structure L_a; and (3) the partial structure S_b matches with the parent structure P_a (see Figure 12.2b). Thus,

$$f(S_a, S_b) = \max \begin{cases} f(S_a, S_{b^p}) + g(a, b) \\ f(L_a, S_b) + |P_a| \cdot u \\ f(P_a, S_b) + |L_a| \cdot u \end{cases} \qquad (12.3)$$

For the third cell type, a is a base pair and b is a base pair. We need to compute four alignment scores because each base pair corresponds to two structures: one child structure and one partial structure. While aligning the child structure L_a with the child structure L_b, it is clear that

$$f(L_a, L_b) = \max \begin{cases} f(S_{a^p}, S_{b^p}) + g(a,b) \\ f(S_{a^p}, L_b) + 2u \\ f(L_a, S_{b^p}) + 2u \end{cases} \tag{12.4}$$

since both a and b are the last components in the respective child structures by precedence order. Equation (12.5) gives the alignment score between the partial structure S_a and the child structure L_b:

$$f(S_a, L_b) = \max \begin{cases} f(S_a, S_{b^p}) + 2u \\ f(P_a, L_b) + |L_a| \cdot u \\ f(L_a, L_b) + |P_a| \cdot u \end{cases} \tag{12.5}$$

In the first case, b is aligned with a gap. If b does not match with a gap, it can be shown that to preserve both precedence and hierarchy constraints, the second and third cases in Eq. (12.5) cover all possible situations. Similarly, we can calculate the score of aligning the child structure L_a and the partial structure S_b:

$$f(L_a, S_b) = \max \begin{cases} f(S_{a^p}, S_b) + 2u \\ f(L_a, P_b) + |L_b| \cdot u \\ f(L_a, L_b) + |P_b| \cdot u \end{cases} \tag{12.6}$$

In aligning the partial structure S_a with the partial structure S_b, there are five possibilities: (1) the parent structure P_a is matched with the parent structure P_b and the child structure L_a is matched with the child structure L_b; (2) the child structure L_a is matched with gaps; (3) the child structure L_b is matched with gaps; (4) the parent structure P_a is matched with gaps; and (5) the parent structure P_b is matched with gaps. Therefore,

$$f(S_a, S_b) = \max \begin{cases} f(P_a, P_b) + f(L_a, L_b) \\ f(P_a, S_b) + |L_a| \cdot u \\ f(S_a, P_b) + |L_b| \cdot u \\ f(L_a, S_b) + |P_a| \cdot u \\ f(S_a, L_b) + |P_b| \cdot u \end{cases} \tag{12.7}$$

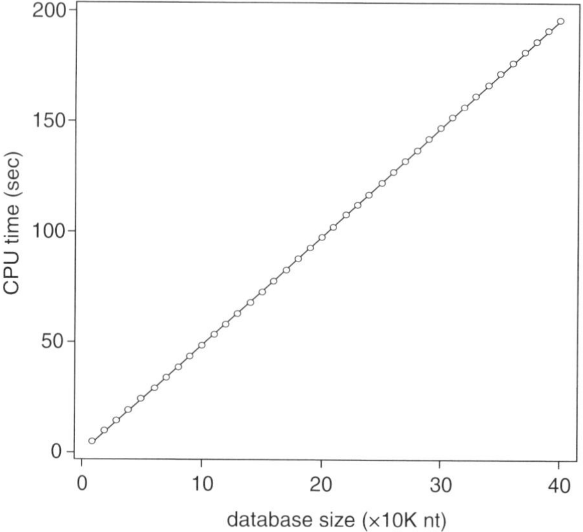

FIGURE 12.3   CPU time versus database size.

12.2.7   Running-Time Analysis

By dynamic programming, the running time of computing an alignment equals the number of writing operations needed to fill the scoring table. Thus, the time complexity of RSmatch is $O(mn)$, where m (n, respectively) is the number of structure components in the query (subject, respectively) RNA structure. To test the scalability, we downloaded the seed sequences for 5S rRNA family from Rfam and randomly selected one annotated structure as the query while folding the rest sequences to prepare the structure database as discussed above. Figure 12.3 shows the RSmatch running time versus the database size. The program was run 10 times and the average values from the 10 runs were plotted. The nearly linear growth of the running time gives an empirical proof that the algorithm's time complexity is bounded by $O(nm)$.

12.3   RSview

The alignment result of RSmatch is displayed in text. We also develop a visualization tool, called RSview, to be used together with RSmatch. Given two RNA molecules, RSview is capable of displaying the output of RSmatch in a colorful and graphical manner. RSview works by integrating RNAview [40] with RSmatch. The programming languages used in implementing RSview include C, Java, and Perl. The original RNAview program accepts RNA structures with three-dimensional

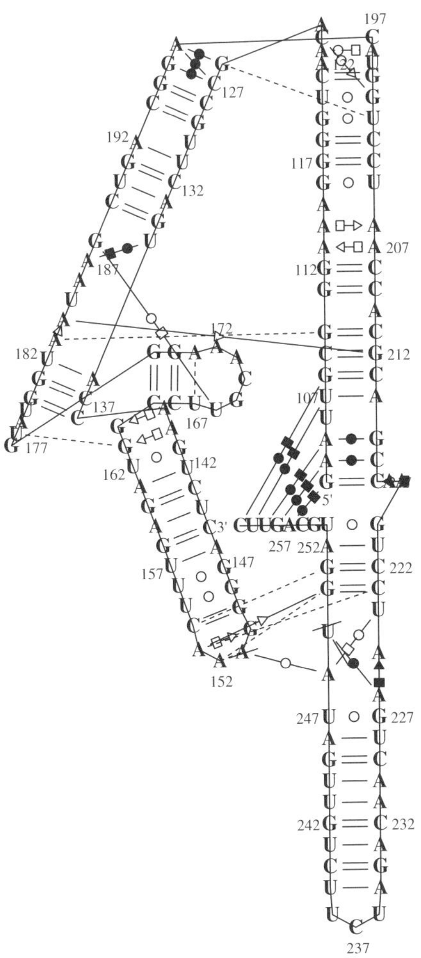

FIGURE 12.4 1GID chain A (P4-P6 RNA ribozyme domain). (*See insert for color representation of figure.*)

coordinate data in PDB, mmCIF, or RNAML format and then produces two-dimensional diagrams of secondary and tertiary RNA structures in Postscript, VRML, or RNAML formats. Figures 12.4 and 12.5 show two example diagrams produced by RNAview. The PDB ID for the structure in Figure 12.4 is 1GID Chain A (P4-P6 RNA ribozyme domain) and the PDB ID for the structure in Figure 12.5 is 1C2X Chain C (5S ribosomal RNA).

Like RNAview, RSview accepts two RNA structures with three-dimensional coordinate data in PDB format. RSview then generates two two-dimensional diagrams in Postscript format, as illustrated in Figures 12.4 and 12.5, together with RNA structures with three-dimensional coordinate data in RNAML format for the input

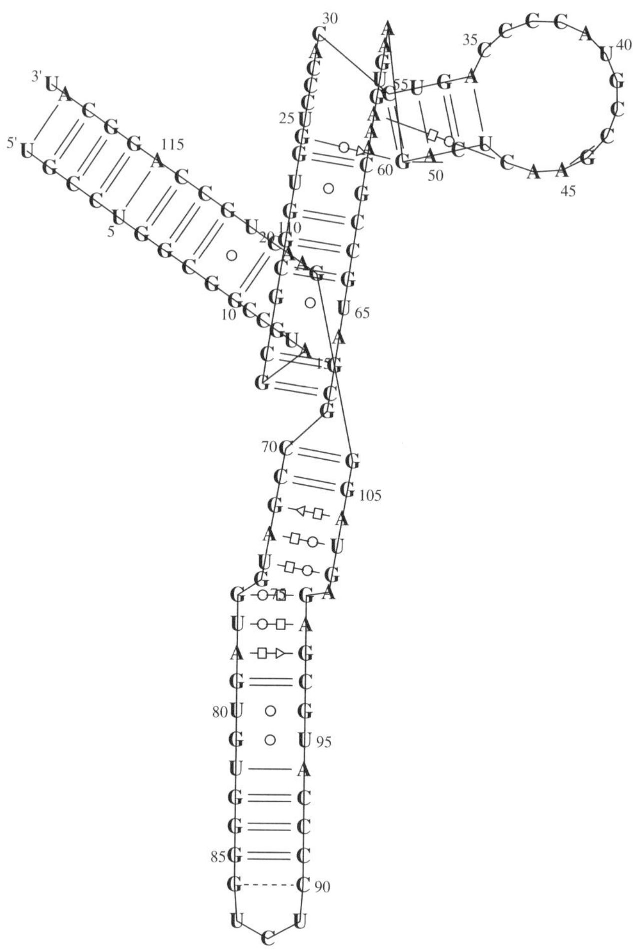

FIGURE 12.5   1C2X chain C (5S ribosomal RNA). (*See insert for color representation of figure.*)

RNA molecules. Next, the sequences of the input RNA molecules are extracted from the RNAML files and are folded into secondary structures using the Vienna RNA package. These secondary structures are then aligned using RSmatch. Finally, RSview combines two simplified two-dimensional diagrams in Postscript format for the RNA molecules with the alignment result obtained from RSmatch. Figure 12.6 shows the output of RSview for the two molecules in Figures 12.4 and 12.5. In Figure 12.6 the nucleotides in cyanine are the unmatched region, and the nucleotides in red are the matched (aligned) region. The blue (starting) line and green (ending) line indicate the best local match with the largest alignment score among all matched (aligned) regions.

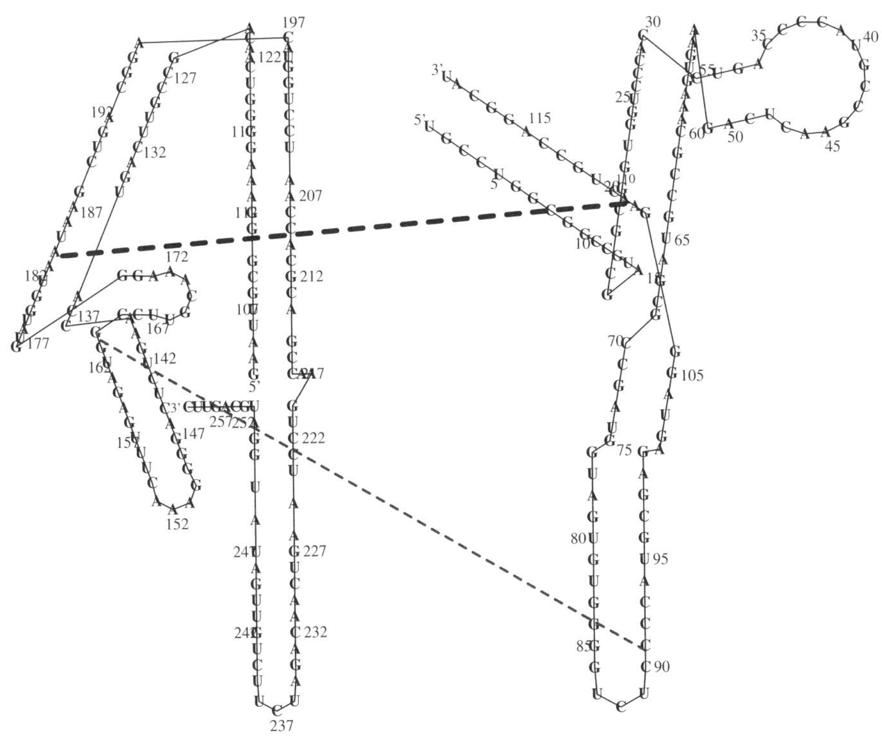

FIGURE 12.6 Output of RSview. (*See insert for color representation of figure.*)

12.4 CONCLUSIONS

We presented a tool, RSmatch, for comparing and aligning RNA secondary structures directly, and a tool, RSview, for graphically displaying the alignment result produced by RSmatch. RSmatch bears similarities to rna_align and RNAforester in that the structural particularities of RNA are either explicitly captured using hierarchical tree or forest structures or implicitly represented using arc-annotated structures. However, RSmatch differs from rna_align and RNAforester in two major aspects. First, RSmatch keeps structural consistency only by allowing single bases matched with single bases and base pairs matched with base pairs, whereas rna_align and RNAforester do not impose this restriction. Second, RSmatch retains the integrity of single-stranded regions by matching one with another instead of breaking a single-stranded region into pieces and aligning them with different single-stranded regions. In addition, RSmatch has less time and space complexities than do the other two tools. The RSmatch and RSview tools will benefit researchers interested in data mining, motif detection, and functional analysis of RNA molecules.

REFERENCES

1. J. T. L. Wang, M. J. Zaki, H. T. T. Toivonen, and D. Shasha, Eds. *Data Mining in Bioinformatics*. Springer-Verlag, New York, 2005.

2. C. Y. Chang, J. T. L. Wang, and R. K. Chang. Scientific data mining: a case study. *Int. J. Software Eng. Knowledge Eng.*, 1998, 8:77–96.

3. J. T. L. Wang, B. A. Shapiro, D. Shasha, K. Zhang, and C. Y. Chang. Automated discovery of active motifs in multiple RNA secondary structures. *Proc. 2nd International Conference on Knowledge Discovery and Data Mining*, 1996, pp. 70–75.

4. S. Griffiths-Jones, A. Bateman, M. Marshall, A. Khanna, and S. R. Eddy. Rfam: an RNA family database. *Nucleic Acids Res.*, 2003, 31:439–441.

5. V. Ambros, B. Bartel, D. P. Bartel, C. B. Berge, J. C. Carrington, X. Chen, G. Dreyfuss, S. R. Eddy, S. Griffiths-Jones, M. Marshall, M. Matzke, G. Ruvkun, and T. Tuschl. A uniform system for microRNA annotation. *RNA*, 2003, 9:277–279.

6. G. Pesole, S. Liuni, G. Grillo, F. Licciulli, F. Mignone, C. Gissi, and C. Saccone. UTRdb and UTRsite: specialized databases of sequences and functional elements of 5′ and 3′ untranslated regions of eukaryotic mRNAs. *Nucleic Acids Res.*, 2002, 30:335–340.

7. B. Mazumder, V. Seshadri, and P. L. Fox. Translational control by the 3′UTR: the ends specify the means. *Trends Biochem. Sci.*, 2003, 28:91–98.

8. S. Kuersten, and E. B. Goodwin. The power of 3′UTR: translational control and development. *Nat. Rev. Genet.*, 2003, 4:626–637.

9. I. L. Hofacker, P. F. Stadler, and R. R. Stocsits. Conserved RNA secondary structures in viral genomes: a survey. *Bioinformatics*, 2004, 20:1495–1499.

10. M. Zuker. Computer prediction of RNA structure. *Methods Enzymol.*, 1989, 180:262–288.

11. P. Schuster, W. Fontana, P. F. Stadler, and I. L. Hofacker. From sequences to shapes and back: a case study in RNA secondary structures. *Proc. Biol. Sci.*, 1994, 255:279–284.

12. I. L. Hofacker. Vienna RNA secondary structure server. *Nucleic Acids Res.*, 2003, 31:3429–3431.

13. E. Rivas and S. R. Eddy. A dynamic programming algorithm for RNA structure prediction including pseudoknots. *J. Mol. Biol.*, 1999, 285:2053–2068.

14. B. Gulko and D. Haussler. Using multiple alignments and phylogenetic trees to detect RNA secondary structure. *Pac. Symp. Biocomput.*, 1996, pp. 350–367.

15. V. R. Akmaev, S. T. Kelley, and G. D. Stormo. A phylogenetic approach to RNA structure prediction. *Proc. Int. Conf. Intell. Syst. Mol. Biol.*, 1999, pp. 10–17.

16. B. Knudsen and J. Hein. Pfold: RNA secondary structure prediction using stochastic context-free grammars. *Nucleic Acids Res.*, 2003, 31:3423–3428.

17. I. L. Hofacker, M. Fekete, and P. F. Stadler. Secondary structure prediction for aligned RNA sequences. *J. Mol. Biol.*, 2002, 319:1059–1066.

18. W. R. Pearson and D. J. Lipman. Improved tools for biological sequence comparison. *Proc. Natl. Acad. Sci. USA*, 1988, 85:2444–2448.

19. S. F. Altschul, W. Gish, W. Miller, E. W. Myers, and D. J. Lipman. Basic local alignment search tool. *J. Mol. Biol.*, 1990, 215:403–410.

20. D. Sankoff. Simultaneous solution of the RNA folding, alignment and protosequence problems. *SIAM J. Appl. Math.*, 1985, 45:810–825.

21. J. Gorodkin, S. L. Stricklin, and G. D. Stormo. Discovering common stem-loop motifs in unaligned RNA sequences. *Nucleic Acids Res.*, 2001, 29:2135–2144.

22. D. H. Mathews and D. H. Turner. Dynalign: an algorithm for finding the secondary structure common to two RNA sequences. *J. Mol. Biol.*, 2002, 317:191–203.

23. O. Perriquet, H. Touzet, and M. Dauchet. Finding the common structure shared by two homologous RNAs. *Bioinformatics*, 2003, 19:108–118.

24. Y. Ji, X. Xu, and G. D. Stormo. A graph theoretical approach for predicting common RNA secondary structure motifs including pseudoknots in unaligned sequences. *Bioinformatics*, 2004, 20:1591–1602.

25. C. Notredame, E. A. O'Brien, and D. G. Higgins. RAGA: RNA sequence alignment by genetic algorithm. *Nucleic Acids Res.*, 1997, 25:4570–4580.

26. J. Kim, J. R. Cole, and S. Pramanik. Alignment of possible secondary structures in multiple RNA sequences using simulated annealing. *Comput. Appl. Biosci.*, 1996, 12:259–267.

27. J. H. Chen, S. Y. Le, and J. V. Maizel. Prediction of common secondary structures of RNAs: a genetic algorithm approach. *Nucleic Acids Res.*, 2000, 28:991–999.

28. B. A. Shapiro and K. Zhang. Comparing multiple RNA secondary structures using tree comparisons. *Comput. Appl. Biosci.*, 1990, 6:309–318.

29. G. H. Lin, B. Ma, and K. Zhang. Edit distance between two RNA structures. *Res. Comput. Mol. Biol.* (RECOMB'01), 2001, pp. 211–220.

30. M. Hochsmann, T. Toller, R. Giegerich, and S. Kurtz. Local similarity in RNA secondary structures. *Proc. Comput. Syst. Bioinf.* (CSB'03), 2003, pp. 159–168.

31. Y. Sakakibara, M. Brown, R. Hughey, I. S. Mian, K. Sjolander, R. C. Underwood, and D. Haussler. Stochastic context-free grammars for tRNA modeling. *Nucleic Acids Res.*, 1994, 22:5112–5120.

32. S. R. Eddy and R. Durbin. RNA sequence analysis using covariance models. *Nucleic Acids Res.*, 1994, 22:2079–2088.

33. T. Lowe and S. R. Eddy. A computational screen for methylation guide snoRNAs in yeast. *Science*, 1999, 283:1168–1171.

34. R. J. Klein and S. R. Eddy. RSEARCH: finding homologs of single structured RNA sequences. *BMC Bioinf.*, 2003, 4:44.

35. I. Holmes and G. M. Rubin. Pairwise RNA structure comparison with stochastic context-free grammars. *Pac. Symp. Biocomput.*, 2002, pp. 163–174.

36. A. Laferriere, D. Gautheret, and R. Cedergren. An RNA pattern matching program with enhanced performance and portability. *Comput. Appl. Biosci.*, 1994, 10:211–212.

37. T. J. Macke, D. J. Ecker, R. R. Gutell, D. Gautheret, D. A. Case, and R. Sampath. RNAMotif, an RNA secondary structure definition and search algorithm. *Nucleic Acids Res.*, 2001, 29:4724–4735.

38. J. A. Jaeger, D. H. Turner, and M. Zuker. Improved predictions of secondary structures for RNA. *Proc. Natl. Acad. Sci. USA*, 1989, 86:7706–7710.

39. M. Zuker. On finding all suboptimal foldings of an RNA molecule. *Science*, 1989, 244: 48–52.

40. H. Yang, F. Jossinet, N. Leontis, L. Chen, J. Westbrook, H. Berman, and E. Westhof. Tools for the automatic identification and classification of RNA base pairs. *Nucleic Acids Res.*, 2003, 31:3450–3460.

41. J. Liu, J. T. L. Wang, J. Hu, and B. Tian. A method for aligning RNA secondary structures and its application to RNA motif detection. *BMC Bioinf.*, 2005, 6:89.

13

INTEGRATIVE ANALYSIS OF YEAST PROTEIN TRANSLATION NETWORKS

Daniel D. Wu and Xiaohua Hu
College of Information Science and Technology, Drexel University, Philadelphia, Pennsylvania

Protein translation is a vital cellular process for any living organism. The maturation of high-throughput technologies and the success of genome projects make it possible to apply computational approaches to the study of biological systems. The availability of interaction databases in particular provides an opportunity for researchers to exploit the immense amount of data *in silico*, such as studying biological networks using network analysis. There has been an extensive effort using computational methods in deciphering transcriptional regulatory networks. However, research on translation regulatory networks has attracted little attention in the bioinformatics and computational biology community, probably due to the nature of available data and the bias of conventional wisdom. In this chapter we present a global network analysis of protein translation networks in yeast, a first step in attempting to facilitate the elucidation of the structures and properties of translation networks. We extract the translation proteome using the MIPS functional category and analyze it in the context of the full protein–protein interaction network. We further derive the individual translation networks from the full interaction network using the extracted proteome. We show that in contrast to the full network, the protein translation networks do not exhibit power-law degree distributions. In addition, we demonstrate the close relationship between the translation networks and other cellular processes, especially transcription and metabolism. We also examine the essentiality and its correlation to connectivity of proteins in the translation networks, the cellular localization of these proteins, and the mapping of these proteins to the kinase-substrate system. These results have potential implications for understanding mechanisms of translational control from a system's perspective.

13.1 PROTEIN BIOSYNTHESIS AND TRANSLATION

The central dogma of molecular biology states that genetic information is transferred from DNA to mRNA through transcription and from mRNA to protein via translation. In every living organism, translation is a vital cellular process in which the information contained in the mRNA sequence is translated into the corresponding protein by complex translation machinery.

There are three major steps in protein biosynthesis: initiation, elongation, and termination. *Initiation* is a series of biochemical reactions leading to the binding of ribosome on the mRNA and the formation of an initiation complex around the start codon. The process involves various regulatory proteins (called *initiation factors*). Eukaryotic protein synthesis exploits various mechanisms to initiate translation, including cap-dependent initiation, reinitiation, and internal initiation. For the majority of mRNAs in the cell, their translation is via the cap-dependent pathway. Although debatable, it is widely believed that some cellular mRNAs contain internal ribosome entry sites (IRESs) and there exists a cap-independent IRES-mediated translation [1]. During *elongation*, codon-specific tRNAs are recruited by the ribosome to grow the polypeptide chain one amino acid at a time while the ribosome moves along the mRNA template (one codon at a time). This process also involves various elongation factors and proceeds in a cyclic manner. In *termination*, the termination codon is recognized by the ribosome. The newly synthesized peptide chain and eventually the ribosomes themselves are released [2].

Recent years have witnessed the breakthrough of high-throughput technologies that have been used in monitoring the various components of the transcription and translation machinery. DNA microarrays enable estimation of the copy number for every mRNA species within a single cell and changes in gene expression temporally or under different physiological conditions [3]. Two-dimensional gel electrophoresis coupled with tandem mass spectrometry makes it possible to measure specific protein levels for thousands of proteins in the cell simultaneously. These high-throughput technologies and the success of several genome projects are rapidly generating an enormous number of data regarding the genes and proteins that govern such cellular processes as transcription and translation. Analyzing these data is providing new insights into the regulatory mechanisms in many cellular systems. One of the major goals in the postgenomic era is to elucidate in a holistic manner the mechanisms by which subcellular processes at the molecular level are manifest at the phenotypic level under physiological and pathological conditions.

The complexity and large size of the transcription and translation machinery make computational approaches attractive and necessary in facilitating our understanding of the design principles and functional properties of these cellular systems. Transcriptional regulation, used by cells to control gene expression, has been a focus in a variety of computational attempts to infer the structure of genetic regulatory networks or to study their high-level properties [4]. However, research on translational regulatory networks has received little attention in the bioinformatics and computational biology community, being either underestimated or neglected. This contrast may be partly due to two factors. First, transcriptional control, other than translational control, has long been regarded by conventional wisdom as the primary control point in gene expression. Second, the success of genome projects and the use of high-throughput technologies provide a tremendous number of data about transcriptional regulation that are readily available for computational analysis. On the contrary, data about translational control are still probably too specialized, so they are consumed primarily by biologists.

Proteins, rather than DNAs or mRNAs, are the executors of the genetic program. They provide the structural framework of a cell and perform a variety of cellular functions, such as serving as enzymes, hormones, growth factors, receptors, and signaling intermediates. Biological and phenotypic complexity eventually derives from changes in protein concentration and localization, posttranslational modifications, and protein–protein interactions. Expression levels of a protein depend not only on transcription rates but also on such control mechanisms as nuclear export and mRNA localization, transcript stability, translational regulation, and protein degradation. Results from biological research have demonstrated that translational regulation is one of the major mechanisms regulating gene expression in cell growth, apoptosis, and tumorigenesis [5]. Therefore, study of protein translation networks, especially from computational systems biology approaches, may provide new insights into our understanding of this important cellular process.

Recently, Mehra and colleagues [6] develop a genome-wide model for the translation machinery in *Escherichia coli* that provides mapping between changes in mRNA levels and changes in protein levels in response to environmental or genetic perturbations. They also propose a mathematical and computational framework [7] that can be applied to the analysis of the sensitivity of a translation network to perturbation in the rate constants and in the mRNA levels in the system. However, toward the goal of understanding how translation machinery functions from a system's perspective, it is imperative that we have a better understanding of the global properties of protein translation networks, especially integrated with functional perspectives. In this chapter we take the first step in pursuing such a goal. We use a graph-theoretic approach to study the protein translation networks in yeast by integrating the protein–protein interaction data, the functional annotations documented in MIPS and GO databases, and some of the recent research results on cellular localization and protein phosphorylation in regard to translation networks.

13.2 METHODS

13.2.1 Graph Notation

We intuitively model a protein translation network as an undirected graph, where vertices represent proteins and edges represent interactions between pairs of proteins. An undirected graph, $G = (V, E)$, is comprised of two sets, vertices V and edges E. An edge e is defined as a pair of vertices (u,v) denoting the direct connection between u and v. The graphs we use in this paper are undirected, unweighted, and simple— meaning that there are no self-loops or parallel edges.

13.2.2 Data Sets

The yeast protein–protein interactions data were downloaded from the General Repository for Interaction Datasets (GRID) [8]. We select GRID because it contains arguably the most comprehensive data. The GRID database includes all published large-scale interaction data sets as well as available curated interactions such as those deposited in BIND [9] and MIPS [10]. The yeast data set we downloaded has 4948 distinct proteins and 18,817 unique interactions. From this network we derive the protein translation networks, which contain all proteins with MIPS functional categories related to protein translation as described in Section 13.3. We also compiled yeast functional annotations and essentiality of proteins from MIPS and GO. Protein phosphorylation data were obtained from ref. 11 and protein localization data from ref. 12.

13.2.3 Analysis of Network Topology

We measure the following basic properties of a translation network: (1) the number of proteins, measured by the number of vertices; (2) the number of interactions, measured by the number of edges; and (3) the size of the largest (or giant) component, measured by the size of the largest connected subgraph. We also measure the following degree-related metrics: (1) the average degree ($\langle k \rangle$), defined as $\langle k \rangle = 2e/n$, where e is the total number of edges and n is the total number of vertices; and (2) the degree distribution, $P(k)$, which measures the frequency of a vertex having degree k. The diameter of a network, $\langle l \rangle$, is defined by the average distance between any two vertices. The distance between two vertices is defined by the number of edges along their shortest path.

For a vertex v we calculate its clustering coefficient by

$$C_v = \frac{E_v}{n(n-1)/2}$$

where n is the number of neighboring vertices of v and E_v is the number of edges among these n neighboring vertices. The average clustering coefficient is calculated by

$$\langle C \rangle = \frac{1}{n} \sum_v C_v$$

Assuming the same degree distribution, we also obtain an average clustering coefficient of an equivalent random network [13], as defined by

$$\langle C_{\text{rand}} \rangle = \frac{(\langle k^2 \rangle - \langle k \rangle)^2}{n \langle k \rangle^3}$$

All statistical analyses are performed using the SPSS software package.

13.3 RESULTS

13.3.1 Global Properties of Protein Translation Networks Within the Full Yeast Interactome

We extract two sets of proteins that are involved in protein biosynthesis from the MIPS functional category database. Table 13.1 shows the functional categories used. The first set of proteins, we name it N1247, belongs to the following categories: 12.04 (translation), 12.04.01 (translation initiation), 12.04.02 (translation elongation), 12.04.03 (translation termination), and 12.07 (translational control). There are a total of 136 unique proteins in this set. The second set, referred to as N12, contains 479 unique proteins in all categories listed in Table 13.1. Therefore, the first set of proteins is actually a subset of the second.

We first study the protein translation networks by using these two sets of proteins in the context of the full yeast interaction network, constructed from yeast protein–protein interaction data from GRID. The basic properties of the full yeast interaction network and the protein translation subnetworks are shown in Table 13.2. One interesting observation of the protein translation networks is the existence of proteins that do not have any interacting partners in the full network. We call them *loner proteins*. This reflects the low coverage of the current interaction database rather than the actual lack of interactions.

Two fundamental network metrics, vertex degree and clustering coefficient, are employed to evaluate the global network characteristics. The vertex degree describes the number of interacting partners for each vertex in the network, whereas the clustering coefficient quantifies how well connected are the neighbors of a vertex in the network. These metrics provide useful insights into the architecture of the underlying network. For the first translation network (N1247), the core set of proteins extracted from MIPS contains 136 proteins, 13 of which do not have interacting partners in the full network. The remaining 123 proteins form a network containing 1100 unique interactions that involve additional 662 proteins. As shown in Table 13.2, the average degree of N1247 is significantly higher than the other larger translation

TABLE 13.1 MIPS Functional Categories Related to Protein Translation

Category	Description	Number of Proteins
12	Protein synthesis	22
12.01	Ribosome biogenesis	63
12.01.01	Ribosomal proteins	245
12.04	Translation	18
12.04.01	Translation initiation	40
12.04.02	Translation elongation	21
12.04.03	Translation termination	9
12.07	Translational control	55
12.10	Aminoacyl-tRNA synthetases	39

TABLE 13.2   Properties of the Full Yeast Interaction Network and the Protein Translation Networks

Network	N1247	N12	Full
Number of interacting proteins	785	1,471	4,948
Number of loner proteins	13	76	0
Number of unique interactions	1,100	2,715	18,817
Average degree $\langle k \rangle$	10.18	8.08	7.61
Average clustering coefficient $\langle C_k \rangle$	0.1096	0.1241	0.1118
C_{rand}	0.0555	0.0304	0.0243

network and the full network. For the second translation network (N12), the number of extracted proteins from MIPS is 479. Again, 76 of them are not shown interacting partners in the data we used. There are additional 1068 proteins interacting with the remaining 403 proteins through 2715 distinct interactions. All three networks show a larger average clustering coefficient than the corresponding C_{rand}, indicating a non-random structure of the underlying networks.

Degree Distributions   The degree distribution is a function describing the probability of a vertex having a specified degree. It is used regularly to classify networks, such as random networks (Poisson distribution) and scale-free networks (power-law distribution). Formally, the degree distribution is defined as [14]

$$P(k) = \sum_{v \in V | d(v) = k} 1$$

where v is a vertex in the vertex set V and $d(v)$ is the degree of vertex v. As reported earlier [15] and shown in Figure 13.1, the degree distribution for the full interaction network displays an approximate power law. However, the two translation networks show only weak power law degree distributions. Regression analysis is performed and the power values with R^2 values are shown in Figure 13.1. We do not pursue other estimation methods (such as that of Goldstein et al. [16]) to get more accurate fitting values for the degree distributions simply because, first, the nature of data we have does not warrant the necessity of doing so, and second, it is not our goal here. However, when we use an alternative approach to evaluate the degree distributions [14], in which the probability $P(k)$ is for all k values greater than or equal to k, we find that the degree distributions for the two translation networks fit better to an exponential regression instead of a power law. It has been proved [14] that the cumulative distribution follows a power law if the original distribution does so, but with a different exponent that is one less than the original exponent. Therefore, considering the nature of the data we have (incomplete and contains noise), the result in Figure 13.2 indicates that the scale-free or non-scale-free topology of the protein translation networks remains an issue to be solved.

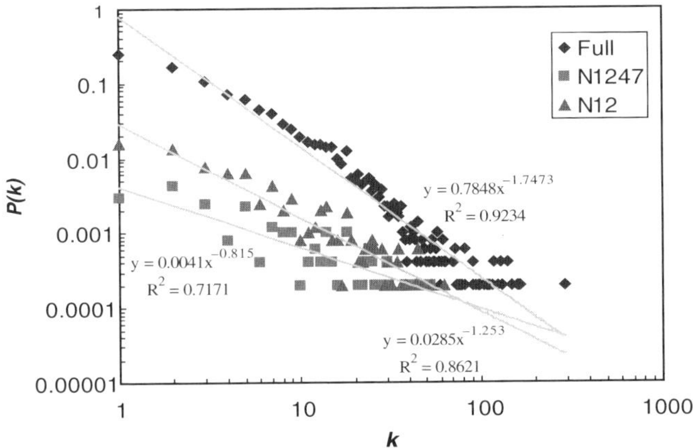

FIGURE 13.1 Degree distributions. The cyan lines show the power-law regression. (*See insert for color representation of figure.*)

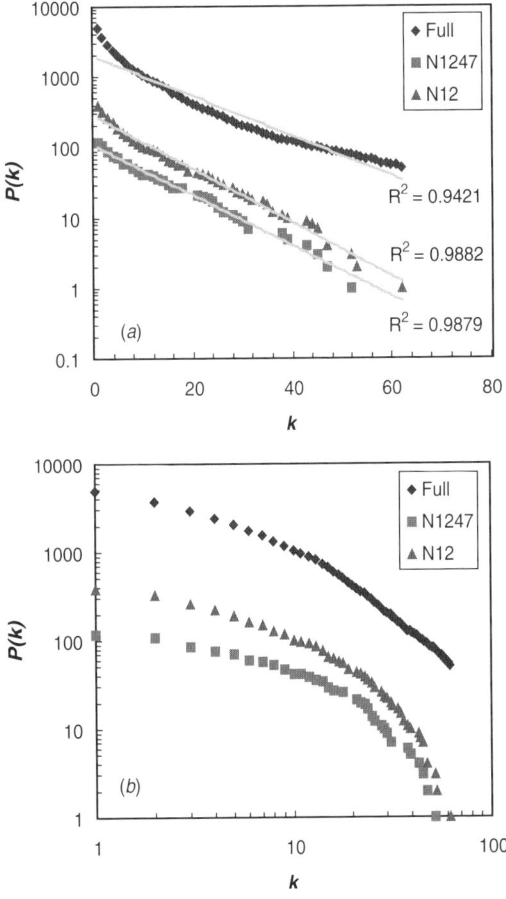

FIGURE 13.2 Cumulative degree distributions: (*a*) semilogarithmic plot with exponential regression; (*b*) log-log plot. (*See insert for color representation of figure.*)

13.3.2 Synthesis of Translation Networks

To further investigate and gain more insights into the organization of protein translation networks, we construct the following translation networks:

1. The first network (named N1247S hereafter, Figure 13.3) contains only the first set of 136 proteins described in Section 3.1. All edges are extracted from the full interaction network. This network only contains interactions between proteins that are both in N1247.

2. The second network (N1247SA, Figure 13.4) is extended from N1247S. It is in fact the isolated N1247 network mentioned in Section 13.3.1, which contains other proteins that interact with those in N1247. Each interaction in the network has at least one of the interacting partners in N1247.

3. The third network [N12S, Figure 13.5(*a*)] contains only proteins in N12. In addition to all of the proteins in N1247, network N12S also contains ribosomal proteins, proteins involved in ribosome synthesis, and proteins with aminoacyl-tRNA transferase activities.

4. The fourth network [N12SA, Figure 13.5(*b*)] is extended from N12S containing proteins either in N12 or interacting with proteins in N12.

The basic properties of networks are listed in Table 13.3. It is noted that all these synthetic translation networks has significantly lowered the average degree due to the existence of the loner proteins, which do not interact with others. Again, no interactions here do not necessarily mean no interactions in reality. Rather, the lack of interactions is due either to the lack of data in our data sets (false negatives) or the lack of qualified interactions (such as restraints posted on N1247S and N12S).

As demonstrated in Figure 13.3, the interactions between proteins within this group are surprisingly low. Most of the interactions exist in clusters corresponding to the three stages of protein translation. The highest connected cluster belongs to the group of proteins (in green) that are involved in translation initiation. This no doubt reflects the belief that translation initiation is one of the most important points in translational control and in active research, leading to high coverage in this area. Figure 13.3 also indicates the same logic for proteins involved in translation elongation (in yellow). Probably the most surprising finding is that there are very few direct interactions of translational control proteins (in blue) either between themselves or with proteins from other clusters (especially those in translation initiation, as one might expect). One possible factor behind this finding may be that the interactions between the control proteins and others are transient such that current technologies, such as high-throughput technologies, are unable to capture them (false negatives). Another factor may relate to the control that may be exerted through indirect (intermediates) interactions. Nonetheless, Figure 13.3 demonstrates the usefulness of network and cluster analyses in helping to delineate the translation process.

By extending the network N1247S to include all interacting partners, N1247SA becomes much more connected, as indicated by the increased size of the giant

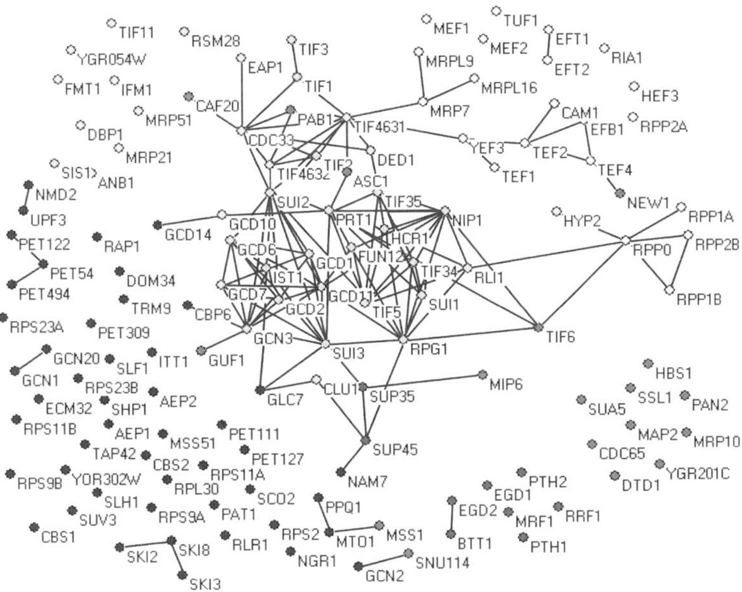

FIGURE 13.3 Synthetic protein translation network N1247S. All proteins are in MIPS functional categories of 12.04 (translation, orange), 12.04.01 (translation initiation, green), 12.04.02 (translation elongation, yellow), 12.04.03 (translation termination, red), and 12.07 (translational control, blue). (*See insert for color representation of figure.*)

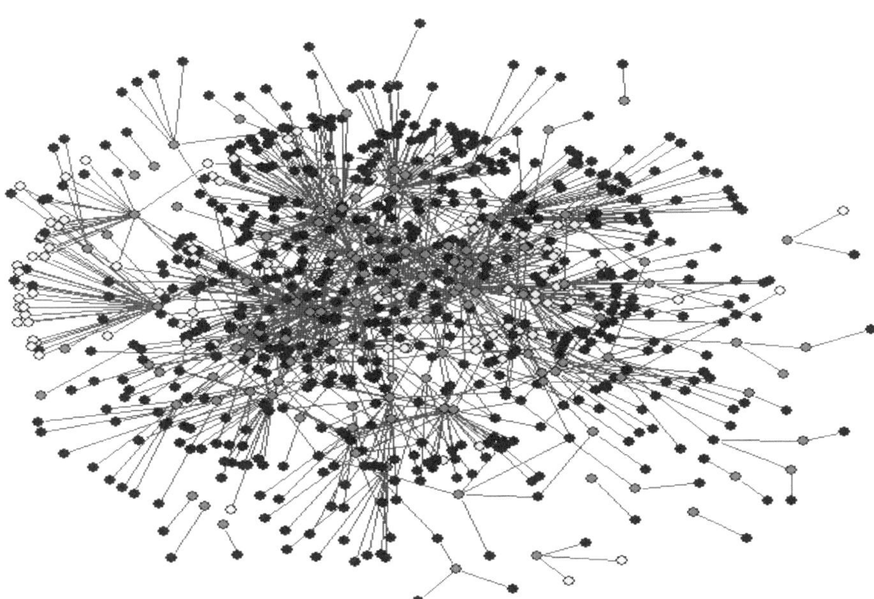

FIGURE 13.4 Synthetic protein translation network N1247SA. At least one of the interacting proteins is in N1247. Proteins in N1247 are indicated by red. Proteins in N12 but not in N1247 are indicated by cyan. All other proteins are shown in black. (*See insert for color representation of figure.*)

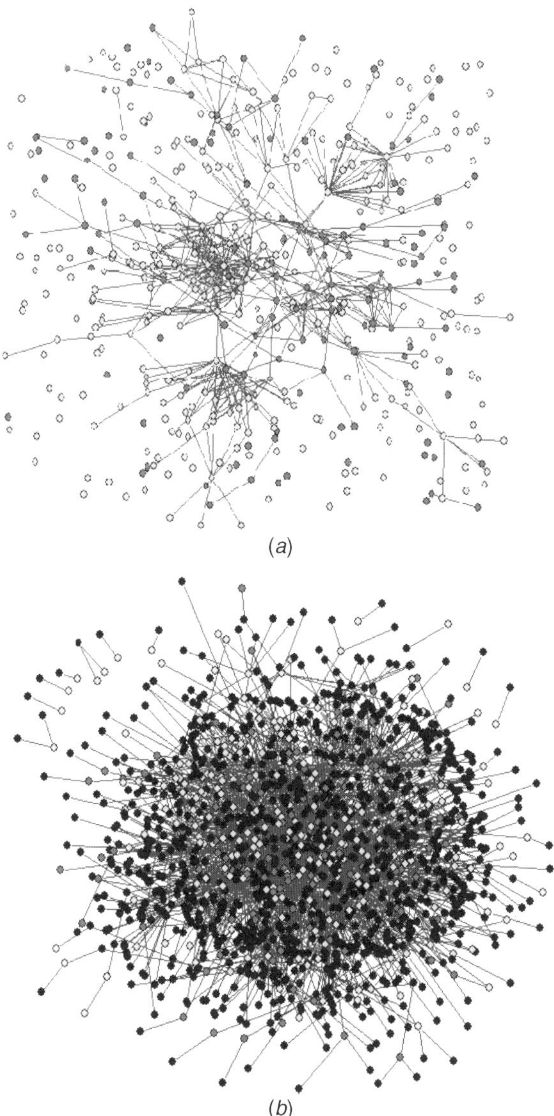

FIGURE 13.5   Synthetic protein translation networks: (*a*) network N12S represents proteins in N12; (*b*) network N12SA contains all proteins that are either in N12 or interacting with proteins in N12. For both networks, proteins in N1247 are in red, remaining N12 proteins are in cyan, all other proteins are in black. (*See insert for color representation of figure.*)

component (from 41% of total proteins to 89%) and the decreased number of loner proteins. There are 662 proteins outside these defined categories which interact with proteins in N1247. A subsequent and natural question to ask is: What are those proteins? The answer is quite intriguing. Table 13.4 lists the top 10 functional categories to which these 662 proteins belong. It should be noted that one protein

TABLE 13.3 Properties of Synthetic Protein Translation Networks

Network	N1247S	N1247SA	N12S	N12SA	Full
Number of proteins	136	798	479	1,547	4,948
Number of unique interactions	152	1,100	543	2,715	18,817
Number of proteins in giant component	56	714	218	1,394	4,857
Number of loner proteins	60	13	230	76	0
Average degree $\langle k \rangle$	2.24	2.76	2.27	3.51	7.61
Diameter $\langle l \rangle$	3.57	4.67	4.36	4.79	4.07

may be listed in multiple categories. We examine these functional categories at different levels (where available) in the function hierarchy used by MIPS. At the top level, more than 46% of these proteins are in the functional category "cell cycle and DNA processing," 42% in "transcription," and more than 37% in "metabolism." Subsequent levels further detail the distributions of these proteins in child categories of top-level parents. This result clearly demonstrates the close relationship between translation and other cellular processes, especially transcription and metabolism.

13.3.3 Essentiality of Proteins in Translation Networks

Network degree (or connectivity) has long been related to protein essentiality [17]. Therefore, we examine here the essentiality of proteins in the translation networks using a gene disruption data set downloaded from MIPS. As shown in Figure 13.6, about 28% of the proteins in the translation networks are lethal to disruption and 70% of them are nonessential. We also examine the essentiality of the loner proteins. As one can expect, the percentage of loner proteins that are essential decreases significantly. Only 15% of the loner proteins are essential. In three networks studied, the average degrees of essential proteins are significantly higher than those of non-essential proteins (Figure 13.7), demonstrating that more connected proteins (with higher degrees) are more likely to be essential.

13.3.4 Cellular Localization of Proteins in Translation Networks

As one may expect, most of the proteins in the translation networks are located in cytoplasm and mitochondria (Figure 13.8). However, since the translation machinery in cells is highly complex and translational control may involve many different mechanisms, we see a variety of distributions of proteins in such locales as nucleolus and nucleus.

13.3.5 Translation Networks and Protein Phosphorylation

Protein phosphorylation is a major regulatory mechanism that controls many basic cellular processes. A phosphorylation map for yeast has been generated by Ptacek and

TABLE 13.4 Functional Categories of Proteins Interacting with Translation Networks

Functional Category		Description	Percent of Proteins
Top level			
	10	Cell Cycle and DNA processing	46.2
	11	Transcription	42.0
	1	Metabolism	37.5
	16	Protein with binding function or cofactor requirement (structural or catalytic)	31.0
	20	Cellular transport, transport facilitation, and transport routes	26.7
	14	Protein fate (folding, modification, destination)	26.1
	42	Biogenesis of cellular components	20.7
	43	Cell type differentiation	11.6
	32	Cell rescue, defense, and virulence	11.3
	34	Interaction with the cellular environment	10.0
Second level			
	10.03	Cell cycle	24.2
	10.01	DNA processing	21.9
	11.04	RNA processing	20.1
	11.02	RNA synthesis	19.6
	20.09	Transport routes	16.0
	16.03	Nucleic acid binding	13.3
	43.01	Fungal/microorganismic cell type differentiation	11.6
	1.05	C-compound and carbohydrate metabolism	11.2
	14.07	Protein modification	10.9
	32.01	Stress response	10.3
Third level			
	11.02.03	mRNA synthesis	17.7
	10.03.01	Mitotic cell cycle and cell cycle control	14.5
	11.04.03	mRNA processing (splicing, 5′-, 3′-end processing)	11.8
	43.01.03	Fungal and other eukaryotic cell type differentiation	11.6
	10.01.05	DNA recombination and DNA repair	10.0
	16.01	Protein binding	9.8
	16.03.03	RNA binding	9.2
	34.11.03	Chemoperception and response	7.6
	01.05.01	C-compound and carbohydrate utilization	7.4
	01.04.01	Phosphate utilization	7.3
Fourth level			
	11.02.03.04	Transcriptional control	10.4
	16.01	Protein binding	9.8
	16.03.03	RNA binding	9.2
	43.01.03.05	Budding, cell polarity, and filament formation	9.2
	11.04.03.01	Splicing	7.6
	10.03.01	Mitotic cell cycle and cell cycle control	7.3
	01.04.01	Phosphate utilization	7.3
	99	Unclassified proteins	7.1
	11.04.01	rRNA processing	6.3
	10.01.05.01	DNA repair	6.2

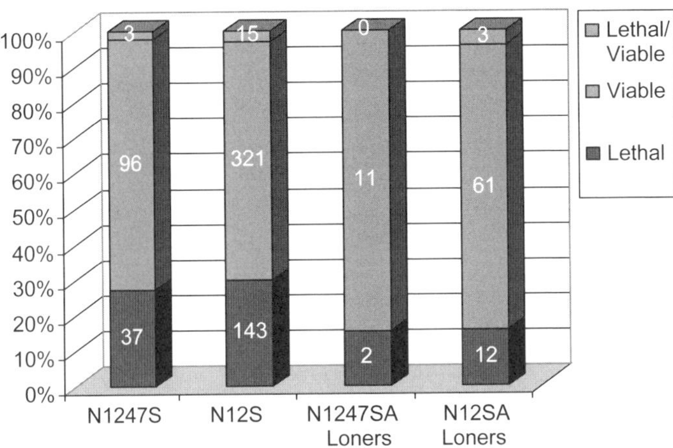

FIGURE 13.6 Essentiality of proteins in translation networks. (*See insert for color representation of figure.*)

colleagues [11]. By using their data, we map the proteins in translation networks to either kinases or substrates for kinases. About 22% of the proteins in translation networks are substrates identified for protein kinases. Even though neither N1247S nor N12S contains any of the 87 kinases testing the map we used, there are 12 proteins in N1247SA and 22 proteins in N12SA that are indeed protein kinases. In addition, 31 proteins in N1247S are substrates for 30 different kinases; 69 kinases can

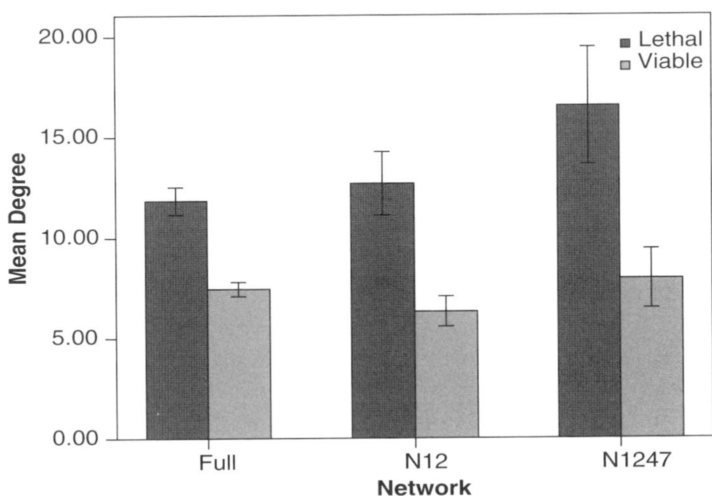

FIGURE 13.7 Essentiality of proteins in translation networks. Error bar at 95% confidence intervals. $p < 0.05$ between lethal and viable proteins in all networks (ANOVA test). (*See insert for color representation of figure.*)

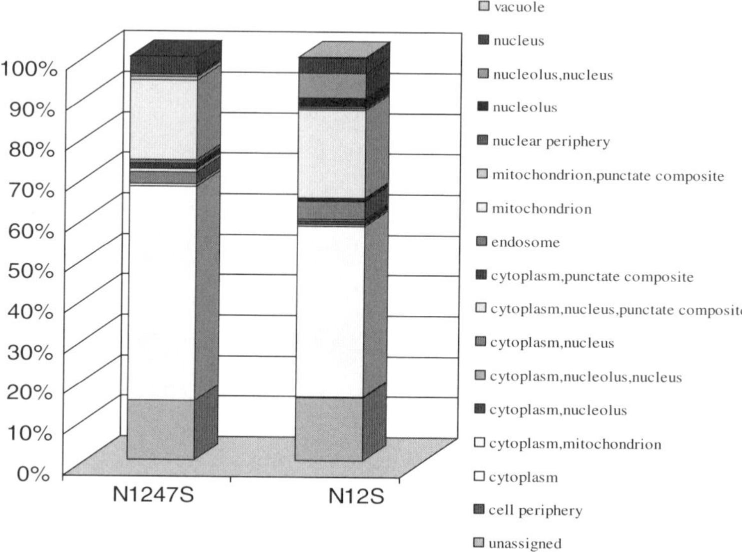

FIGURE 13.8   Cellular localization of proteins in a translation network. (*See insert for color representation of figure.*)

phosphorylate 190 proteins in N1247SA; 56 kinases can phosphorylate 105 proteins in N12S, and 361 proteins are substrates for 78 kinases (i.e., almost 90% of the 87 kinases in the yeast kinase-substrate map).

13.4   CONCLUSIONS

In this chapter we present a systematic global analysis of protein translation networks in yeast. As far as we know, this is the first report of this type of study. We first construct the full protein–protein interaction network and examine the translation proteome in the context of this full network. We define the translation proteome by using the MIPS functional category. The average degree of the protein interaction network containing the major translation-related proteins is significantly higher than an expanded translation network and the full network. Although the full network is scale-free, the degree distributions of translation networks do not display clear power-law behavior. The clustering coefficients of the translation networks indicate non-random or hierarchical structures of underlying networks. Reconstruction and analysis of the translation networks clearly demonstrate (1) the existence of clusters corresponding to different stages of the translation process; (2) the close relationship between translation machinery and other cellular processes, especially transcription and metabolism; and (3) the relationship between the translation networks and protein phosphorylation.

This work is the first step in our effort to elucidate the structure and properties of the protein translation networks. Such an effort may facilitate the computational dissection of translation networks and provide new insights into mechanisms of translational control from a system perspective.

Acknowledgments

This work was supported in part by National Science Foundation career grant IIS 0448023 and NSF CCF 0514679, Pennsylvania Department of Health Tobacco Settlement Formula grants 240205 and 240196, and Pennsylvania Department of Health grant 239667.

REFERENCES

1. W. C. Merrick (2004). Cap-dependent and cap-independent translation in eukaryotic systems. *Gene*, 332: 1–11.
2. V. M. Pain (1996). Initiation of protein synthesis in eukaryotic cells. *Eur. J. Biochem.*, 236: 747–767.
3. D. J. Lockhart and E. A. Winzeler (2000). Genomics, gene expression and DNA arrays. *Nature*, 405: 827–836.
4. H. de Jong (2002). Modeling and simulation of genetic regulatory systems: a literature review. *J. Comput. Biol.*, 9: 67–103.
5. E. C. Holland (2004). Regulation of translation and cancer. *Cell Cycle*, 3: 452–455.
6. A. Mehra, K. H. Lee, and V. Hatzimanikatis (2003). Insights into the relation between mRNA and protein expression patterns, I: Theoretical consideration. *Biotechnol. Bioeng.*, 84: 822–833.
7. A. Mehra and V. Hatzimanikatis (2006). An algorithmic framework for genome-wide modeling and analysis of translation networks. *Biophys. J.*, 90: 1136–1146.
8. B.-J. Breitkreutz, C. Stark, and M. Tyers (2003). The GRID: the General Repository for Interaction Datasets. *Genome Biol.*, 4: R23.
9. G. D. Bader, D. Betel, and C. W. Hogue (2003). BIND: the Biomolecular Interaction Network Database. *Nucleic Acids Res.*, 31(1): 248–250.
10. H. W. Mewes, D. Frishman, U. Guldener, G. Mannhaupt, K. Mayer, M. Mokrejs, B. Morgenstern, M. Munsterkotter, S. Rudd, and B. Weil (2002). MIPS: a database for genomes and protein sequences. *Nucleic Acids Res.*, 30: 31–34.
11. J. Ptacek et al. (2005). Global analysis of protein phosphorylation in yeast. *Nature*, 438: 679–684.
12. W. K. Huh et al. (2003). Global analysis of protein localization in budding yeast. *Nature*, 425: 686–691.
13. M. E. J. Newman (2003). Random graphs as models of networks, in S. Bornholdt and H. G. Schuster, Eds., *Handbook of Graphs and Networks: From the Genome to the Internet*. Wiley-VCH, Berlin, pp. 35–68.
14. M. E. J. Newman (2003). The structure and function of complex networks. *SIAM Rev.*, 45: 167–256.

15. A. L. Barabasi and R. Albert (1999). Emergence of scaling in random networks. *Science*, 286: 509–512.

16. M. L. Goldstein, S. A. Morris, and G. G. Yen (2004). Problems with fitting to the power-law distribution. *Eur. Phys. J. B*, 41: 255–258.

17. H. Jeong, S. P. Mason, A. L. Barabasi, and Z. N. Oltvai (2001). Lethality and centrality in protein networks. *Nature*, 411: 41–42.

14

IDENTIFICATION OF TRANSMEMBRANE PROTEINS USING VARIANTS OF THE SELF-ORGANIZING FEATURE MAP ALGORITHM

MARY QU YANG

National Human Genome Research Institute, National Institutes of Health, Bethesda, Maryland

JACK Y. YANG

Harvard Medical School, Harvard University, Boston, Massachusetts

CRAIG W. CODRINGTON

Department of Physics, Purdue University, West Lafayette, Indiana

Membrane proteins account for roughly one-third of all proteins and play a crucial role in processes such as cell-to-cell signaling, transport of ions across membranes, and energy metabolism [4,5,34] and are a prime target for therapeutic drugs [5,13,14,32]. One important subfamily of membrane proteins are the transmembrane proteins, of which there are two main types:

1. α-helical proteins, in which the membrane-spanning regions are made up of α-helixes
2. β-barrel proteins, in which the membrane-spanning regions are made up of β-strands

β-Barrel proteins are found mainly in the outer membrane of gram-negative bacteria, and possibly in eukaryotic organelles such as mitochondria, whereas α-helical proteins are found in eukaryotes and the inner membranes of bacteria [1].

Given the obvious biological and medical significance of transmembrane proteins, it is of tremendous practical importance to identify the location of transmembrane regions and the overall orientation of the protein in the membrane (i.e., whether or not the N-terminus is located on the cytoplasmic side). There are difficulties in obtaining the three-dimensional structure of membrane proteins using experimental techniques:

- Membrane proteins are difficult to crystallize, due to their amphipathic nature; because they have both a hydrophilic part and a hydrophobic part, they are not entirely soluble in either aqueous or organic solvents. As a result they are difficult to analyze using x-ray crystallography, which requires that the protein be crystallized.

- Membrane proteins tend to denature upon removal from the membrane, making their three-dimensional structure nearly impossible to analyze.

This has led to a surge of interest in applying techniques from machine learning and bioinformatics to infer secondary structure from primary structure in proteins. These include discriminant analysis [22], decision trees [33], neural networks [3,6,21,30], support vector machines [2,19,23,25,40], and hidden Markov models [35,37].

The emphasis in this chapter is on the prediction of α-helical transmembrane regions in proteins using variants of the self-organizing feature map algorithm [17]. As pointed out in [5], identifying α-helical transmembrane regions is easier than identifying β-strand transmembrane regions, because the former tend to have a high proportion of hydrophobic amino acid residues; nevertheless, it is by no means a solved problem.

To identify features that are useful for this task, we have conducted a detailed analysis of the physiochemical properties of transmembrane and intrinsically unstructured proteins; the results of this analysis are given in Section 14.1.

14.1 PHYSIOCHEMICAL ANALYSIS OF PROTEINS

We are interested in constructing a classifier to distinguish transmembrane segments from nontransmembrane segments. We therefore have a two-class classification problem, in which we take class 1 to be transmembrane and class 2 to be nontransmembrane. Due to the different side chains, each amino acid has different physiochemical properties. In this section we analyze several of these properties to determine which properties are most effective in discriminating transmembrane segments from nontransmembrane segments. Specifically, the properties that we analyze are (1) hydropathy, (2) polarity, (3) flexibility, (4) polarizability, (5) van der Waals volume, (6) bulkiness, and (7) electronic effects. Certain properties, such as hydropathy and polarity, can be measured in different ways; this results in different scales. We are also interested in determining which scale is most effective in discriminating transmembrane segments from nontransmembrane segments.

Two methods were used to assess the effectiveness of a given feature X in discriminating transmembrane segments from nontransmembrane segments:

1. The *degree of overlap* between the two distributions $p_X(x|\text{class}1)P\{\text{class}1\}$ and $p_X(x|\text{class}2)P\{\text{class}2\}$ is proportional to the Bayes error, the smallest probability of error attainable by any classifier. Thus, the smaller the overlap, the more easily the two classes can be discriminated. Here

class 1	transmembrane
class 2	nontransmembrane
$p_X(x|\text{class } i)$	probability density of feature X,
	restricted to class i instances
$P\{\text{class } i\}$	prior probability of class i instances

    We refer to $p_X(x|\text{class } i)P\{\text{class } i\}$ as a joint distribution because

    $$p_X(x|\text{class } i)P\{\text{class } i\} = p(x, \text{class } i)$$

2. The *overlap ratio*, a quantitative measure of how well the two classes can be discriminated, was calculated as follows: A graph is constructed such that:

    (a) The horizontal axis, which corresponds to the feature X, is divided into bins.

    (b) The y-value associated with the bin corresponding to X values between x_a and x_b is the fraction of all instances in the training set that belong to class 1 and have a value for feature X in the range $[x_a, x_b)$.

    The graph represents an approximation to the function $P\{\text{class}1|x\}$. We define the complementary function $P\{\text{class}2|x\}$ using

    $$P\{\text{class } 2|x\} = 1 - P\{\text{class } 1|x\}$$

    The overlap ratio is then defined as

    $$\text{overlap ratio} = \frac{\text{area under both } P\{\text{class } 1|x\} \text{ and } P\{\text{class } 2|x\}}{\text{area under} P\{\text{class } 1|x\} + \text{area under} P\{\text{class } 2|x\}}$$

    The smaller the overlap ratio, the more easily the two classes can be discriminated.

Given a sequence of amino acids, the "pointwise" feature value associated with a particular position in the sequence depends only on which of the 20 amino acids occurs at that position. To increase the robustness of our classifier, we work with average rather than pointwise feature values. The average of a given feature associated with a particular amino acid A in the sequence is the average of the pointwise feature values associated with the amino acids contained in a sliding window of width L centered at A. Note that in the preceding discussion, x is taken to be an average feature value over a window, as opposed to being a pointwise feature value, so the degree to which the two classes can be separated depends not only on the particular features chosen, but also on the window size L; it is therefore important to determine the optimal window size for each feature. In our analysis we consider all odd feature window sizes in the range 9 to 31.

14.1.1 Hydropathy

Hydropathy provides a measure of the relative hydrophobicity of amino acids. There are a number of hydropathy scales in common use, including those proposed by Kyte and Doolittle [18], Eisenberg [10], Engelman et al. [11], and Liu and Deber [20]; these are listed in Table 14.1 under the headings H_{KD}, H_{Ei}, H_{En}, and H_{LD}, respectively. On the basis of the overlap ratios listed in Tables 14.2 and 14.3, and the fact that the joint probability distributions plotted in Figure 14.1 are reasonably well separated, we conclude that whereas all four scales can be used to discriminate transmembrane segments for nontransmembrane segments in transmembrane proteins, the Liu–Deber scale is best suited to this task.

TABLE 14.1 Physiochemical Properties of Amino Acids^a

AA	H_{KD}	H_{Ei}	H_{En}	H_{LD}	P_G	P_Z	B	F	V	Z	E
I	4.5	1.38	3.1	4.41	5.2	0.13	21.4	0.46	124	4.3	0.08
V	4.2	1.08	2.6	3.02	5.9	0.13	21.57	0.39	105	2.3	0.02
L	3.8	1.06	2.8	4.76	4.9	0.13	21.4	0.37	124	4.2	0
F	2.8	1.19	3.7	5	5.2	0.35	19.8	0.31	135	8	−0.47
C	2.5	0.29	2	2.5	5.5	1.48	13.46	0.35	86	2.7	0.03
M	1.9	0.64	3.4	3.23	5.7	1.43	16.2	0.3	124	5.1	−0.39
A	1.8	0.62	1.6	0.16	8.1	0	11.5	0.36	67	1.1	0.09
G	−0.4	0.48	1	−3.31	9	0	3.4	0.54	48	0.03	0
T	−0.7	−0.05	1.2	−1.08	8.6	1.66	15.77	0.44	93	2.7	−0.5
S	−0.8	−0.18	0.6	−2.85	9.2	1.67	9.47	0.51	73	1.6	−0.45
W	−0.9	0.81	1.9	4.88	5.4	2.1	21.67	0.31	163	12	−0.21
Y	−1.3	0.26	−0.7	2	6.2	1.61	18.03	0.42	141	8.8	−0.49
P	−1.6	0.12	−0.2	−4.92	8	1.58	17.4	0.51	90	4.3	0.08
H	−3.2	−0.4	−3	−4.63	10.4	51.6	13.69	0.32	118	6.3	−0.43
E	−3.5	−0.74	−8.2	−1.5	12.3	49.9	13.57	0.5	109	4.1	0.07
N	−3.5	−0.78	−4.8	−3.79	11.6	3.38	12.82	0.46	96	3.7	−0.8
Q	−3.5	−0.85	−4.1	−2.76	10.5	3.53	14.45	0.49	114	4.8	−0.47
D	−3.5	−0.9	−9.2	−2.49	13	49.7	11.68	0.51	91	3	0
K	−3.9	−1.5	−8.8	−5	11.3	49.5	15.7	0.47	135	5.2	−0.65
R	−4.5	−2.53	−12.3	−2.77	10.5	52	14.28	0.53	148	8.5	−0.56

^{a}H_{KD}, H_{Ei}, H_{En}, and H_{LD} indicate the Kyte–Doolittle, Eisenberg–Schwarz–Komaromy–Wall, Engelman–Steitz–Goldman, and Liu–Deber hydropathy scales, respectively; P_G and P_Z indicate the Grantham and Zimmerman–Eliezer–Simha polarity scales, respectively; $B =$ bulkiness, $F =$ flexibility, $V =$ van der Waals volume, $Z =$ polarizability, $E =$ electronic effects, and AA $=$ amino acid.

TABLE 14.2 Overlap Ratios for Discriminating Transmembrane Segments from Nontransmembrane Segments in Membrane Proteins as a Function of Window Length (W.L.)[a]

W.L.	H_{KD}	H_{Ei}	H_{En}	H_{LD}	P_G	P_Z	B	F	E
31	0.249	0.221	0.26	0.198	0.249	0.211	0.423	0.294	0.504
29	0.232	0.197	0.241	0.183	0.233	0.223	0.397	0.278	0.499
27	0.231	0.203	0.213	0.194	0.232	0.232	0.412	0.266	0.462
25	0.238	0.198	0.227	0.178	0.215	0.269	0.393	0.269	0.411
23	0.217	0.204	0.219	0.177	0.208	0.233	0.385	0.258	0.434
21	0.209	0.204	0.215	0.166	0.216	0.197	0.37	0.252	0.379
19	0.214	0.222	0.22	0.199	0.224	0.235	0.415	0.259	0.389
17	0.201	0.252	0.218	0.199	0.219	0.206	0.393	0.259	0.442
15	0.191	0.195	0.201	0.214	0.224	0.193	0.356	0.283	0.456
13	0.216	0.203	0.217	0.178	0.203	0.189	0.325	0.283	0.50
11	0.21	0.199	0.228	0.185	0.204	0.168	0.346	0.277	0.493
9	0.231	0.205	0.222	0.2	0.232	0.28	0.396	0.299	0.562

[a]H_{KD}, H_{Ei}, H_{En}, H_{LD} indicate the Kyte–Doolittle, Eisenberg–Schwarz–Komaromy–Wall, Engelman–Steitz–Goldman, and Liu–Deber hydropathy scales, respectively, P_G and P_Z indicate the Grantham and Zimmerman–Eliezer–Simha polarity scales, respectively, B = bulkiness, F = flexibility, and E = electronic effects.

TABLE 14.3 Overlap Ratios for Discriminating Intrinsically Unstructured Segments from Intrinsically Structured Segments in Membrane Proteins as a Function of Window Length (W.L.)[a]

W.L.	H_{KD}	H_{Ei}	H_{En}	H_{LD}	P_G	P_Z	B	F
31	0.318	0.163	0.17	0.243	0.22	0.134	0.349	0.227
29	0.221	0.229	0.167	0.249	0.138	0.161	0.351	0.238
27	0.222	0.15	0.164	0.23	0.17	0.142	0.221	0.263
25	0.216	0.234	0.162	0.241	0.175	0.142	0.364	0.272
23	0.253	0.143	0.16	0.253	0.163	0.157	0.238	0.254
21	0.182	0.139	0.144	0.267	0.176	0.159	0.323	0.271
19	0.285	0.142	0.149	0.257	0.172	0.251	0.337	0.291
17	0.29	0.199	0.148	0.266	0.183	0.307	0.353	0.279
15	0.32	0.17	0.155	0.274	0.182	0.183	0.338	0.361
13	0.264	0.18	0.165	0.284	0.194	0.254	0.358	0.34
11	0.31	0.228	0.195	0.281	0.22	0.446	0.345	0.358
9	0.372	0.23	0.226	0.325	0.269	0.251	0.416	0.401

[a]H_{KD}, H_{Ei}, H_{En}, H_{LD} indicate the Kyte–Doolittle, Eisenberg–Schwarz–Komaromy–Wall, Engelman–Steitz–Goldman, and Liu–Deber hydropathy scales, respectively, P_G and P_Z indicate the Grantham and Zimmerman–Eliezer–Simha polarity scales, respectively; B = bulkiness, and F = flexibility.

14.1.2 Polarity

There are a number of polarity scales in common use, including those proposed by Grantham [12] and Zimmerman et al. [41]; these are listed in Table 14.2 under the

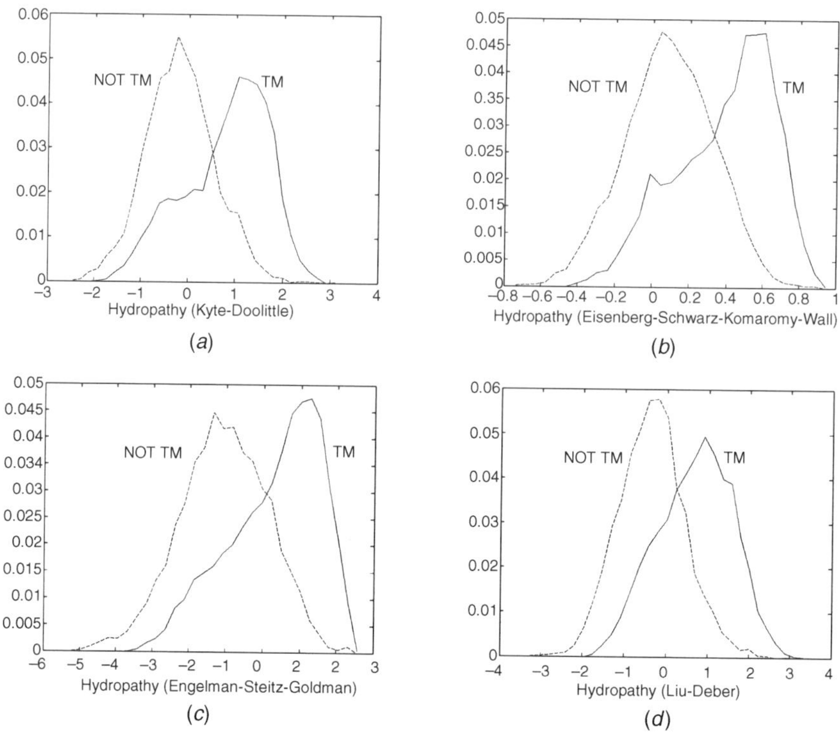

FIGURE 14.1 Joint probability distributions $p(x, \text{TM})$ and $p(x, \text{NOT TM})$, where x is hydropathy as measured by the (*a*) Kyte–Doolittle, (*b*) Eisenberg–Schwarz–Komaromy–Wall, (*c*) Engelman–Steitz–Goldman, and (*d*) Liu–Deber scales. TM, transmembrane segment; NOT TM, nontransmembrane segment. The overlap between the two distributions gives the Bayes error.

headings P_G and P_Z, respectively. On the basis of the overlap ratios listed in Tables 14.3 and 14.4 and the fact that the joint probability distributions plotted in Figure 14.2 are reasonably well separated, we conclude that although both scales can be used for discriminating transmembrane segments for nontransmembrane segments, the Grantham scale is slightly better for this task.

14.1.3 Flexibility

Proteins are dynamic molecules that are in constant motion. The structural flexibility that enables this motion has been associated with various biological process, such as molecular recognition and catalytic activity [7,36,39]; it follows that flexibility can help to predict protein function. On the basis of the overlap ratios listed in Tables 14.2 and 14.3 (under the heading "F") and the fact that the joint probability distributions in Figure 14.3 are reasonably well separated, we conclude that flexibility may be useful for discriminating transmembrane segments from nontransmembrane segments.

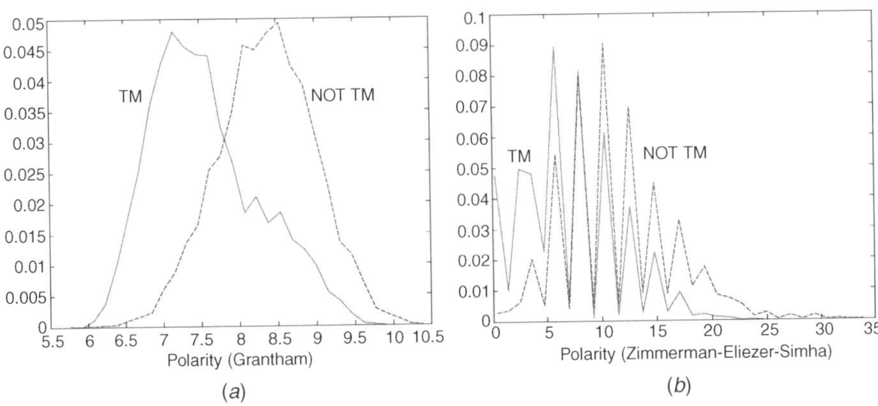

FIGURE 14.2   Joint probability distributions $p(x, \text{TM})$ and $p(x, \text{NOT TM})$, where x is polarity as measured by the (*a*) Grantham and (*b*) Zimmerman–Eliezer–Simha scales. TM, transmembrane segment, and NOT TM, nontransmembrane segment. The overlap between the two distributions gives the Bayes error.

14.1.4   Electronic Polarizability and van der Waals Volume

Due to the highly overlapping probability distributions in Figure 14.3 and the relatively large overlap ratios for electronic polarizability and van der Waals volume listed in Table 14.1 under the headings P and V, respectively, we conclude that neither of these properties can be used reliably to discriminate transmembrane from nontransmembrane segments.

14.1.5   Bulkiness

Bulkiness, a measurement of the volume occupied by an amino acid, is correlated with hydrophobicity [24]. On the basis of the relatively large overlap ratios listed in Tables 14.2 and 14.3 (under the heading B) and the relatively high degree of overlap between the joint probability distributions in Figure 14.3, we conclude that on its own, bulkiness cannot be used reliably to discriminate transmembrane segments from nontransmembrane segments. However, it may be useful in combination with other features.

14.1.6   Electronic Effects

Dwyer [8] proposed that electronic properties of amino acids may play a role in protein folding and stability. He analyzed electron distributions and polarizability using quantum mechanical calculations and derived a scale that takes into account electronic effects such as steric, inductive, resonance, and field effects [9]. On the basis of the relatively large overlap ratios listed in Table 14.3 (under the heading E) and the relatively high degree of overlap between the joint probability distributions in Figure 14.3, we conclude that on their own, electronic effects cannot be used reliably

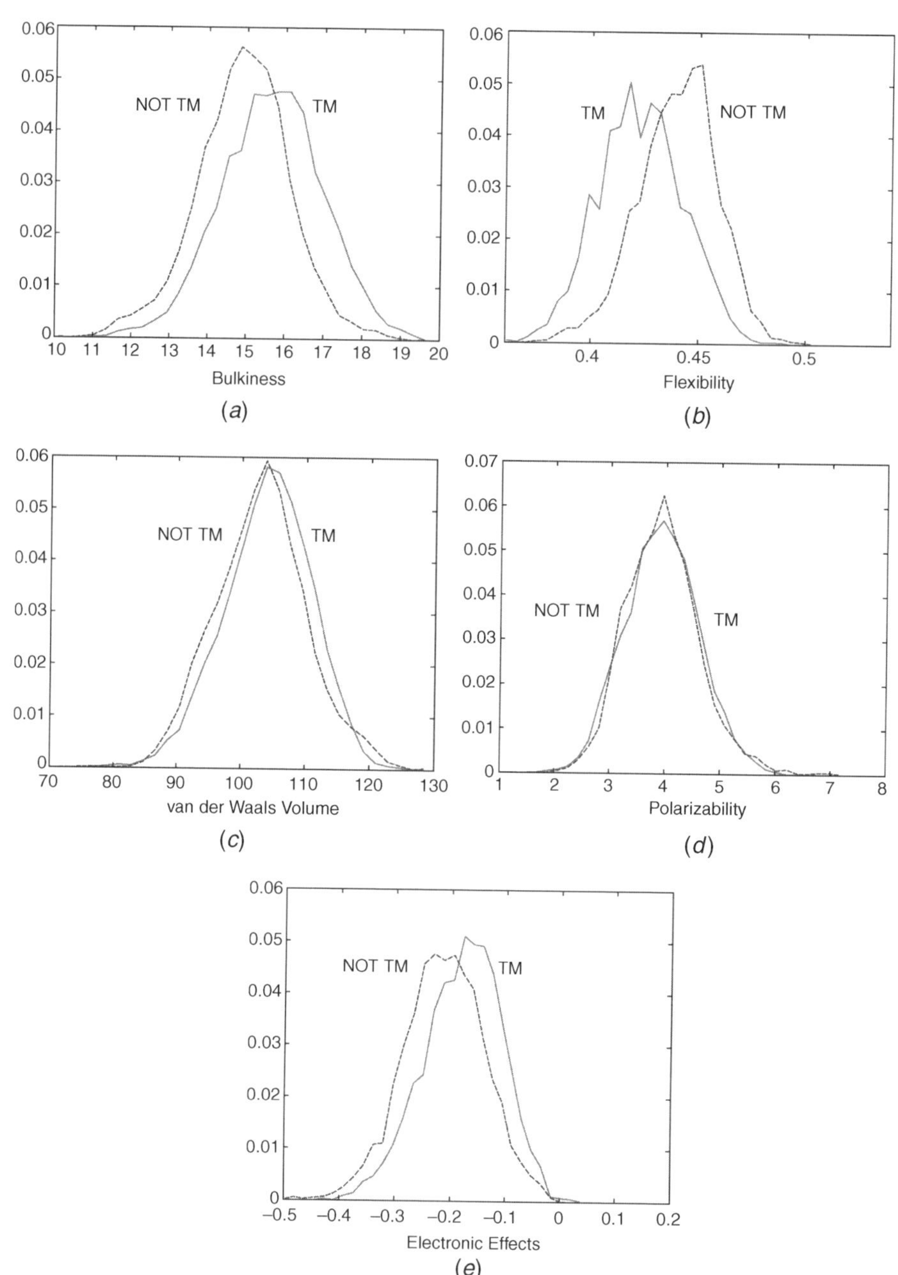

FIGURE 14.3 Joint probability distributions $p(x, \text{TM})$ and $p(x, \text{NOT TM})$, where x is (a) bulkiness, (b) flexibility, (c) van der Waals volume, (d) polarizability, and (e) electronic effects. TM, transmembrane segment; NOT TM, nontransmembrane segment. The overlap between the two distributions gives the Bayes error.

to discriminate transmembrane segments from nontransmembrane segments. However, they may be useful in combination with other features.

14.1.7 Summary

On the basis of the analysis above, we decided to base our transmembrane classifier on three features:

1. Hydropathy (Liu–Deber scale)
2. Polarity (Grantham scale)
3. Flexibility

Furthermore, we observed tendencies for certain properties in transmembrane and nontransmembrane segments (Table 14.1). Our results are in agreement with previous work that found that transmembrane segments tend to be more hydrophobic than nontransmembrane segments, due to the fact that transmembrane α-helices require a stretch of 12 to 35 hydrophobic amino acids to span the hydrophobic region inside the membrane [18].

14.2 VARIANTS OF THE SOM ALGORITHM

Our approach to discriminating transmembrane segments from nontransmembrane segments is based on the self-organized global ranking (SOGR) algorithm [31], which itself was inspired by Kohonen's self-organizing map (SOM) algorithm [17]. In the *SOM algorithm*, each neuron has associated with it a topological neighborhood, and the algorithm is such that neighboring neurons in the topological space tend to arrange themselves over time into a grid in feature space that mimics the neighborhood structure in the topological space. The *SOGR algorithm* differs from the SOM algorithm by dropping the topological neighborhood and replacing it with the concept of a global neighborhood generated by ranking. We consider several variants of the SOGR algorithm:

- The first variant modifies the way that neurons are initialized in the feature space; we call this variant SOGR-I.
- The SOGR algorithm updates the weights after each new instance is presented to the network. Because of this, the results may be affected by the order in which instances are presented to the network. The second variant, which we call SOGR-IB ("B" stands for "batch update"), removes the dependence on the order in which instances are presented only by updating the weights after each cycle, where a cycle involves presenting the entire training set to the network, one instance at a time. This variant also uses the modified initialization procedure described above.

Before we describe our modifications to the self-organized global ranking (SOGR) algorithm [31], we describe the SOGR algorithm itself.

14.2.1 The SOGR Algorithm

We assume that m neurons (or, in SOGR terminology, codebook vectors) are used. Let the initial position of neuron j at time $t = 0$ be $\vec{W}_j(0)$, and assume that the training set consists of instances $(\vec{x}_i, y_i)$, $i = 1, \ldots, n$.

1. *Initialization.* Choose initial positions $\vec{W}_j(0)$ in feature space for the m neurons by assigning the neurons random positions in feature space.

2. Present the instances in the training set to the network, one at a time. As each instance is presented to the network, the time index t is increased by 1. For each instance $(\vec{x}_i, y_i)$ in the training set, the positions of one or more neurons are adjusted as follows:

   (a) *Identifying winning neurons.* Find the R closest neurons to the feature vector $\vec{x}_i$; that is, find the R neurons with the smallest value of $\|\vec{x}_i - \vec{W}_j(t)\|$. These R neurons constitute the "neighborhood" of the input vector. Let Γ be the set of indices of the R winning neurons.

   (b) *Updating weights.* Adjust the positions of each of the R closest neurons using the update rule

   $$\vec{W}_j(t + 1) = \vec{W}_j(t) + \eta_t(\vec{x}_i - \vec{W}_j(t)) \qquad \text{for } j \in \Gamma$$

   where η_t is the learning rate. The learning rate is chosen to decrease with time to force convergence of the algorithm. In [31] it is suggested that the learning rate be decreased at an exponential rate and that it should be smaller for larger neighborhood sizes R.

3. *Assigning classes to neurons.* Associated with each neuron j is a count of the number of instances belonging to each class that are closer to neuron j than any other neuron. This count is calculated as follows:

   (a) For each neuron, initialize the counts to zero.

   (b) For each instance $(\vec{x}_i, y_i)$ in the training set, find the closest neuron to the feature vector $\vec{x}_i$; that is, find the neuron with the index j^*, where

   $$j* = \arg \min_j \|\vec{x}_i - \vec{W}_j(t)\|$$

   and increment the count in neuron $j*$ corresponding to class y_i by 1.

   (c) After all instances in the training set have been considered, each neuron is assigned to the class corresponding to the largest count for that neuron.

After the training process has been completed, a test instance can be classified by assigning it the class label of the nearest neuron.

14.2.2 Variants of the SOGR Algorithm

SOGR-I   The first variant we consider is SOGR-I; this variant modifies the initialization scheme of SOGR. Specifically, assume that the feature space is d dimensional, so that the feature vectors $\vec{x}_i$ belong to R^d. For each feature k, we find the largest and smallest value of that feature over the entire training set, which are, respectively, L_k and U_k:

$$L_k = \min_i x_{ik}$$

$$U_k = \max_i x_{ik}$$

where x_{ik} is the kth element of the feature vector $\vec{x}_i$. Then the initial positions of the m neurons are chosen as

$$W_{jk}(0) = L_k + \frac{j-1}{m-1}(U_k - L_k) \qquad \begin{array}{l} j = 1, \ldots, m \\ k = 1, \ldots, d \end{array}$$

Thus, the m neurons are evenly distributed along the line connecting $(L_1, L_2, \ldots, L_d)$ to $(U_1, U_2, \ldots, U_d)$.

This approach has several advantages over other initialization methods:

- It guarantees that the neurons will be in some sense evenly distributed throughout the feature space. Random initialization, on the other hand, does not guarantee this. If one has a large feature space, say of 60 dimensions, and comparatively few neurons, say 50, then with random initialization those neurons will with high probability not be evenly distributed throughout the feature space.

- Even a small number of neurons can be used to populate the feature space. If we consider an alternative initialization procedure in which one populates the feature space with a d-dimensional grid of neurons, and there are q grid points along each feature space axis, the total number of neurons required to populate this grid is q^d. For example, if $q = 3$ and the feature space has 60 dimensions, the number of neurons required is

$$q^d = 3^{60} \approx 4.239 \times 10^{28}$$

  which is clearly infeasible.

To illustrate how the SOGR-I algorithm works, we applied it to an artificial data set with two features. We generated two clusters of points, with each cluster corresponding to a class:

- Class 1 consisted of 53 training and 22 testing instances drawn from a bivariate normal distribution with mean vector $\vec{M}_1$ and covariance matrix C_1, where

$$\vec{M}_1 = \begin{bmatrix} 1 \\ 1 \end{bmatrix} \qquad \vec{C}_1 = \begin{bmatrix} 0.4 & 0 \\ 0 & 0.4 \end{bmatrix}$$

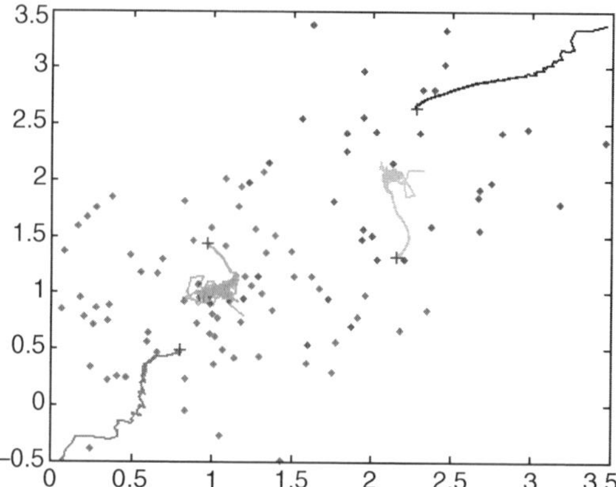

FIGURE 14.4   SOGR-I trajectories for a two-class problem with four neurons.

- Class 2 consisted of 53 training and 22 testing instances drawn from a bivariate normal distribution with mean vector $\vec{M}_2$ and covariance matrix C_2, where

$$\vec{M}_2 = \begin{bmatrix} 2 \\ 2 \end{bmatrix} \qquad \vec{C}_2 = \begin{bmatrix} 0.4 & 0 \\ 0 & 0.4 \end{bmatrix}$$

Figure 14.4 shows the trajectory for each neuron (as a function of the time index t) when the SOGR-I algorithm is applied to these data. Note how the trajectories start out from initial positions that are evenly distributed along the diagonal (which is how neurons are initialized in SOGR-I) and end up at final positions marked by a "$+$". The final arrangement of neurons appears reasonable in that neurons are positioned at locations in the feature space where the density of instances is high.

SOGR-IB   The second variant, SOGR-IB, addresses several problems with the original SOGR algorithm:

- Because the SOGR algorithm updates the weights after each new instance is presented to the network, the results may be affected by the order in which instances are presented to the network. This represents an additional source of variability that may increase the mean square prediction error.
- Because the SOGR algorithm updates the weights after each new instance is presented to the network, the trajectories of the neurons can oscillate wildly. This effect is independent of the order-dependent effect discussed above, for it occurs even when the order in which the training instances are cycled through is fixed from one iteration over the training data to the next. Figure 14.7(c) shows this "jitter" in the trajectories of the neurons for the SOGR algorithm for the case of 30 neurons, with a neighborhood size of 6 and an iteration limit of 1500.

- The SOGR algorithm specifies that the learning rate should be decreased during the course of training: for example, at an exponential rate. The problem is that if the learning rate is decreased too rapidly, the neurons may get stuck before they have reached their optimal positions.

SOGR-IB ("B" stands for "batch update") addresses these problems in two ways:

- It uses a *batch update strategy* for updating the positions of the neurons in feature space. This eliminates the dependence of the results on the order in which instances are presented to the network and stabilizes the trajectories of the neurons, which is expected to reduce the variability of the results.
- It uses a fixed, but small learning rate η_t, which eliminates the problem of the weights getting stuck because the learning rate η_t was decreased too quickly. We give a convergence proof below that demonstrates that the fixed learning rate strategy will converge to a neighborhood of the optimal solution, provided that a sufficiently small learning rate is used.

To distinguish the SOGR strategy of updating the weights after each instance from the batch update strategy discussed above, we shall refer to the former as a *stochastic gradient update strategy*. The SOGR-IB algorithm is described below.

1. *Initialization.* Choose initial positions $\vec{W}_j(0)$ in feature space for the m neurons using the SOGR-I initialization strategy. Set $t = 0$.
2. Repeat the following until the *energy*, defined by

$$Q(t) = \frac{1}{2nR} \sum_{\text{instances } i} \sum_{\text{neurons } j} m_{ij} \|\vec{x}_i - \vec{W}_j(t)\|^2 \qquad (14.1)$$

   does not reach a new minimum over a number of iterations through the training set, where n is the number of training instances and R is the number of neurons neighboring a given training instances that will be updated (m_{ij} will be defined later). After each pass through the training set, the time index t is incremented by 1.

   (a) Let $\vec{Z}_j$ be the accumulator corresponding to neuron j. Initialize $\vec{Z}_j$ to 0 for all neurons j.
   (b) Present the instances $(\vec{x}_i, y_i)$ in the training set to the network one at a time. After each instance is presented, the accumulators are updated as follows:

   - *Identifying winning neurons.* Find the R closest neurons to the feature vector $\vec{x}_i$; that is, find the R neurons with the smallest value of $\|\vec{x}_i - \vec{W}_j(t)\|$. These R neurons constitute the neighborhood of the input vector. Let Γ be the set of indices of the R winning neurons.

- *Updating accumulators.* Adjust the accumulators corresponding to each of the R closest neurons using the update rule

$$\vec{Z}_j = \vec{Z}_j + \frac{1}{nR}\eta_t(\vec{x}_i - \vec{W}_j(t)) \qquad \text{for neurons } j \in \Gamma$$

- where η_t is the learning rate.

(c) *Updating neurons.* After all instances in the training set have been presented to the network, update the neurons using the update rule

$$\vec{W}_j(t+1) = \vec{W}_j(t) + \vec{Z}_j \qquad \text{for all neurons } j$$

where n is the number of instances in the training set.

3. *Assigning classes to neurons.* Same as step 3 in the SOGR algorithm above.

There are a couple of points worth noting:

- Note the different role of the time index t in the SOGR and SOGR-I algorithms compared to the SOGR-IB algorithm. In the former, t is incremented after each training instance is presented, whereas in the latter, t is incremented after each pass through the training set.
- One iteration through steps (a), (b), and (c) above is equivalent to updating the weights using the formula

$$\vec{W}_j(t+1) = \vec{W}_j(t) + \frac{1}{nR}\sum_{i=1}^{n}\eta_t m_{ij}(\vec{x}_i - \vec{W}_j(t)) \qquad \text{for all neurons } j \qquad (14.2)$$

- where m_{ij} is 1 if neuron j is one of the R closest neurons to the feature vector $\vec{x}_i$, and is zero otherwise. In this formula, all neurons are updated simultaneously.

To analyze the SOGR-IB algorithm, it is helpful to think of each instance $(\vec{x}_i, y_i)$ as defining a quadratic *error surface* of the form

$$E_i(\vec{W}) = \tfrac{1}{2}\|\vec{x}_i - \vec{w}\|^2$$

Then for the SOGR algorithm, the change in weights after the instance $(\vec{x}_i, y_i)$ has been presented to the network can be thought of as a move "downhill" on the error surface $E_i(\vec{W})$. However, each instance has its own error surface, so the downhill direction varies from one instance to the next, which can lead to a very erratic trajectory for the neuron weights. Thus, the gradient is not constant at a given value of the weight $\vec{W}$, but rather, varies from instance to instance. This is why the SOGR and SOGR-I update rules are called stochastic gradient update rules.

By contrast, the batch update rule descends on an error surface that represents a weighted sum of per-instance error surfaces. This can be seen by noting that by (14.2), the batch update rule has the form

$$
\begin{aligned}
\vec{W}_j(t+1) &= \vec{W}_j(t) + \vec{Z}_j \\
&= \vec{W}_j(t) + \frac{1}{nR} \sum_{i=1}^{n} \eta_t m_{ij}(\vec{x}_i - \vec{W}_j(t)) \\
&= \vec{W}_j(t) + \eta_t \frac{1}{nR} \sum_{i=1}^{n} m_{ij}(\vec{x}_i - \vec{W}_j(t)) \\
&= \vec{W}_j(t) - \eta_t \frac{1}{nR} \sum_{i=1}^{n} m_{ij} \nabla_{\vec{W}_j} \left[\frac{1}{2} \|\vec{x}_i - \vec{W}_j(t)\|^2 \right]
\end{aligned}
$$

Here $\nabla_{\vec{W}_j}$ is the vector gradient operator, whose kth component is $\partial/\partial W_{jk}(t)$, where W_{jk} is the kth component of the vector $\vec{W}_j$. Thus,

$$
\begin{aligned}
\vec{W}_j(t+1) &= \vec{W}_j(t) - \eta_t \nabla_{\vec{W}_j} \frac{1}{nR} \sum_{i=1}^{n} m_{ij} \underbrace{\left[\frac{1}{2} \|\vec{x}_i - \vec{W}_j(t)\|^2 \right]}_{E_i(\vec{W}_j(t))} \\
&= \vec{W}_j(t) - \eta_t \nabla_{\vec{W}_j} \left[\frac{1}{nR} \sum_{i=1}^{n} m_{ij} E_i(\vec{W}_j(t)) \right] \qquad (14.3)
\end{aligned}
$$

Equation (14.3) provides the justification that the batch update rule amounts to gradient descent on a weighted sum of per-instance error surfaces.

We now show that SOGR-IB can be viewed as an stepwise procedure to minimize the objective function

$$
Q(t) = \frac{1}{2nR} \sum_{\text{instances } i} \sum_{\text{neurons } j} m_{ij} \|\vec{x}_i - \vec{W}_j(t)\|^2 \qquad (14.4)
$$

where n is the number of training instances, and for each instance i, $m_{ij} = 1$ for R neurons (indexed by j), and is zero for the other neurons.

- Given that the weights are fixed, $Q(t)$ is minimized by setting $m_{ij} = 1$ for the R neurons that are closest to instance i [i.e., those with the smallest value of $\|\vec{x}_i - \vec{W}_j(t)\|^2$] and setting $m_{ij} = 0$ for the other neurons.
- Given that the m_{ij} are fixed, we can reduce $Q(t)$ by shifting the weights of each neuron k in the direction of the negative gradient; that is,

$$
\vec{W}_k(t+1) = \vec{W}_k(t) - \eta_t \nabla_{\vec{W}_k} Q(t)
$$

- where η_t is the learning rate and the gradient of $Q(t)$ with respect to the weight vector $\vec{W}_k$ of neuron k is

$$\nabla_{\vec{W}_k} Q(t) = \frac{1}{nR} \sum_{\text{instances } i} m_{ik}(\vec{W}_k(t) - \vec{x}_i)$$

- resulting in the following update equation for neuron k:

$$\vec{W}_k(t+1) = \vec{W}_k(t) - \eta_t \frac{1}{nR} \sum_{\text{instances } i} m_{ik}[\vec{W}_k(t) - \vec{x}_i]$$

This is exactly the SOGR-IB update equation (14.2).

- Since for the Euclidean norm, the per-instance error surfaces $E_i(\vec{W}_j)$ are convex, it follows that the weighted sum of error surfaces over which the minimization is performed is also convex, and hence that each step (14.2) reduces the energy $Q(t)$ provided that the learning rate η_t is sufficiently small.
- Thus, each time the m_{ij} or the weights $\vec{W}_k(t)$ are adjusted using the SOGR-IB procedure, the value of the energy function $Q(t)$ is decreased. Since $Q(t) \geq 0$, $Q(t)$ must have a minimum, which implies that the SOGR-IB algorithm must converge.

The batch update rule has several advantages over the stochastic gradient update rule:

- Because the neuron weights are updated only after all instances in the training set have been presented to the network, the results are independent of the order in which instances are presented to the network.
- Because the batch update performs gradient descent on a weighted sum of error surfaces, the trajectories resulting from the batch update rule will be much smoother than those resulting from the stochastic update rule; this is borne out by the trajectories shown in Figures 14.5 to 14.8. Because the trajectories are smoother, it is expected that there will be less variability in the results obtained using the batch update rule than in those obtained using the stochastic update rule.
- The SOGR algorithm specifies that the learning rate should be decreased during the course of training: for example, at an exponential rate. The problem is that if the learning rate is decreased too rapidly, the neurons may get stuck before they have reached their optimal positions. The solution to this problem is to use a small but fixed learning rate. The problem with this approach is that when a fixed learning rate is used with the stochastic update rule, the trajectories can be very erratic if the learning rate is too large, as can be seen from Figures 14.5(c), 14.6(c), and 14.8(b). However, when the batch update rule is used in conjunction with a fixed learning rate, the trajectories are stable and smooth, as can be seen from Figures 14.5(d), 14.6(d), and 14.8(c).

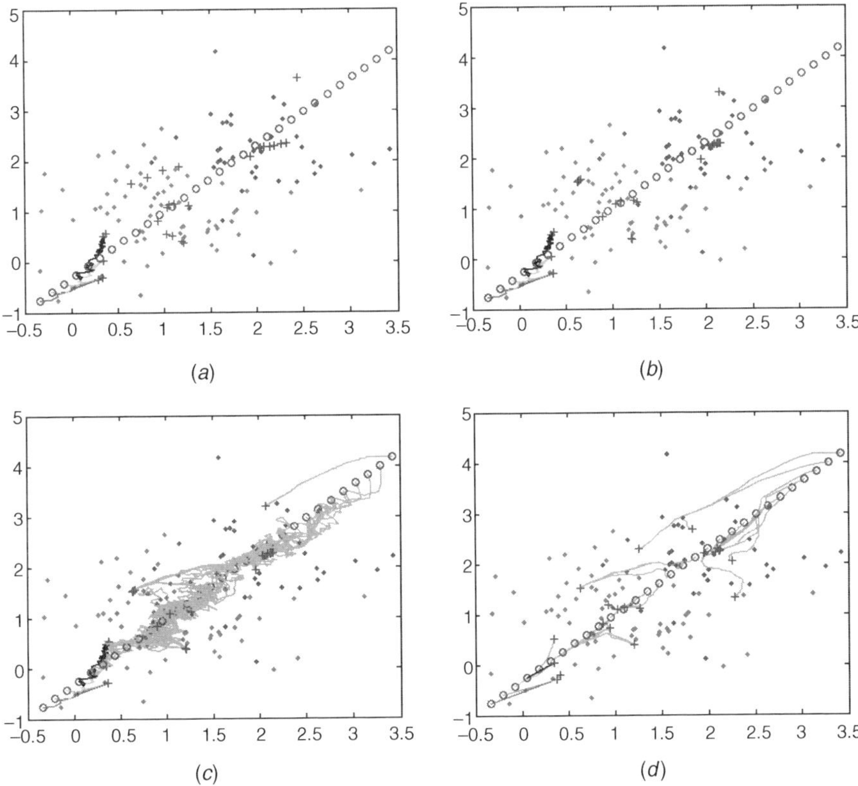

FIGURE 14.5 Neuron trajectories for artificial data for the case of 30 neurons, using $R = 10$ neighbors. Positions of neurons after (*a*) 100; (*b*) 1000; (*c*) 1500 iterations of the stochastic gradient update rule with learning rate $\eta_t = 0.1/(1 + t)$; (*d*) neuron trajectories after 1500 iterations of the batch update rule with constant learning rate $\eta_t = 0.2$.

14.2.3 Improving Classifier Performance by Filtering

If each residue in a contiguous sequence of residues is classified as either transmembrane (TM) or nontransmembrane (NOT TM), classification accuracy can often be improved by filtering the classifications themselves. This strategy exploits the observation that the class labels (TM or NOT TM) of nearby residues tend to be highly correlated. We have developed two filtering strategies, which we refer to as filters I and II; these are described below.

Filter I Filter I depends on two parameters, a filter window length M and a filter threshold θ, where $0.5 \leq \theta \leq 1$. The filter operates on a given amino acid residue A as follows:

- A window of length M is centered on A.

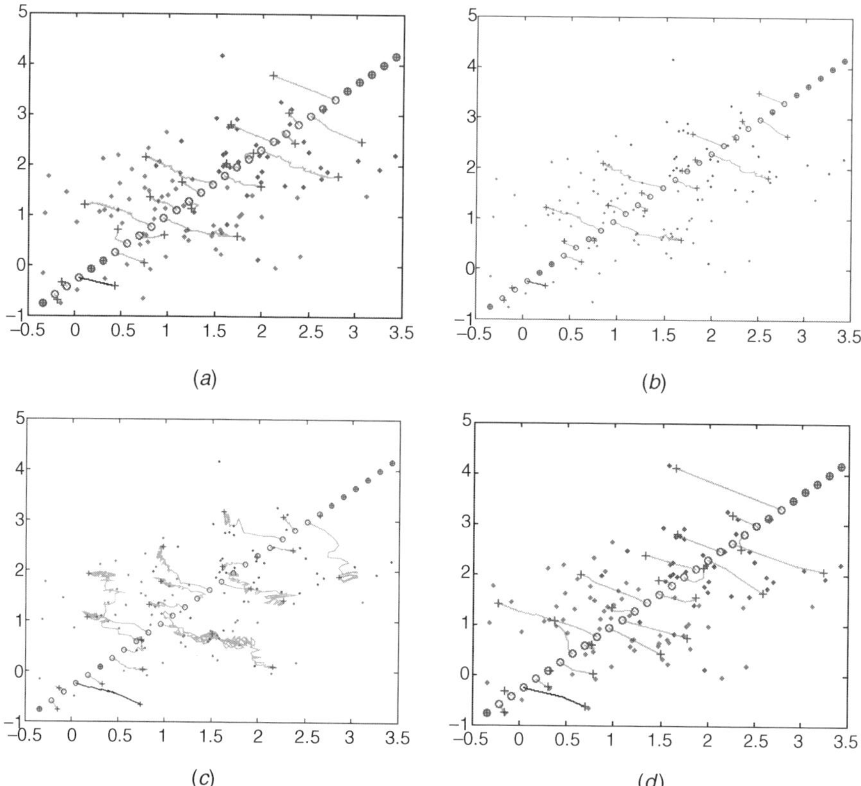

FIGURE 14.6 Trajectories and positions of neurons for the case of 30 neurons, using $R = 1$ neighbor, after 1500 iterations of (*a*) the stochastic gradient update rule with learning rate $\eta_t = 0.1/(1 + t)$; (*b*) the stochastic gradient update rule with exponentially decreasing learning rate $\eta_t = 0.1e^{-0.5t}$; (*c*) the stochastic gradient update rule with constant learning rate $\eta_t = 0.2$; (*d*) the batch update rule with constant learning rate $\eta_t = 0.2$.

- If the fraction of residues in the window having a class label L different from that of residue A exceeds θ, the class label of residue A is changed to L; otherwise, the class label of residue A is left unchanged.

In this filtering step, the updated class labels are not carried forward to subsequent filtering steps, so one can conceive of the filter as operating on all residues in parallel.

Filter II   In classification problems we are often interested in maximizing the true positive rate (also called the *sensitivity*), as this rate reflects the ability of the classifier to detect the "signal." For example, if we want to design a classifier to indicate whether or not a given person has cancer (in this case the signal is "having cancer"), the cost of saying that he does not have cancer when in fact he does (the false negative rate) is much higher than the cost of saying that the person has cancer when in fact he

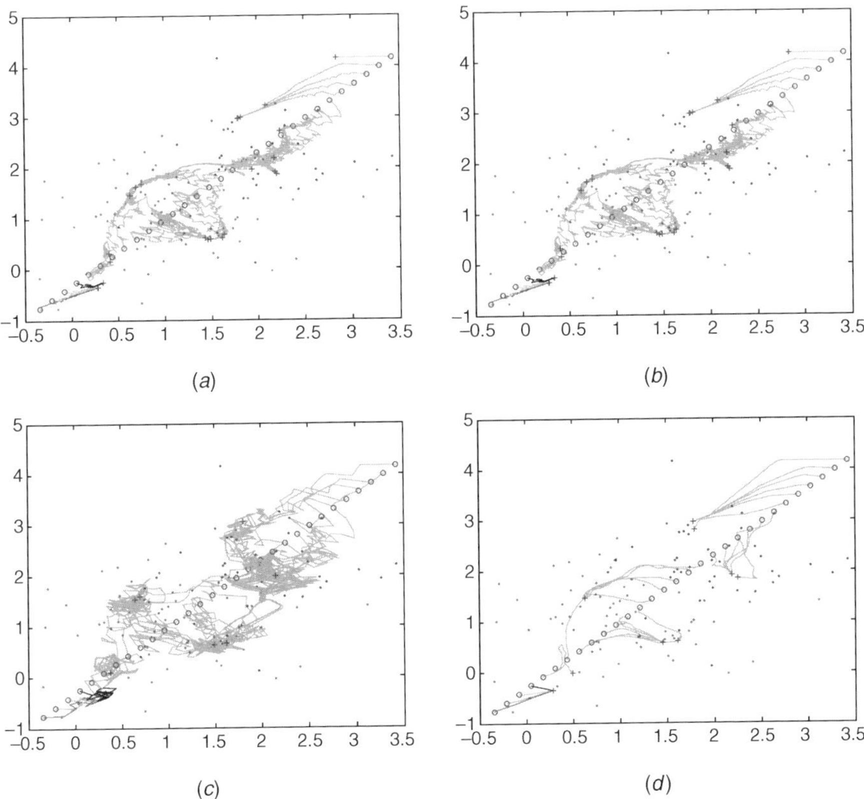

FIGURE 14.7 Trajectories and positions of neurons for the case of 30 neurons, using $R = 6$ neighbor, after 1500 iterations of (*a*) the stochastic gradient update rule with learning rate $\eta_t = 0.1/(1 + t)$; (*b*) the stochastic gradient update rule with exponentially decreasing learning rate $\eta_t = 0.1e^{-0.5t}$; (*c*) the stochastic gradient update rule with constant learning rate $\eta_t = 0.2$; (*d*) the batch update rule with constant learning rate $\eta_t = 0.2$.

does not (the false positive rate). Thus, it is more important to make the false negative rate small than it is to make the false positive rate small. Since

$$\text{true positive rate} = 1 - \text{false negative rate}$$
$$\text{true negative rate} = 1 - \text{false positive rate}$$

it is desirable in many applications to make the true positive rate (i.e., the sensitivity) large at the expense of the true negative rate (i.e., the specificity). In our case, a true positive corresponds to the case of correctly classifying a residue whose true class is transmembrane.

Filter II, designed based on the observation that transmembrane segments tend to be between 12 and 35 residues long, turns out to have the desirable property that it can

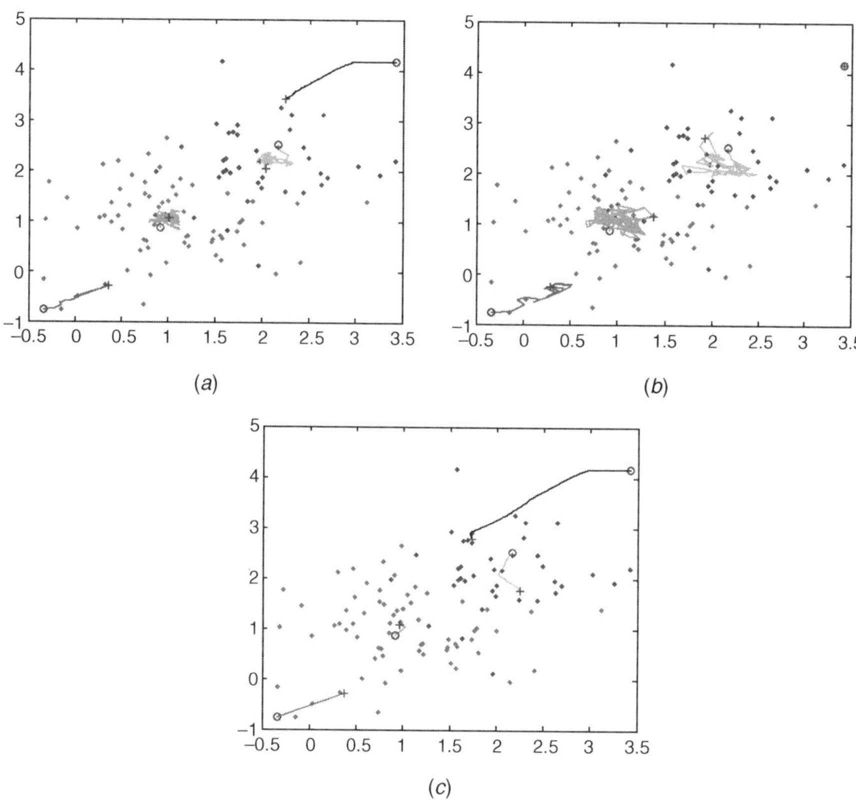

FIGURE 14.8   Neuron trajectories for artificial data for the case of four neurons using $R = 1$ neighbors. Positions of neurons and trajectories after 1500 interations of (*a*) the stochastic gradient update rule with decreasing learning rate; (*b*) the stochastic gradient update rule with a fixed learning rate of 0.2; (*c*) the batch update rule with a fixed learning rate of 0.2.

be used to enhance sensitivity at the expense of specificity. Further details may be found in [38].

14.3   RESULTS

Our data set consisted of 51 transmembrane proteins with multiple transmembrane segments obtained from Swiss-Prot, totaling 7638 transmembrane and 10368 non-transmembrane residues.

The SOGR-I and SOGR-IB classifiers require that several parameters be specified: the number of neurons, the neighborhood size R, and the length of the window over which features are extracted. In addition, filter I depends on two parameters, the filter threshold θ and the filter window length, while filter II depends on three parameters (see [38]).

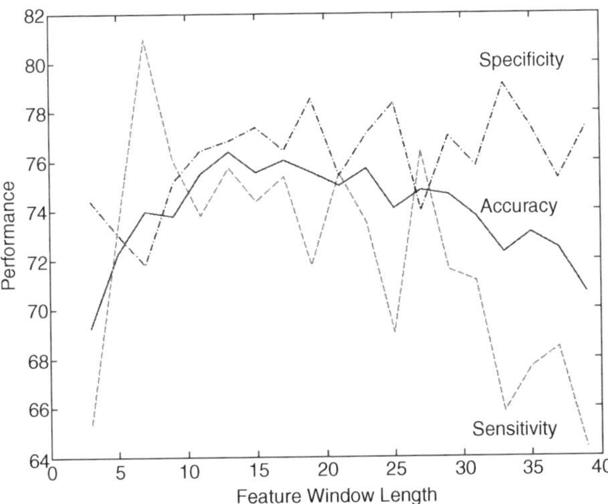

FIGURE 14.9 Performance of the SOGR-I classifier as a function of the length of the window used to extract features, based on threefold cross-validation (fixed learning rate $\eta_t = 0.05$, neighborhood size $R = 2$, number of neurons $= 16$, filter I threshold $\theta = 0.65$, filter I window length $= 11$).

The length of the window over which features are extracted is a significant factor in determining the performance of a classifier. Figure 14.9 shows the performance of the SOGR-I classifier as a function of the feature window length. These results suggest that the optimum window length for distinguishing between transmembrane and nontransmembrane segments is approximately 13. This is close to the lower limit of the length of tranmembrane segments, which tend to be between 12 and 35 residues long [18]. Note that the sensitivity is maximized for a feature window length of approximately 7. Based on a series of experiments, we settled on a feature window length of 11, a network size of 16 neurons, and a neighborhood size of $R = 2$.

Figure 14.10 shows the effect of the filter I threshold θ on performance. Whereas the accuracy, sensitivity, and specificity all peak at around $\theta = 0.5$, the specificity remains approximately constant for $\theta > 0.5$.

Figure 14.11 shows the effect of the filter I window length on performance. The accuracy, sensitivity, and specificity all reach a maximum near a filter window length of about 23. The accuracy and sensitivity decay fairly rapidly for filter window lengths greater than 25, while the specificity drops off more slowly.

The SOGR-I and SOGR-IB classifiers were benchmarked against C4.5 [27], a decision tree classifier, and SVM^{light} version 6.01 [16], a support vector machine classifier. Table 14.5 shows the results for the case of the three features identified by the physiochemical analysis given in Section 14.1—hydropathy (Lie–Deber scale), polarity (Grantham scale), and flexibility—while Table 14.6 shows the results for a subset of these features—polarity (Grantham scale) and flexibility. For the SOGR-I and SOGR-IB classifiers, results are shown both before and after using filter II. Tenfold cross-validation was used.

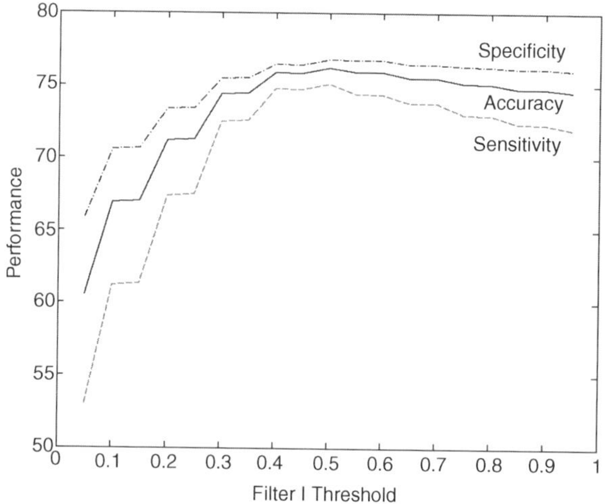

FIGURE 14.10   Performance of the SOGR-I classifier as a function of the filter I threshold θ, based on threefold cross-validation (fixed learning rate $\eta_t = 0.05$, neighborhood size $R = 2$, number of neurons = 16, feature window length = 11, filter I window length = 11).

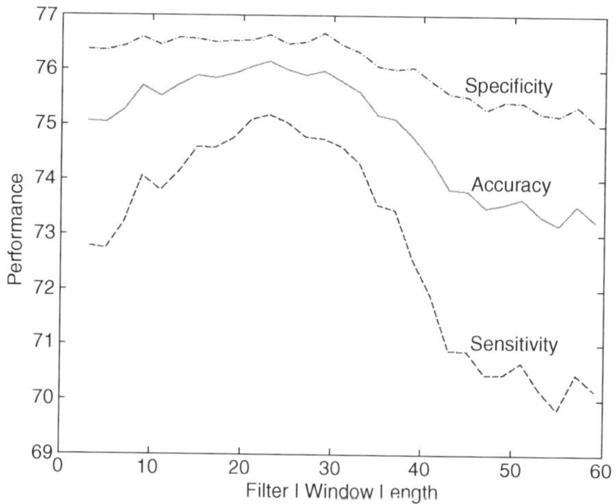

FIGURE 14.11   Performance of the SOGR-I classifier as a function of the filter I window length, based on threefold cross-validation (fixed learning rate $\eta_t = 0.05$, neighborhood size $R = 2$, number of neurons = 16, feature window length = 11, filter I threshold $\theta = 0.65$).

14.4   DISCUSSION AND CONCLUSIONS

As is evident from Tables 14.4 and 14.5, the four classifiers—SOGR-I, SOGR-IB, C4.5, and SVM^{light} version 6.01—performed comparably. Mean accuracy ranged from 74.3 to 76.4%. For comparison, accuracies of 95% have been reported in the literature [30], but it is difficult to compare results across different data sets.

TABLE 14.4 Accuracy Comparison of the SOGR-I and SOGR-IB Classifiers (in Combination with Filter II), a Decision Tree Classifier (C4.5), and a Support Vector Machine Classifier (SVM^{light} Version 6.01)^a

| | SOGR-I | | SOGR-IB | | C4.5 | | |
Fold	Before Filtering	After Filtering	Before Filtering	After Filtering	Before Pruning	After Pruning	SVM
1	72.2311	73.7149	72.2311	72.1251	72.4960	72.5490	72.9730
2	69.0476	67.2746	67.1733	65.9574	67.8318	67.6798	67.3252
3	77.1277	77.3404	76.9149	78.8298	77.5532	77.6596	77.7660
4	81.8913	85.4326	84.5875	87.7666	83.7827	83.7827	83.4608
5	79.3146	83.6499	78.4889	81.1313	78.3237	78.4476	78.1586
6	81.4600	84.5114	83.6230	86.6744	82.8119	83.1595	82.0780
7	75.9410	75.9410	76.8266	75.6458	75.6458	75.9410	76.3100
8	78.3488	79.7704	79.2783	79.5517	79.8797	79.9891	79.2783
9	64.1365	65.9471	65.0418	67.8273	64.5543	64.5543	64.7632
10	67.2325	67.4539	65.6089	67.2325	66.8635	67.0111	67.6753
Mean	74.7	76.1	75.0	76.3	75.0	75.1	75.0
Std. dev.	6.2	7.4	7.2	7.9	6.8	6.8	6.5

^{a}Based on tenfold cross-validation. Three features were used: hydropathy (Lie–Deber scale), polarity (Grantham scale), and flexibility.

The SOGR-IB classifier in combination with filter II performed best overall, but the margins were very slim. As expected, the SOGR-IB classifier performed slightly better than the SOGR-I classifier. In many cases filter II improved the predictions of the SOGR-I and SOGR-IB classifiers by 1 to 2% or more, but this was not true in all

TABLE 14.5 Accuracy Comparison of the SOGR-I and SOGR-IB Classifiers (in Combination with Filter II), a Decision Tree Classifier (C4.5), and a Support Vector Machine Classifier (SVM^{light} Version 6.01)^a

| | SOGR-I | | SOGR-IB | | C4.5 | | |
Fold	Before Filtering	After Filtering	Before Filtering	After Filtering	Before Pruning	After Pruning	SVM
1	71.7541	70.6942	72.0721	72.4430	72.3900	72.6020	72.6550
2	65.1469	64.4884	65.8561	64.9443	66.1601	66.1601	67.0719
3	77.1277	77.7660	78.4043	80.2128	76.3830	77.5532	77.4468
4	83.0986	86.5594	85.0302	85.1509	83.7827	83.7827	83.0181
5	77.2502	78.9017	77.6631	79.2320	76.4244	76.4244	79.1082
6	81.9235	84.1251	83.2368	85.8633	82.8505	82.8119	82.1166
7	75.5720	73.7269	76.6052	77.9336	75.7934	75.8672	75.9410
8	79.4423	82.5588	79.4423	82.5041	79.7704	79.4970	79.4423
9	64.1365	65.8078	64.3454	66.5042	64.2061	64.2061	64.4150
10	67.4539	67.1587	67.5277	69.0037	67.0849	67.0849	67.0849
Mean	74.3	75.2	75.0	76.4	74.5	74.6	74.8
Std. dev.	6.8	8.0	7.2	7.7	6.9	6.9	6.7

^{a}Based on ten-fold cross-validation. Two features were used polarity (Grantham scale) and flexibility.

TABLE 14.6 Intrinsically Unstructured Segments and Transmembrane Segments Tend to Have Opposite Physiochemical Properties

| | Segment Type | |
Property	Transmembrane	Intrinsically Unstructured
Hydropathy	High	Low
Polarity	Low	High
Bulkiness	High	Low
Flexibility	Low	High
Electronic effects	High	Low

cases. Filter II has the further advantage that it can be used to enhance sensitivity at the expense of specificity.

We close by drawing a connection to our previous work involving intrinsically unstructured proteins [38]. These are proteins that in contrast to most proteins, do not need to assume a particular three-dimensional configuration to carry out their function; instead of folding into specific three-dimensional structures, these proteins exist as dynamic ensembles in their native state [15,26,28,29]. Specifically, we observed the following:

- Intrinsically unstructured segments and transmembrane segments tend to have opposite properties, as summarized in Table 14.6.

- Transmembrane proteins appear to be much richer than other proteins in intrinsically unstructured segments; about 70% of transmembrane proteins contain intrinsically unstructured regions, compared to about 35% of other proteins.

- In approximately 70% of transmembrane proteins that contain intrinsically unstructured segments, the intrinsically unstructured segments are close to transmembrane segments.

These observations suggest a relationship between intrinsically unstructured segments and transmembrane segments; a deeper understanding of this relationship may provide insight into the structure and function of transmembrane proteins and may lead to improved protein secondary structure prediction.

REFERENCES

1. P. G. Bagos, T. D. Liakopoulos, and S. J. Hamodrakas. Evaluation of methods for predicting the topology of β-barrel outer membrane proteins and a consensus prediction method. *Bioinformatics*, 6(7), 2005.

2. Y.-D. Cai, G.-P. Zhou, and K.-C. Chou. Support vector machines for predicting membrane protein types by using functional domain composition. *Biophys. J.*, 84(5):3257–3263, 2003.

3. R. Casadio, P. Fariselli, C. Taroni, and M. Compiani. A predictor of transmembrane α-helix domains of proteins based on neural networks. *Eur. Biophys. J.*, 24(3):165–178, Feb. 1996.

4. R. Chapman, C. Sidrauski, and P. Walter. Intracellular signaling from the endoplasmic reticulum to the nucleus. *Annu. Rev. Cell Dev. Biol.*, 14:459–485, 1993.

5. C. P. Chen and B. Rost. State-of-the-art in membrane protein prediction. *Appl. Bioinf.*, 1(1):21–35, 2002.

6. G. W. Dombi and J. Lawrence. Analysis of protein transmembrane helical regions by a neural network. *Protein Sci.*, 3(4):557–566, 1994.

7. A. K. Dunker and Z. Obradovic. The protein trinity: linking function and disorder. *Nat. Biotechnol.*, 19(9):805–806, Sept. 2001.

8. D. S. Dwyer. Electronic properties of the amino acid side chains contribute to the structural preferences in protein folding. *J. Biomol. Struct. Dyn.*, 18(6):881–892, June 2001.

9. D. S. Dwyer. Electronic properties of amino acid side chains: quantum mechanics calculation of substituent effects. *BMC Chem. Biol.*, 5(2):1–11, Aug. 2005.

10. D. Eisenberg, E. Schwarz, M. Komaromy, and R. Wall. Analysis of membrane and surface protein sequences with the hydrophobic moment plot. *J. Mol. Biol.*, 179(1):125–142, 1984.

11. D. M. Engelman, T. A. Steitz, and A. Goldman. Identifying nonpolar transbilayer helices in amino acid sequences of membrane proteins. *Annu. Rev. Biophys. Biophys. Chem.*, 15:321–353, 1986.

12. R. Grantham. Amino acid difference formula to help explain protein evolution. *Science*, 185(4154):862–864, Sept. 1974.

13. T. Gudermann, B. Nurnberg, and G. Schultz. Receptors and g proteins as primary components of transmembrane signal transduction, 1: g-protein-coupled receptors: structure and function. *Mol. Med.*, 73:51–63, 1995.

14. C. Heusser and P. Jardieu. Therapeutic potential of anti-IgE antibodies. *Curr. Opin. Immunol.*, 9:805–813, 1997.

15. L. M. Iakoucheva, P. Radivojac, C. J. Brown, T. R. O'Connor, J. G. Sikes, Z. Obradovic, and A. K. Dunker. The importance of intrinsic disorder for protein phosphorylation. *Nucleic Acids Res.*, 32(3):1037–1049, Feb. 2004.

16. T. Joachims. Making large-scale SVM learning practical, in B. Schölkopf, C. Burges, and A. Smola, Ed., *Advances in Kernel Methods: Support Vector Learning*. MIT Press, Cambridge, MA, 1999.

17. T. Kohonen. Self-organizing formation of topologically correct feature maps. *Biol. Cybern.*, 43(1):59–69, 1982.

18. J. Kyte and R. Doolittle. A simple method for displaying the hydropathic character of a protein. *J. Mol. Biol.*, 157:105–132, 1982.

19. H. H. Lin, L. Y. Han, C. Z. Cai, Z. L. Ji, and Y. Z. Chen. Prediction of transporter family from protein sequence by support vector machine approach. *Proteins: Struct. Funct. Bioinf.*, 62(1):218–231, 2006.

20. L.-P. Liu and C. M. Deber. Guidelines for membrane protein engineering derived from de novo designed model peptides. *Biopolymers (Peptide Sci.)*, 5(47):41–62, 1998.

21. R. Lohmann, G. Schneider, D. Behrens, and P. Wrede. A neural network model for the prediction of membrane-spanning amino acid sequences. *Protein Sci.*, 3(9):1597–1601, 1994.

22. E. N. Moriyama and J. Kim. Protein family classification with discriminant function analysis, in J. P. Gustafson, R. Shoemaker, and J. W. Snape, Ed., *Genome Exploitation: Data Mining the Genome*. Springer-Verlag, New York, 2005.

23. Navjyot~K. Natt, Harpreet Kaur, and G. P. S. Raghava. Prediction of transmembrane regions of β-barrel proteins using ANN- and SVM-based methods. *Proteins: Struct. Funct. Bioinf.*, 56(1):11–18, May 7, 2004.

24. F. Ortolani, M. Raspanti, and M. Marchini. Correlations between amino acid hydrophobicity scales and stain exclusion capacity of type 1 collagen fibrils. *J. Electron Microsc.*, 43:32–38, 1994.

25. K.-J. Park, M. M. Gromiha, P. Horton, and M. Suwa. Discrimination of outer membrane proteins using support vector machines. *Bioinformatics*, 21(23):4223–4229, 2005.

26. K. Peng, S. Vucetic, P. Radivojac, C. J. Brown, A. K. Dunker, and Z. Obradovic. Optimizing long intrinsic disorder predictors with protein evolutionary information. *J. Bioinf. Computat. Biol.*, 3(1):35–60, Feb. 2005.

27. J. R. Quinlan. *C4.5: Programs for Machine Learning*. Morgan Kaufmann, San Francisco, CA, 1993.

28. P. Radivojac, Z. Obradovic, D. K. Smith, G. Zhu, S. Vucetic, C. J. Brown, J. D. Lawson, and A. K. Dunker. Protein flexibility and intrinsic disorder. *Protein Sci.*, 13(1):71–80, Jan. 2004.

29. P. Romero and A. K. Dunker. Intelligent data analysis for protein disorder prediction. *Artif. Intell. Rev.*, 14, 2000.

30. B. Rost, R. Casadio, P. Fariselli, and C. Sander. Transmembrane helices predicted at 95% accuracy. *Protein Sci.*, 4(3):521–533, 1995.

31. M. I. Saglam, O. Ersoy, and I. Erer. Self-organizing global ranking algorithm and its applications, in *Intelligent Engineering Systems Through Artificial Neural Networks*, Vol. 14, pp. 893–898, 2004.

32. H. U. Saragovi and K. Gehring. Development of pharmacological agents for targeting neurotrophins and their receptors. *Trends Pharmacol. Sci.*, 21:93–98, 2000.

33. S. Shimozono, A. Shinohara, T. Shinohara, S. Miyano, S. Kuhara, and S. Arikawa. Knowledge acquisition from amino acid sequences by machine learning system BONSAI. *Trans. Inf. Process. Soc. Jpn.*, 35:2009–2018, 1994.

34. B. J. Soltys and R. S. Gupta. Mitochondrial proteins at unexpected cellular locations: export of proteins from mitochondria from an evolutionary perspective. *Int. Rev. Cytol.*, 194:133–196, 2000.

35. E. L. L. Sonnhammer, G. von Heijne, and A. Krogh. A hidden Markov model for predicting transmembrane helices in protein sequences. *Proc. 6th International Conference on Intelligent Systems for Molecular Biology* (ISMB), pp. 175–182, AAAI Press, Menlo Park, CA, 1998.

36. J. A. Tainer, E. D. Getzoff, H. Alexander, R. A. Houghten, A. J. Olson, R. A. Lerner, and W. A. Hendrickson. The reactivity of anti-peptide antibodies is a function of the atomic mobility of sites in a protein. *Nature*, 312:127–134, 1984.

37. H. Viklund and A. Elofsson. Best α-helical transmembrane protein topology predictions are achieved using hidden Markov models and evolutionary information. *Protein Sci.*, 13(7):1908–1917, 2004.

38. M. Q.-X. Yang. Predicting protein structure and function using machine learning methods. Ph.D. dissertation, Purdue University, West Lafayette, IN, 2005.

39. Z. Yuan, J. Zhao, and Z. X. Wang. Flexibility analysis of enzyme active sites by crystallographic temperature factors. *Protein Eng.*, 16:109–114, 2003.

40. Z. Yuan, J. S. Mattick, and R. D. Teasdale. SVMtm: support vector machines to predict transmembrane segments. *J. Comput. Chem.*, 25(5):632–636, 2004.

41. J. M. Zimmerman, N. Eliezer, and R. Simha. The characterization of amino acid sequences in proteins by statistical methods. *J. Theor. Biol.*, 21(2):170–201, 1968.

15

triCluster: MINING COHERENT CLUSTERS IN THREE-DIMENSIONAL MICROARRAY DATA

LIZHUANG ZHAO AND MOHAMMED J. ZAKI

Department of Computer Science, Rensselaer Polytechnic Institute, Troy, New York

Traditional clustering algorithms work in the full-dimensional space; i.e., they consider the value of each point in all the dimensions, and try to group the similar points together. Biclustering [7], however, does not have such a strict requirement. If some points are similar in several dimensions (a subspace), they will be clustered together in that subspace. This is very useful, especially for clustering in a high-dimensional space where often only some dimensions are meaningful for a subset of points. Biclustering has proved of great value for finding interesting patterns in the microarray expression data [8], which record the expression levels of many genes (the rows/points) for different biological samples (the columns/dimensions). Biclustering is able to identify the coexpression patterns of a subset of genes that might be relevant to a subset of the samples of interest.

Besides biclustering along the gene–sample dimensions, there has been a lot of interest in mining gene expression patterns across time [4]. The proposed approaches are also mainly two-dimensional (i.e., finding patterns along the gone–time dimensions). In this chapter we are interested in mining triclusters [i.e., mining coherent clusters along the gene–sample–time (temporal) or gene–sample–region (spatial) dimensions].

There are several challenges in mining microarray data for bi- and triclusters. First, biclustering itself is known to be a NP-hard problem [7], and thus many proposed algorithms of mining bicusters use heuristic methods or probabilistic approximations, which as a trade-off decrease in the accuracy of the final clustering results. Extending these methods to triclustering will be even harder. Second, microarray data are inherently susceptible to noise, due to varying experimental conditions; thus, it is essential that the methods be robust to noise. Third, given that we do not understand the complex gene regulation circuitry in the cell, it is important that clustering methods allow overlapping clusters that share subsets of genes, samples, or time courses/spatial regions. Furthermore, the methods should be flexible enough to mine several (interesting) type of clusters and should not be too sensitive to input parameters.

In this chapter we present a novel, efficient, deterministic, triclustering method called *triCluster* that addresses the foregoing challenges. Following are the key features of our approach:

1. We mine only the maximal triclusters satisfying certain homogeneity criteria.

2. The clusters can be positioned arbitraily anywhere in the input data matrix, and they can have arbitrary overlapping regions.

3. We use a flexible definition of a cluster that can mine several types of triclusters, such as triclusters having identical or approximately identical values for all dimensions or a subset of the dimensions, and triclusters that exhibita scaling or shifting expression values (where one dimension is an approximately constant multiple of or is at an approximately constant offset from another dimension, respectively).

4. triCluster is a deterministic and complete algorithm which utilizes the inherent unbalanced property (number of genes being a lot more than the number of samples or time slices) in microarray data for efficient mining.

5. triCluster can optionally merge or delete triclusters that have large overlaps and can relax the similarity criteria automatically. It can thus tolerate some noise in the data set and lets users focus on the most important clusters.

6. We present a useful set of metrics to evaluate the clustering quality, and we show that triCluster can find substantially significant triclusters in the real microarray data sets.

15.1 PRELIMINARY CONCEPTS

Let $G = \{g_0, g_1, \ldots, g_{n-1}\}$ be a set of n genes, let $S = \{s_0, s_1, \ldots, s_{m-1}\}$ be a set of m biological samples (e.g., different tissues or experiments), and let $T = \{t_0, t_1, \ldots, t_{l-1}\}$ be a set of l experimental time points. A three-dimensional microarray data set is a real-valued $n \times m \times l$ matrix $D = G \times S \times T = \{d_{ijk}\}$ (with $i \in [0, n-1], j \in [0, m-1], k \in [0, l-1]$), whose three dimensions correspond to genes, samples, and times, respectively (note that the third dimension can also be a spatial region of interest but without loss of generality (w.l.o.g.), we as consider time as the third dimension). Each entry d_{ijk} records the (absolute or relative) expression level of gene g_i in sample s_j at time t_k. For example, Figure 15.1*a* shows a data set with

Time t_0

	s_0	s_1	s_2	s_3	s_4	s_5	s_6
g_0	3.6	1.0	1.0		1.0	1.0	1.0
g_1	3.0	2.5			2.0		1.0
g_2		5.0			5.0		5.0
g_3	6.6	5.5					2.0
g_4	9.0	7.5			6.0		3.0
g_5	6.6				4.4		2.0
g_6		3.0			3.0		3.0
g_7		8.0	8.0		8.0	8.0	
g_8	6.0	5.0			4.0		2.0
g_9		4.0	4.0		4.0	4.0	4.0

Time t_1

	s_0	s_1	s_2	s_3	s_4	s_5	s_6
g_0		0.5	0.5		0.5	0.5	0.5
g_1		3.0			2.4		1.2
g_2		2.5			2.5		2.5
g_3		5.5					2.0
g_4		9.0			7.2		3.6
g_5					4.4		2.0
g_6		1.5			1.5		1.5
g_7		4.0	4.0		4.0	4.0	
g_8		6.0			4.8		2.4
g_9		2.0	2.0		2.0	2.0	2.0

(a)

Time t_0

	s_3	s_0	s_6	s_4	s_1	s_5	s_2
g_5		6.6	2.0	4.4			
g_2			5.0	5.0	5.0		
g_6			3.0	3.0	3.0		
g_0		3.6	1.0	1.0	1.0	1.0	1.0
g_9			4.0	4.0	4.0	4.0	4.0
g_7				8.0	8.0	8.0	8.0
g_3		6.6	2.0		5.5		
g_4		9.0	3.0	6.0	7.5		
g_8		6.0	2.0	4.0	5.0		
g_1		3.0	1.0	2.0	2.5		

Time t_1

	s_3	s_0	s_6	s_4	s_1	s_5	s_2
g_5							
g_2			2.5	2.5	2.5		
g_6			1.5	1.5	1.5		
g_0			0.5	0.5	0.5	0.5	0.5
g_9			2.0	2.0	2.0	2.0	2.0
g_7				4.0	4.0	4.0	4.0
g_3							
g_4			3.6	7.2	9.0		
g_8			2.4	4.8	6.0		
g_1			1.2	2.4	3.0		

(b)

FIGURE 15.1 (*a*) Example of a microarray data set; (*b*) some clusters.

10 genes, seven samples, and two time points. For clarity, certain cells have been left blank; we assume that these are filled by some random expression values.

A *tricluster* C is a submatrix of the data set D, where $C = X \times Y \times Z = \{c_{ijk}\}$, with $X \subseteq G$, $Y \subseteq S$, and $Z \subseteq T$ provided that certain conditions of homogeneity are satisfied. For example, a simple condition might be that all values $\{c_{ijk}\}$ are identical or approximately equal. If we are interested in finding common gene coexpression patterns across different samples and times, we can find clusters that have similar values in the G dimension, but can have different values in the S and T dimensions. Other homogeneity conditions can also be defined, such as similar values in the S dimension, an order-preserving submatrix, and so on [16].

Let $C = X \times Y \times Z = \{c_{ijk}\}$ be a tricluster, and let $C_{2,2} = \begin{bmatrix} c_{ia} & c_{ib} \\ c_{ja} & c_{jb} \end{bmatrix}$ be any arbitrary 2×2 submatrix of C [i.e., $C_{2,2} \subseteq X \times Y$ (for some $z \in Z$) or $C_{2,2} \subseteq X \times Z$ (for some $y \in Y$) or $C_{2,2} \subseteq Y \times Z$ (for some $x \in X$)]. We all C a *scaling cluster* if we have $c_{ib} = \alpha_i c_{ia}$ and $c_{jb} = \alpha_j c_{ja}$, and further, $|\alpha_i - \alpha_j| \le \varepsilon$ [i.e., the expression values differ by an approximately (within ε) constant *multiplicative factor* α]. We call C a *shifting cluster* iff we have $c_{ib} = \beta_i + c_{ia}$ and $c_{jb} = \beta_j + c_{ja}$, and further, $|\beta_i - \beta_j| \leq \varepsilon$ [i.e., the expression values differ by a approximately (within ε) constant *additive factor β*].

We say that cluster $C = X \times Y \times Z$ is a subset of $C' = X' \times Y' \times Z'$ iff $X \subseteq X', Y \subseteq Y'$, and $Z \subseteq Z'$. Let $\mathcal{B}$ be the set of all triclusters that satisfy the given homogeneity conditions; then $C \in \mathcal{B}$ is called a *maximal tricluster* iff there does not exit another cluster $C' \in \mathcal{B}$ such that $C \subset C'$. Let $C = X \times Y \times Z$ be a tricluster, and let $C_{2,2} = \begin{bmatrix} c_{ia} & c_{ib} \\ c_{ja} & c_{jb} \end{bmatrix}$ be an arbitrary 2×2 submatrix of C [i.e., $C_{2,2} \subseteq X \times Y$ (for some $z \in Z$) or $C_{2,2} \subseteq X \times Z$ (for some $y \in Y$) or $C_{2,2} \subseteq Y \times Z$ (for some $x \in X$)]. We call C a *valid cluster* iff it is a maximal tricluster satisfying the following properties:

1. Let $r_i = |c_{ib}/c_{ia}|$ and $r_j = |c_{jb}/c_{ja}|$ be the ratio of two column values for a given row (i or j). We require that $\max(r_i r_j / \min(r_i, r_j) - 1 \leq \varepsilon$, where ε is a maximum ratio value.

2. If $c_{ia} \times c_{ib} < 0$, then $\mathrm{sign}(c_{ia}) = \mathrm{sign}(c_{ja})$ and $\mathrm{sign}(c_{ib}) = \mathrm{sign}(c_{jb})$, where $\mathrm{sign}(x)$ returns $-1/1$ if x is negative/nonnegative (expression values of zero are replaced with a small random positive correction value in the preprocessing step). This allows us easily to mine data sets having negative expression values. (It also prevents us from reporting that, for example, expression ratio $-5/5$ is equal to $5/-5$.)

3. We require that the cluster satisfy maximum range thresholds along each dimension. For any $c_{i_1 j_1 k_1} \in C$ and $c_{i_2 j_2 k_2} \in C$, let $\delta = |c_{i_1 j_1 k_1} - c_{i_2 j_2 k_2}|$. We require the following conditions: (a) If $j_1 = j_2$ and $k_1 = k_2$, then $\delta \leq \delta^x$, (b) if $i_1 = i_2$ and $k_1 = k_2$, then $\delta \leq \delta^y$, and (c) if $i_1 = i_2$ and $j_1 = j_2$, then $\delta \leq \delta^z$, where δ^x, δ^y, and δ^z represent the maximum range of expression values allowed along the gene, sample, and time dimensions.

4. We require that $|X| \geq mx, |Y| \geq my$, and $|Z| \geq mz$, where mx, my, and mz denote minimum cardinality thresholds for each dimension.

Lemma 15.1 (*Symmetry Property*) Let $C = X \times Y \times Z$ be a tricluster, and let $\begin{bmatrix} c_{ia} & c_{ib} \\ c_{ja} & c_{jb} \end{bmatrix}$ an arbitrary 2×2 submatrix of $X \times Y$ (for some $z \in Z$) or $X \times Z$ (for some $y \in Y$) or $Y \times Z$ (for some $x \in X$). Let $r_i = |c_{ib}/c_{ia}|$, $r_j = |c_{jb}/c_{ja}|$, $r_a = |c_{ja}/c_{ia}|$, and $r_b = |c_{jb}/c_{ib}|$ then

$$\frac{\max(r_i, r_j)}{\min(r_i, r_j)} - 1 \leq \varepsilon \iff \frac{\max(r_a, r_b)}{\min(r_a, r_b)} - 1 \leq \varepsilon$$

Proof: W.l.o.g. assume that $r_i \geq r_j$; then $|c_{ib}/c_{ia} \geq |c_{jb}/c_{ja}| \iff |c_{ja}/c_{ia}| \geq |c_{jb}/c_{ib}| \iff r_a \geq r_b$. We now have

$$\frac{\max(r_i, r_j)}{\min(r_i, r_j)} = \frac{r_i}{r_j} = \frac{|c_{ib}/c_{ia}|}{|c_{jb}/c_{ja}|} = \frac{|c_{ja}/c_{ia}|}{|c_{jb}/c_{ja}|} = \frac{r_a}{r_b} = \frac{\max(r_a, r_b)}{\min(r_a, r_b)}$$

Thus,

$$\frac{\max(r_i, r_j)}{\min(r_i, r_j)} - 1 \le \varepsilon \iff \frac{\max(r_a, r_b)}{\min(r_a, r_b)} - 1 \le \varepsilon \quad \blacksquare$$

The symmetric property of our cluster definition allows for very efficient cluster mining. The reason is that we are now free to mine clusters searching over the dimensions with the least cardinality. For example, instead of searching for subspace clusters over subsets of the genes (which can be large), we can search over subsets of samples (which are typically very few) or over subsets of time courses (which are also not large).

Note that by definition, a cluster represents a scaling cluster (if the ratio is 1.0, it is a uniform cluster). However, our definition allows for the mining of shifting clusters as well, as indicated by the following lemma.

Lemma 15.2 (*Shifting Cluster*) Let $C = X \times Y \times Z = \{c_{xyz}\}$ be a maximal tricluster. Let $e^c = \{e^{c_{xyz}}\}$ be the tricluster obtained by applying the exponential function (base e) to each value in C. If e^c is a (scaling) cluster, then C is a shifting cluster.

Proof: Let $C_{2,2} = \begin{bmatrix} c_{ia} & c_{ib} \\ c_{ja} & c_{jb} \end{bmatrix}$ be an arbitrary 2×2 submatrix of C. Assume that e^c is a valid scaling cluster. Then, by definition, $e^{c_{ib}} = \alpha_i e^{c_{ia}}$. But this immediately implies that $\ln(e^{c_{ib}}) = \ln(\alpha_i e^{c_{ia}})$, which gives us $c_{ib} = \ln(\alpha_i) + c_{ia}$. Similarly, we have $c_{jb} = \ln(\alpha_j) + c_{ja}$. Setting $\beta_i = \ln(\alpha_i)$ and $\beta_j = \ln(\alpha_j)$, we have that C is a shifting cluster. $\blacksquare$

Note that the clusters can have arbitrary positions anywhere in the data matrix, and they can have arbitrary overlapping regions (although triCluster can optionally merge or delete overlapping clusters under certain scenarios). We impose the minimum size constraints (i.e., mx, my, and mz to mine large enough clusters). Typically, $\varepsilon \approx 0$, so that the ratios of the values along one dimension in the cluster are similar (by Lemma 1, this property is also applicable for the other two dimensions), i.e., the ratios can differ by at most ε. Further, different choices of dimensional range thresholds (δ^x, δ^y, and δ^z) produce different types of clusters:

1. If $\delta^x = \delta^y = \delta^z = 0$, we obtain a cluster that has identical values along all dimensions.
2. If $\delta^x = \delta^y = \delta^z \approx 0$, we obtain clusters with approximately identical values.
3. If $\delta^x \approx 0, \delta^y \ne 0$, and $\delta^z \ne 0$, we obtain a cluster ($X \times Y \times Z$), where each gene $g_i \in X$ has similar expression values across the different samples Y and the different times Z, and different genes' expression values cannot differ by more than the threshold δ^x. Similarly, we can obtain other cases by setting (a) $\delta^x \ne 0, \delta^y, \approx 0$, and $\delta^z \ne 0$ or (b) $\delta^x \ne 0, \delta^y \ne 0$, and $\delta^z \approx 0$.
4. If $\delta^x \approx 0, \delta^y \approx 0$, and $\delta^z \ne 0$, we obtain a cluster with similar values for genes and samples, but the time courses are allowed to differ by an arbitrary scaling factor. Similar cases are obtained by setting (a) $\delta^x \approx 0, \delta^y \ne 0$, and $\delta^z \approx 0$, and (b) $\delta^x \ne 0, \delta^y \approx 0$, and $\delta^z \approx 0$.

5. If $\delta^x \neq 0$, $\delta^y \neq 0$, and $\delta^z \neq 0$, we obtain a cluster that exhibits scaling behavior on genes, samples, and times, and the expression values are bounded by δ^x, δ^y, and δ^z, respectively.

Note also that triCluster also allows different ε values for different pairs of dimensions. For example, we may use one value of ε to constrain the expression values for, say the gene–sample slice, but we may then relax the maximum ratio threshold for the temporal dimension to capture more interesting (and big) changes in expression as time progresses.

For example, Figure 15.1b shows some examples of different clusters that can be obtained by permuting some dimensions. Let $mx = my = 3$, $mz = 2$, and $\varepsilon = 0.01$. If we let $\delta^x = \delta^y = \delta^z = \infty$ (i.e., if they are unconstrained), $C_1 = \{g_1, g_4, g_8\} \times \{s_1, s_4, s_6\} \times \{t_0, t_1\}$ is an example of a scaling cluster (i.e., each point values along one dimension is some scalar multiple of another point values along the same dimension). We also discover two other maximal overlapping clusters, $C_2 = \{g_0, g_2, g_6, g_9\} \times \{s_1, s_4, s_6\} \times \{t_0, t_1\}$ and $C_3 = \{g_0, g_7, g_9\} \times \{s_1, s_2, s_4, s_5\} \times \{t_0, t_1\}$. Note that if we set $my = 2$, we would find another maximal cluster $C_4 = \{g_0, g_2, g_6, g_7, g_9\} \times \{s_1, s_4\} \times \{t_0, t_1\}$, which is subsumed by C_2 and C_3. We shall see later that triCluster can optionally delete such a cluster in the final steps. If we set $\delta^x = 0$ and let δ^y and δ^z be unconstrained, we will not find cluster C_1, whereas all other clusters will remain valid. This is because if $\delta^x = 0$, the values for each gene in the cluster must be identical; however, since δ^y and δ^z are unconstrained, the cluster can have different coherent values along the samples and times. Since ε is symmetric for each dimension, triCluster first discovers all unconstrained clusters rapidly and then prunes unwanted clusters if δ^x, δ^y, or δ^z are constrained. Finally, it optionally deletes or merges mined clusters if certain overlapping criteria are met.

15.2 RELATED WORK

Although there has been work on mining gene expression patterns across time, to the best of our knowledge there is no previous method that mines triclusters. On the other hand, there are many full-space and biclustering algorithms designed to work with microarray data sets, such as feature-based clustering [1,2,27], graph-based clustering [13,24,28], and pattern-based clustering [5,7,15,18,26]. Below we review briefly some of these methods and refer the reader to an excellent recent survey on biclustering [16] for more details. We begin by discussing time-series expression clustering methods.

15.2.1 Time-Based Microarray Clustering

Jiang et al. [14] gave a method to analyze the gene–sample–time microarray data. It treats the gene–sample–time microarray data as a gene×sample matrix with each as a vector of the values along the time dimension. For any two such vectors, it uses their *Pearson's correlation coefficient* as the distance. Then for each gene, it groups similar time vectors together to form a sample subset. After that, it enumerates the subset of all the genes to find those subsets of genes whose corresponding sample subsets result in a considerable intersection set. Jiang et al. discussed two methods: grouping samples first and grouping genes first. Although the paper dealt with three-dimensional microarray data, it considers the time dimension in full space (i.e., all the values along the time dimension) and is thus unable to find temporal trends that are applicable to only a subset of the times, and as such, it casts the three-dimensional problem into a biclustering problem.

In general, most previous methods apply traditional full space clustering (with some improvements) to the gene time-series data. Thus, these methods are not capable of mining coherent subspace clusters (i.e., these methods sometimes will miss important information obscured by the data noise). For example, Erdal et al. [9] extract a 0/1 vector for each gene, such that there is a 1 whenever there is a big change in its expression from one time to the next. Using longest common subsequence length as similarity, they perform a full-dimensional clustering. The subspaces of time points are not considered, and the sample space is ignored. Moller et al. [17] present another time-series microarray clustering algorithm. For any two time vectors $[x_1(t_1), x_2(t_2), \ldots, x_k(t_k)]$ and $[y_1(t_1), y_2(t_2), \ldots, y_k(t_k)]$ they calculate

$$\mathrm{sim}\,(x, y) = \sum_{k=1}^{n} \frac{(x_{k+1} - x_k) - (y_{k+1} - y_k)}{t_{k+1} - t_k}$$

Then they use a full-space repeated fuzzy clustering algorithm to partition the time-series clusters. Ramoni et al. [20] present a Bayesian method for model-based gene expression clustering. It represents gene expression dynamics as autoregressive equations and uses an agglomerative method to search for the clusters. Feng et al. [10] proposed a time-frequency-based full-space algorithm using a measure of functional correlation set between time-course vectors of different genes. Filkov et al. [11] addressed the analysis of short-term time-series gene microarray data by detecting the period in a predominantly cycling data set and the phase between phase-shifted cyclic data sets. It, too, is a full-space clustering method for gene time-series data. For a more comprehensive look at time-series gene expression analysis, see the paper by Bar-Joseph [4], who divides the computational challenges into four analysis levels: experimental design, analysis, pattern recognition, and gene networks. Bar-Joseph discusses the computational and biological problems at each level, reviews some methods proposed to deal with these issues, and highlights some open problems.

15.2.2 Feature- and Graph-Based Clustering

PROCLUS [1] and ORCLUS [2] use projective clustering to partition a data set into clusters occurring in possibly different subsets of dimensions in a high-dimensional data set. PROCLUS seeks to find axis-aligned subspaces by partitioning the set of points and then uses a hill-climbing technique to refine the partitions. ORCLUS finds arbitrarily oriented clusters by using ideas related to singular-value decomposition.

Other subspace clustering methods include CLIQUE [3] and DOC [19]. These methods are not designed to mine coherent patterns from microarray data sets.

CLIFF [27] iterates between feature (genes) filtering and sample partitioning. It first calculates k best features (genes) according to their intrinsic discriminability using current partitions. Then it partitions the samples with these features by keeping the minimum normalized weights. This process iterates until convergence. COSA [12] allows traditional clustering algorithms to cluster on a subset of attributes rather than on all of them. Principal component analysis for gene expression clustering has also been proposed [30].

HCS [13] is a full-space clustering algorithm. It cuts a graph into subgraphs by removing some edges, and repeats until all the vertices in each subgraph are similar enough. MST [28] is also a full-space clustering method. It uses greedy method to construct a minimum spanning tree and splits the current tree(s) repeatedly until the average edge length in each subtree is below some threshold. Then each tree is a cluster. SAMBA [24] uses a bipartite graph to model and implement the clustering. It repeatedly finds the maximal highly connected subgraph in the bipartite graph. Then it performs local improvement by addign or deleting a single vertex until no further improvement is possible. Other graph-theoretic clustering algorithms include CLICK [21] and CAST [6].

There are some common drawbacks concerning the algorithms above applied to microarray data sets. First, some of them are radomized methods based on shrinking and expansion, which sometimes results in incomplete cluster. Second, none of them can deal properly with overlapped clusters. Third, the greedy methods will lead to a local optimum that may miss some important (part of) clusters. Moreover, full-space clustering is even not biclustering and will compromise the important clusters by considering irrelevant dimensions. In general, none of them are deterministic and thus cannot guarantee that all valid (overlapped) clusters are found.

15.2.3 Patern-Based Clustering

δ-Biclustering [7] uses the mean-squared residue of a submatrix $(X \times Y)$ to find biclusters. If a submatrix with enough size has a mean-squared residue less than threshold δ, it is a δ-bicluser. Initially, the algorithm starts with the entire data matrix and repeatedly adds or deletes a row or column from the current matrix in a greedy way until convergence. After having found a cluster, it replaces the submatrix with random values and continues to find the next-best cluster. This process iterates until no additional clusters can be found. One limitation of δ-biclustering is that it may converge to a local optimum and cannot guarantee that all clusters will be found. It can also easily miss overlapping clusters, due to the random value substitutions that it makes. Another move-based δ-biclustering method was proposed in [29]. However, it, too, is an iterative improvement-based method.

pCluster [26] defines a cluster C as a submatrix of the original data set such that for any 2×2 submatrix $\begin{bmatrix} c_{xa} & c_{xb} \\ c_{ya} & c_{yb} \end{bmatrix}$ of C, $|(c_{xa} - c_{ya}) - (c_{xb} - c_{yb})| < \delta$, where δ is a threshold. The algorithm first scans the data set to find all column-pair and row-pair maximal clusters called MDS. Then it does the pruning in turn using the row-pair

MDS and the column-pair MDS. It then mines the final clusters based on a prefix tree. pCluster is symmetric (i.e., it treats rows and columns equally) and it is capable of finding clusters similar to those found by triCluster, but it does not merge or prune clusters and is not robust to noise. Further, we show that it runs much slower than triCluster on real microarray data sets.

xMotif [18] requires that all gene expressions in a bicluster be similar across all the samples. It randomly picks a seed sample s and a sample subset d (called a *discriminating set*), and then finds all such genes that are conserved across all the samples in d. xMotif uses a Monte Carlo method to find the clusters that cover all genes and samples. However, it cannot guarantee to find all the clusters because of its random sampling process.

Another stochastic algorithm OPSM [5] has a drawback similar to that of xMotif, but uses a different cluster definition. It defines a cluster as a submatrix of the original data matrix after row and column permutation, where the row values are in a nondecreasing pattern. Another method using this definition is OP-Cluster [15]. Gene clustering methods using self-organizing maps [23] and iterated two-way clustering [25] have also been proposed; a systematic comparison of these and other biclustering methods may be found in ref. 16.

15.3 THE triCluster ALGORITHM

As outlined above, triCluster mines arbitrarily positioned and overlapping scaling and shifting patterns from a three-dimensional data set as well as several specializations. Typically, three-dimensional microarray data sets have more genes than samples, and perhaps an equal number of time points and samples (i.e., $|G| \geq |T| \approx |S|$). Due to the symmetric property, triCluster always transposes the input three-dimensional matrix such that the dimension with the largest cardinality (say, G) is a first dimension; we then make S the second and T the third dimension. triCluster has three principal steps: (1) For each $G \times S$ time slice matrix, find the valid ratio ranges for all pair of samples and construct a range multigraph; (2) mine the maximal biclusters from the range multigraph; (3) construct a graph based on the mined biclusters (as vertices) and get the maximal triclusters; and (4) optionally, delete or merge clusters if certain overlapping criteria are met. We look at each step below.

15.3.1 Constructing a Range Multigraph

Given a data set D, the minimum-size thresholds, mx, my, and mz, and the maximum ratio threshold ε, let s_a and s_b be any two sample columns in some time t of D and let $r_x^{ab} = d_{xa}/d_{xb}$ be the ratio of the expression values of gene g_x in columns s_a and s_b, where $x \in [0, n-1]$. A *ratio range* is defined as an interval of ratio values, $[r_l, r_u]$, with $r_l \leq r_u$. Let $\mathcal{G}_{ab}([r_l, r_u]) = \{g_x : r_x^{ab} \in [r_l, r_u]\}$ be the set of genes, called the *gene set*, whose ratios with respect to columns s_a and s_b lie in the given ratio range, and if $r_x^{ab} < 0$, all the values in the same column have the same signs (negative/ nonnegative).

s_0/s_6	3.0	3.0	3.0	3.3	3.3	3.6
Row	g_1	g_4	g_8	g_3	g_5	g_0

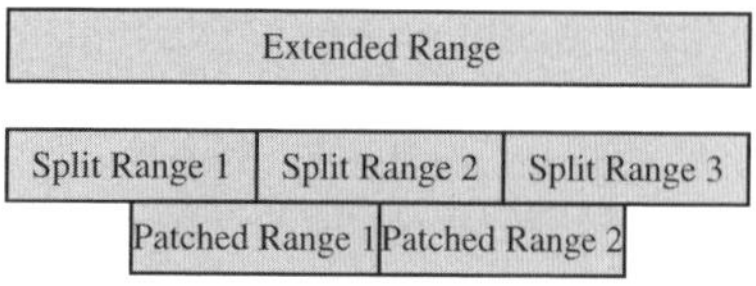 Extended Range		

Split Range 1	Split Range 2	Split Range 3
Patched Range 1	Patched Range 2	

        (*a*)   (*b*)

FIGURE 15.2 (*a*) Sorted ratios of column s_0/s_6 in Figure 15.1; (*b*) split and patched ranges.

In the first step, triCluster quickly tries to summarize the valid ratio ranges that can contribute to some bicluster. More formally, we call a ratio range *valid* iff (1) $\max(|r_u|, |r_l|)/\min(|r_u|, |r_l|) - 1 \le \varepsilon$ (i.e., the range satisfies the maximum ratio threshold imposed in our cluster definition); (2) $|\mathcal{G}_{ab}([r_l, r_u])| \ge mx$ [i.e., there are enough (at least mx) genes in the gene set; this is imposed since our cluster definition requires any cluster to have at least mx genes]; (3) if there exists a $r_x^{ab} < 0$, all the values $\{d_{xa}\}/\{d_{xb}\}$ in the same column have same signs (negative/nonnegative); and (4) $[r_l, r_u]$ is maximal with respect to ε (i.e., we cannot add another gene to $\mathcal{G}_{ab}([r_l, r_u])$ and yet preserve the ε bound). Intuitively, we want to find all the maximal ratio ranges that satisfy the ε threshold and span at least mx genes. Note that there can be multiple ranges between two columns and also that some genes may not belong to any range.

Figure 15.2 shows the ratio values for different genes using columns s_0/s_6 at time t_0 for our running example in Figure 15.1. Assume that $\varepsilon = 0.01$ and $mx = 3$; there is only one valid ratio range, [3.0, 3.0], and the corresponding gene set is $\mathcal{G}_{s_0 s_6}([3.0, 3.0]) = \{g_1, g_4, g_8\}$. Using a sliding window approach (with window size: $r_x^{.06} \times \varepsilon$ for each gene g_x) over the sorted ratio values, triCluster find all valid ratio ranges for all pairs of columns $s_a, s_b \in S$. If at any stage there are more than mx rows within the window, a range is generated. It is clear that different ranges may overlap. For instance, if we let $\varepsilon = 0.1$, we would obtain two valid ranges, [3.0, 3.3] and [3.3, 3.6], with overlapping gene sets $\{g_1, g_4, g_8, g_3, g_5\}$ and $\{g_3, g_5, g_0\}$, respectively. If there are consecutive overlapping valid ranges, we merge them into an extended range, even though the maximum ratio threshold ε is exceeded. If the extended range is too wide, say more than 2ε, we split the extended range into several blocks of range at most 2ε *(split ranges)*. To avoid missing any potential clusters, we also add some overlapping *patched ranges*. This process is illustrated in Figure 15.2*b*. Note that an added advantage of allowing extended ranges is that it makes the method more robust to noise, since often the users may set a stringent ε condition, whereas the data might require a larger value.

Given the set of all valid ranges, as well as the extended split or patched ranges, across any pairs of columns s_a and s_b with $a < b$, given as $\mathcal{R}^{ab} = \{R_i^{ab} = [r_{l_i}^{ab}, r_{u_i}^{ab}] : s_a, s_b \in S\}$, we construct a weighted, directed range multigraph $M = (V, E)$, where $V = S$ (the set of all samples), and for each $R_i^{ab} \in \mathcal{R}^{ab}$ there exists a weighted, directed edge $(s_a, s_b) \in E$ with weight $w = r_{u_i}^{ab}/r_{l_i}^{ab}$. In addition, each edge in the range multigraph has associated with it the gene set corresponding to the range on that edge. For example, suppose that $my = 3$, $mx = 3$, and $\varepsilon = 0.01$. Figure 15.3 shows the range multigraph constructed from Figure 15.1 for time t_0. Another range multigraph is obtained for time t_1.

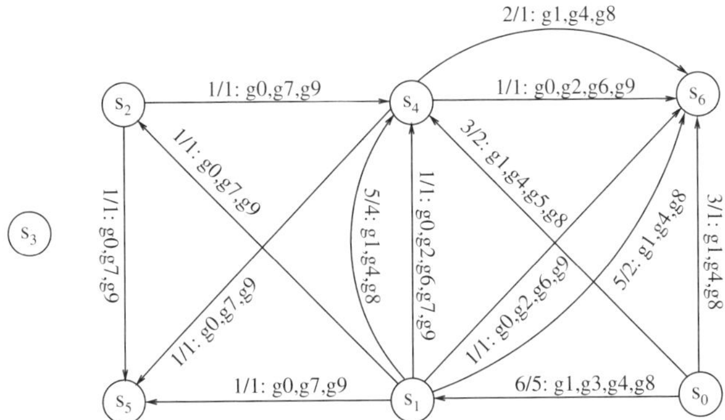

FIGURE 15.3 Weighted, directed range multigraph.

15.3.2 Mining Biclusters from a Range Multigraph

The range multigraph represents in a compact way all the valid ranges that can be used to mine potential biclusters corresponding to each time slice, and thus filters out most of the unrelated data. biCluster uses a depth-first search (DFS) on the range multigraph to mine all the biclusters, as shown in pseudocode in Figure 15.4. It takes as input the set of parameter values $\varepsilon, mx, my, \delta^x$, and δ^y, the range graph M^t for a given time point t, and the set of genes all G and samples S. It will output the final set of all biclusters C^t for that time course. biCluster is a recursive algorithm that at each call accepts a current *candidate* bicluster $C = X \times Y$ and a set of not yet expanded samples P. The initial call is made with a cluster $C = G \times \emptyset$ with all genes G, but no samples, and with $P = S$, since we have not yet processed any samples. Before passing C to the recursive call, we make sure that $|C \cdot X| \geq mx$ (which is certainly true in the initial call, and also at line 16). Line 2 checks if the cluster meets the the maximum gene and sample range thresholds δ^x and δ^y, and also the minimum sample cardinality my (line 3). If so, we next check if C is not already contained in some maximal cluster $C' \in C^t$ (line 3). If not, we add C to the set of final clusters C^t (line 6), and we remove any cluster $C'' \in C$ already subsumed by C (line 5).

Lines 7 to 17 generate a new candidate cluster by expanding the current candidate by one more sample, constructing the appropriate gene set for the new candidate, before making a recursive call. biCluster begins by adding to the current cluster C each new sample $s_b \in P$ (line 7) to obtain a new candidate C^{new} (lines 8 and 9). Samples already processed are removed from P (line 10). Let s_a be all samples added to C since the previous recursive call. If no previous vertex s_a exists (which will happen during the initial call, when $C \cdot Y = \emptyset$), we simply call biCluster with the new candidate. Otherwise, biCluster tries all combinations of each qualified range edge R_i^{ab} between s_a and s_b for all $s_a \in C \cdot Y$ (line 14), obtains their gene-set intersection $\cap_{\text{all } s_a \in C \cdot Y} \mathcal{G}(R_i^{ab})$, and intersects with $C \cdot X$ to obtain the valid genes in the new cluster C^{new} (line 15). If the new cluster has at least mx genes, another recursive call to biCluster is made (lines 16 and 17).

Input   : parameters: $\varepsilon, mx, my, \delta^x, \delta^y$, range graph M_t, set of genes G and samples S

Output   : bicluster set $\mathcal{C}^t$

Initialization: $\mathcal{C}^t = \emptyset$, call $\textsc{biCluster}(C = G \times \emptyset, S)$

1   $\textsc{biCluster}(C = X \times Y, P)$:

2   **if** C *satisfies* δ^x, δ^y **then**

3   **if** $|C.Y| \geq my$ **then**

4   **if** $\nexists C' \in \mathcal{C}^t$, *such that* $C \subset C'$ **then**

5   Delete any $C'' \in \mathcal{C}$, if $C'' \subset C$

6            $\mathcal{C}^t \longleftarrow \mathcal{C}^t + C$

7   **foreach** $s_b \in P$ **do**

8      $C^{new} \longleftarrow C$

9      $C^{new}.Y \longleftarrow C^{new}.Y + s_b$

10     $P \longleftarrow P - s_b$

11   **if** $C.Y = \emptyset$ **then**

12        $\textsc{biCluster}(C^{new}, P)$

13   **else**

14   **forall** $s_a \in C.Y$ *and each* $R_i^{ab} \in \mathcal{R}^{ab}$ *satisfying* $|\mathcal{G}(R_i^{ab}) \cap C.X| \geq mx$ **do**

15           $C^{new}.X \longleftarrow (\bigcap_{all\ s_a \in C.Y} \mathcal{G}(R_i^{ab})) \cap C.X$

16   **if** $|C^{new}.X| \geq mx$ **then**

17             $\textsc{biCluster}(C^{new}, P)$

FIGURE 15.4   biCluster algorithm.

For example, let's consider how the clusters are mined from the range graph M^{t_0} shown in Figure 15.3. Let $mx = 3$, $my = 3$, $\varepsilon = 0.01$, as before. Initially, biCluster starts at vertex s_0 with the candidate cluster $\{g_0 \ldots, g_9\} \times \{s_0\}$. We next process vertex s_1; since there is only one edge, we obtain a new candidate $\{g_1, g_3, g_4, g_8\} \times \{s_0, s_1\}$. From s_1 we process s_4 and consider both the edges: For $w = 5/4, \mathcal{G} = \{g_1, g_4, g_8\}$, we obtain the new candidate $\{g_1, g_4, g_8\} \times \{s_0, s_1, s_4\}$, but the other edge, $w = 1/1, \mathcal{G} = \{g_0, g_2, g_6, g_7, g_9\}$, will not have enough genes. We then further process s_6. Of the two edges between s_4 and s_6, only one (with weight 2/1) will yield a candidate cluster $\{g_1, g_4, g_8\} \times \{s_0, s_1, s_4, s_6\}$. Since this is maximal and meets all parameters, at this point we have found one (C_1) of the three final clusters shown in Figure 15.1b. Similarly, when we start from s_1, we find the other two clusters, $C_3 = \{g_0, g_7, g_9\} \times \{s_1, s_2, s_4, s_5\}$ and $C_2 = \{g_0, g_2, g_6, g_9\} \times \{s_1, s_4, s_6\}$. Intuitively, we are searching for maximal cliques (on samples), with cardinality at least my, that also satisfy the minimum number of genes constraint mx.

Input   : parameters: $\varepsilon, mx, my, mz, \delta^x, \delta^y, \delta^z$, bicluster sets $\{\mathcal{C}^t\}$ of all times, set of genes G, samples S and times T

Output   : cluster set $\mathcal{C}$

Initialization: $\mathcal{C} = \emptyset$, call $\text{TRICLUSTER}(C = G \times S \times \emptyset, T)$

1 $\text{TRICLUSTER}\ (C = X \times Y \times Z, P)$;

2 **if** C *satisfies* $\delta^x, \delta^y, \delta^z$ **then**

3   **if** $|C.Z| \geq mz$ **then**

4   **if** $\not\exists C' \in \mathcal{C}$, *such that* $C \subset C'$ **then**

5   Delete any $C'' \in \mathcal{C}$, if $C'' \subset C$

6              $\mathcal{C} \longleftarrow \mathcal{C} + C$

7 **foreach** $t_b \in P$ **do**

8      $C^{new}.Z \longleftarrow C.Z + t_b$

9      $P \longleftarrow P - t_b$

10   **forall** $t_a \in C.Z$ *and each bicluster* $c_i^{t_a} \in \mathcal{C}^{t_a}$ *satisfying* $|c_i^{t_a}.X \cap C.X| \geq mx$ *and* $|c_i^{t_a}.Y \cap C.Y| \geq my$ **do**

11          $C^{new}.X \longleftarrow (\bigcap_{all\ t_a \in C.Z} c_i^{t_a}.X) \cap C.X$

12          $C^{new}.Y \longleftarrow (\bigcap_{all\ t_a \in C.Z} c_i^{t_a}.Y) \cap C.Y$

13   **if** $|C^{new}.X| \geq mx$ *and* $|C^{new}.Y| \geq my$ *and the ratios at time* t_b, t_a *are coherent* **then**

14              $\text{TRICLUSTER}(C^{new}, P)$

FIGURE 15.5   triCluster algorithm.

15.3.3 Getting Triclusters from a Bicluster Graph

After having got the maximal bicluster set $\mathcal{C}^t$ for each time slice t, we use them to mine the maximal triclusters. This is accomplished by enumerating the subsets of the time slices as shown in Figure 15.5, using a process similar to the biCluster clique mining (Figure 15.4). For example, from Figure 15.1 we can get the biclusters from the two time points t_0 and t_1 as shown in Figure 15.6. Since the clusters are identical, to illustrate our tricluster mining method, let's assume that we also obtain other biclusters at times points t_3 and t_8. Assume that the minimum-size threshold is $mx \times my \times mz = 3 \times 3 \times 3$. TriCluster starts from time t_0, which contains three biclusters. Let's begin with cluster C_1 at time t_0, given as $C_1^{t_0}$. For each bicluster C^{t_1}, only $C_1^{t_1}$ can be used for extension since $C_1^{t_0} \cap C_1^{t_1} = \{g_1, g_4, g_8\} \times \{s_0, s_1, s_4, s_6\}$, which can satisfy the cardinality constraints (Figure 15.4, line 15). We continue by processing time t_3, but the cluster cannot be extended. So we try t_8, and we may find that we can extend it by mean of $C_1^{t_8}$. The final result of this path is $\{g_1, g_4, g_8\} \times \{s_0, s_1, s_4, \} \times \{t_0, t_1, t_8\}$. Similarly, we try all such paths and keep

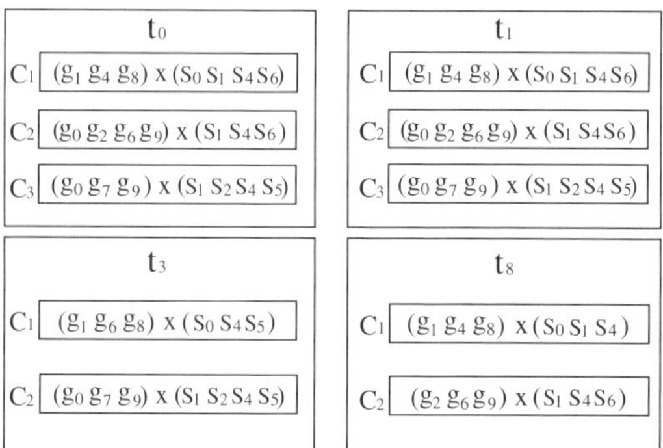

FIGURE 15.6 Tricluster example.

maximal triclusters only. During this process we also need to check the coherent property along the time dimension, as the tricluster definition requires, between the new time slice and the preceding one. For example, for the three biclusters in Figure 15.1, the ratios between t_1 and t_0 are 1.2 (for C_1) and 0.5 (for C_2 and C_3), respectively. If the extended bicluster has no such coherent values in the intersection region, triCluster will prune it.

The complexity of this part (along the time dimension) is the same as that of biclusters generation (biCluster) for one time slice. But since biCluster needs to run $|T|$ times, the total running time is $|T| \times$ [time (multigraph) + time (biCluster)] + time (triCluster).

15.3.4 Merging and Pruning Clusters

After mining the set of all clusters, triCluster optionally merges or deletes certain clusters with large overlap. This is important, since real data can be noisy, and the users may not know the correct values for different parameters. Furthermore, many clusters having large overlaps only make it harder for users to select the important ones.

Let $A = X_A \times Y_A \times Z_A$ and $B = X_B \times Y_B \times Z_B$ be any two mined clusters. We define the *span* of a cluster $C = X \times Y \times Z$, to be the set of gene–sample–time tuples that belong to the cluster, given as $L_C = \{(g_i, s_j, t_k)|g_i \in X, s_j \in Y, t_k \in Z\}$. Then we can define the following derived spans:

- $L_{A \cup B} = L_A \cup L_B$
- $L_{A-B} = L_A - L_B$
- $L_{A+B} = L_{(X_A \cup X_B)} \times (Y_A \cup Y_B) \times (Z_A \cup Z_B)$

If any of the following three overlap conditions are met, triCluster either deletes or merges the clusters involved:

1. For any two clusters A and B, if $L_A > L_B$, and if $|L_{B-A}|/|L_B| < \eta$, then delete B. As illustrated in Figure 15.7a (for clarity, we use two-dimensional figures here),

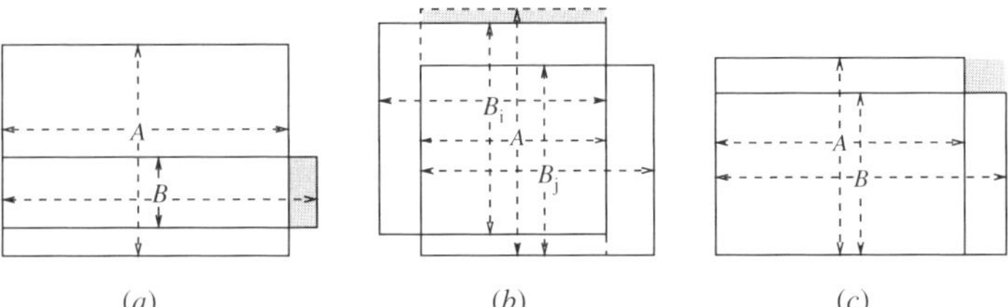

FIGURE 15.7 Three pruning or merging cases.

this means that if the cluster with the smaller span *(B)* has only a few extra elements, delete the smaller cluster.

2. This is a generalization of case 1. For a cluster A, if there exists a set of clusters $\{B_i\}$, such that $|L_A - L \cup_i B_i|/|L_A| < \eta$, delete cluster A. As shown in Figure 15.7b, A is mostly covered by the $\{B_i\}$'s. Therefore, it can be deleted.

3. For two clusters A and B, if $|L_{(A+B)-A-B}|/|L_{A+B}| < \gamma$, merge A and B into one cluster, $(X_A \cup X_B) \times (Y_A \cup Y_B) \times (Z_A \cup Z_B)$. This case is shown in Figure 15.7c. Here η and γ are user-defined thresholds.

15.3.5 Complexity Analysis

Since we have to evaluate all pairs of samples, compute their ratios, and find the valid ranges over all the genes, the range multigraph construction step takes time $O(|G||S|^2|T|)$. The bicluster mining step and tricluster mining step correspond to constrained maximal clique enumeration (i.e., cliques satisfying the *mx*, *my*, *mz*, δ^x, δ^y, δ^z parameters) from the range multigraph and bicluster graph. Since in the worst case there can be an exponential number of clusters, these two steps are the most expensive. The precise number of clusters mined depends on the data set and the input parameters. Nevertheless, for microarray data sets, triCluster is likely to be very efficient for the following reasons: First, the range multigraph prunes away much of the noise and irrelevant information. Second, the depth of the search is likely to be small, since microarray data sets have far fewer samples and times than genes. Third, triCluster keeps intermediate gene sets for all candidate clusters, which can prune the search the moment the input criteria are not met. The merging and pruning steps apply only to those pairs of clusters that actually overlap, which can be determined in $O(|\mathcal{C}| \log(|\mathcal{C}|))$ time.

15.4 EXPERIMENTS

Unless noted otherwise, all experiments were done on a Linux/Fedora virtual machine (Pentium-M, 1.4 GHz, 448M memory) over Windows XP through middleware VMware. We used both synthetic and real microarray data sets to evaluate the triCluster algorithm. For the real data set we used yeast cell–cycle regulated genes [22] (http://genome-www.stanford.edu/cellcycle). The goal of the study was to identify all genes whose mRNA levels are regulated by the cell cycle.

Synthetic data sets allow us to embed clusters and then to test how triCluster performs for varying input parameters. We generate synthetic data using the following steps: The input parameters to the generator are the total number of genes, samples, and times; number of clusters to embed; percentage of overlapping clusters; dimensional ranges for the cluster sizes; and the amount of noise for the expression values. The program randomly picks cluster positions in the data matrix, ensuring that no more than the required number of clusters overlap. The cluster sizes are generated uniformly between each dimensional ranges. For generating the expression values within a cluster, we generate at random, base values (v_i, v_j, and v_k) for each dimension in the cluster. Then the expression value is set as $d_{ijk} = v_i \cdot v_j \cdot v_k \cdot (1 + \rho)$, where ρ doesn't exceed the random noise level. Once all clusters are generated, the noncluster regions are assigned random values.

15.4.1 Results from Synthetic Data Sets

We first wanted to see how triCluster behaves with varying input parameters in isolation. We generated synthetic data with the following default parameters: data matrix size $4000 \times 30 \times 20(G \times S \times T)$, number of clusters 10, cluster size $150 \times 6 \times 4(X \times Y \times Z)$, percentage overlap 20%, and noise level 3%. For each experiment we keep all default parameters except for the varying parameter. We also choose appropriate parameter values for triCluster so that all embedded clusters were found. Figure 15.8a–f shows triCluster's sensitivity to different parameters. We found that the time increases approximately linearly with the number of genes in a cluster (a). This is because the range multigraph is constructed on the samples, not on the genes; more genes lead to longer gene sets (per edge), but the intersection time is essentially linear in gene-set size. The time is exponential with the number of samples (b), since we search over the sample subset space. Finally, the time for increasing time slices is also linear for the range shown (c), but in general the dependence will be exponential, since triCluster searches over subsets of time points after mining the biclusters for each time slice. The time is linear with respect to number of clusters (d), whereas the overlap percentage does not seem to have much impact on the time (e). Finally, as we add more noise, the more time that is required to mine the clusters (f), since there is more chance that a random gene or sample can belong to a cluster.

15.4.2 Results from Real Microarray Datasets

We define several metrics to analyze the output from different biclustering algorithms. If C is the set of all clusters output, then:

1. *Cluster#:* cardinality $|C|$
2. *Element-Sum:* sum of the spans of all clusters (i.e., $Element\_Sum = \sum_{C \in C} |L_C|$)

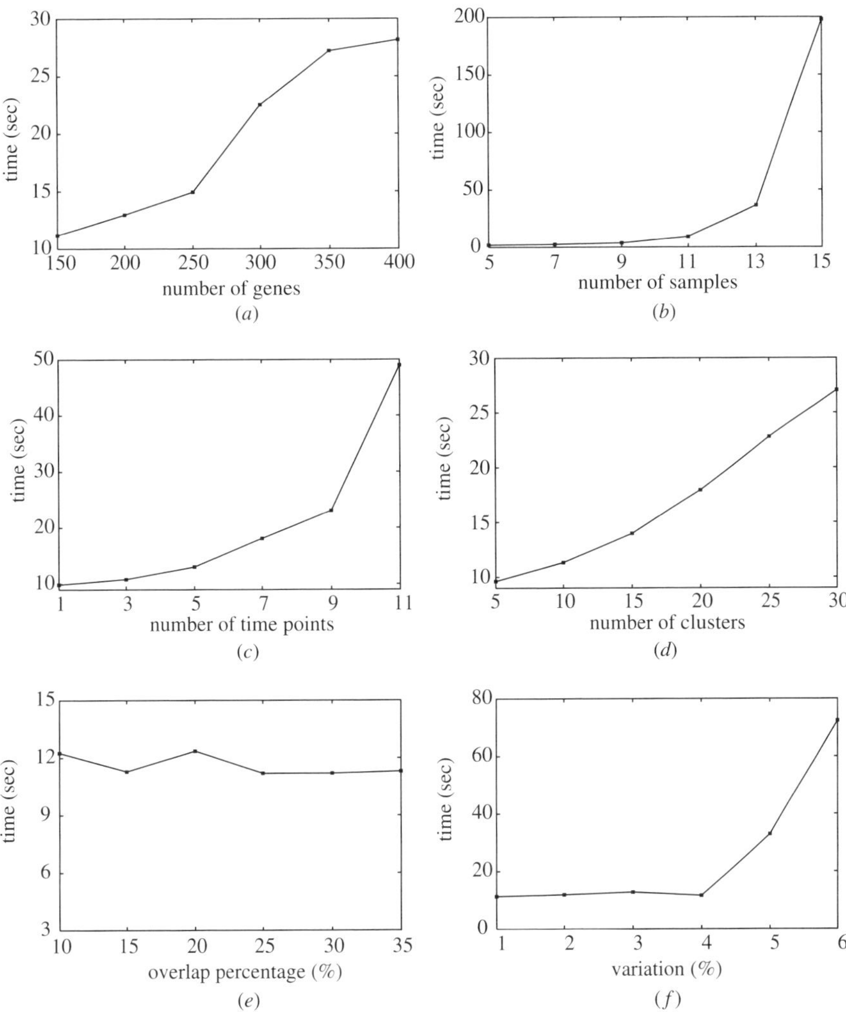

FIGURE 15.8 Evaluation of triCluster on synthetic data sets.

3. *Coverage:* span of the union of all clusters (i.e., $Coverage = |L_{\cup_{C \in c} c}|$)

4. *Overlap:* given as $\frac{Element\_Sum - Coverage}{Coverage}$

5. *Fluctuation:* average variance across a given dimension across all clusters

For the yeast cell cycle data we looked at the time slices for the elutritration experiments. There are a total of 7679 genes whose expression value is measured from time 0 to 390 minutes at 30-minute intervals. Thus, there are 14 total time points. Finally, we use 13 of the attributes of the raw data as the samples (e.g., the raw values for the average and normalized signal for Cy5 and Cy3 dyes, and the ratio of those values). Thus, we obtain a three-dimensional expression matrix of size:

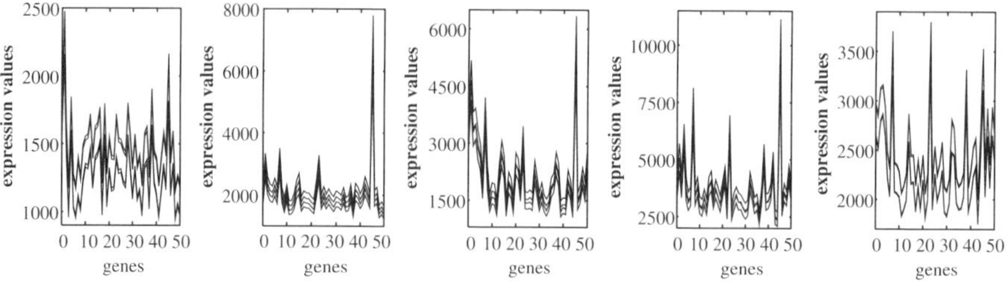

FIGURE 15.9 Sample curves.

$T \times S \times G = 14 \times 13 \times 7679$. We mined these data looking for triclusters with minimum size at least $mx = 50$ (genes), $my = 4$ (samples), $mz = 5$ (time points), and we set $\varepsilon = 0.003$ (however, we relax the ε threshold along the time dimension). The per dimension thresholds $\delta^x, \delta^y, \delta^z$ were left unconstrained. triCluster output five clusters in 17.8 s, with the following metrics:

Number of clusters	5
Number of elements	6520
Coverage	6520
Overlap	0.00%
Fluctuation	T:626.53, S:163.05, G:407.3

We can see that none of the five clusters was overlapping. The values for the total span across the clusters was 6520 cells, and the variances along each dimension are also shown. To see a mined tricluster visually, we plot various two-dimensional views of one of the clusters (C_0) in Figures 15.9, 15.10, and 15.11. Figure 15.9 shows how the expression values for the genes (x-axis) changes across the samples (y-axis) for different time points (the different sun plots). Figure 15.10 shows how the gene expression (x-axis) changes across the different time slices (y-axis) for different samples (the different subplots). Finally, Figure 15.11 shows what happens at different times (x-axis) for different genes (y-axis) across different samples (the different subplots). These figures show that triCluster is able to mine coherent clusters across any combination of the gene–sample–time dimensions.

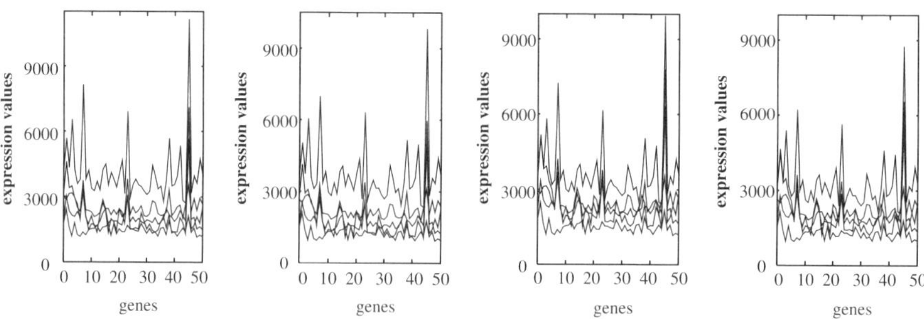

FIGURE 15.10 Time curves.

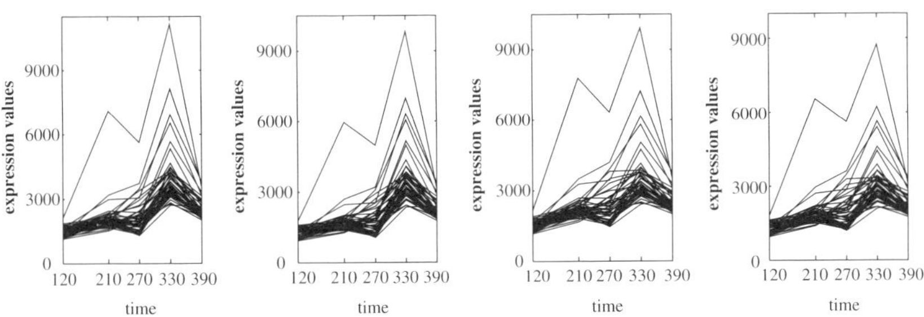

FIGURE 15.11 Gene curves.

The gene ontology (GO) project (www.geneontology.org) aims at developing three structured, controlled vocabularies (ontologies) that describe gene products in terms of their associated biological processes, cellular components, and molecular functions in a species-independent manner. We used the yeast genome gene ontology term finder (www.yeastgen-ome.org) to verify the biological significance of triCluster's result. We obtained a hierarchy of GO terms for each gene within each cluster for each of the three categories: processes, cellular components, and gene functions. Table 15.1 shows the significant shared GO terms (or parents of GO terms) used to describe the set of genes in each cluster. The table shows the number of genes in each cluster and the significant GO terms for the process, function, and component ontologies. Only the most significant common terms are shown. For example, for cluster C_0, we find that the genes are involved primarily in the ubiquitin cycle. The tuple $(n = 3, p = 0.00346)$ means that of the 51 genes, three belong to this process, and the statistical significance is given by the p-value of 0.00346. Within each category, the terms are given in descending order of significance (i.e., increasing p-values). Further, only p-values lower than 0.01 are shown; the other genes in the cluster share other terms, but at a lower significance. From the table it is clear that the clusters are distinct along each category. For example, the most significant process for C_0 is the ubiquitin cycle, for C_1 it is G1/S transition of mitotic cell cycle, for C_2 it is lipid transport, for C_3 it is physiological process/organelle organization and biogenesis, and for C_4 it is pantothenate biosynthesis. Looking at the function, we find the most significant terms to be protein phosphatase regulator activity for C_1, oxidoreductase activity for C_2, MAP kinase activity for C_3, and ubiquitin conjugating enzyme activity for C_4. Finally, the clusters also differ in terms of the cellular component: C_2 genes belong to the cytoplasm, C_3 genes to the membrane, and C_4 to the Golgi vesicle.

These results indicate that triCluster can find potentially biologically significant clusters in genes or samples or times or any combinations of these three dimensions. Since the method can mine coherent subspace clusters in any three-dimensional data set, triCluster will also prove useful in mining temporal and/or spatial dimensions. For example, if one of the dimensions represents genes, another the spatial region of interest, and the third dimension the time, triCluster can find interesting expression patterns in different regions at different times.

TABLE 15.1 Significant Shared GO Terms (Process, Function, Component) for Genes in Various Clusters

Cluster	No. of Genes	Process	Function	Cellular Component
C_0	51	Ubiquitin cycle ($n = 3$, $p = 0.00346$), protein polyubiquitination ($n = 2$, $p = 0.00796$), carbohydrate biosynthesis ($n = 3$, $p = 0.00946$)		
C_1	52	G1/S transition of mitotic cell cycle ($n = 3$, $p = 0.00468$), mRNA polyadenylylation ($n = 2$, $p = 0.00826$)	Protein phosphatase regulator activity ($n = 2$, $p = 0.00397$), phosphatase regulator activity ($n = 2$, $p = 0.00397$)	
C_2	57	Lipid transport ($n = 2$, $p = 0.0089$)	Oxidoreductase activity ($n = 7$, $p = 0.00239$), lipid transporter activity ($n = 2$, $p = 0.00627$), antioxidant activity ($n = 2, p = 0.00797$)	Cytoplasm ($n = 41$, $p = 0.00052$), microsome ($n = 2$, $p = 0.00627$), vesicular fraction ($n = 2$, 0.00627), microbody ($n = 3$, $p = 0.00929$), peroxisome ($n = 3, p = 0.00929$)
C_3	97	Physiological process ($n = 76$, $p = 0.0017$), organelle organization and biogenesis ($n = 15$, $p = 0.00173$), localization ($n = 21, p = 0.00537$)	MAP kinase activity ($n = 2$, $p = 0.00209$), deaminase activity ($n = 2$, $p = 0.00804$), hydrolase activity, acting on carbon–nitrogen, but not peptide, bonds ($n = 4$, $p = 0.00918$), receptor signaling protein serine/threonine kinase activity ($n = 2$, $p = 0.00964$)	Membrane ($n = 29, p = 9.36\text{e-}06$), cell ($n = 86, p = 0.0003$), endoplasmic reticulum ($n = 13$, $p = 0.00112$), vacuolar membrane ($n = 6$, $p = 0.0015$), cytoplasm ($n = 63$, $p = 0.00169$) intracellular ($n = 79$, $p = 0.00209$), endoplasmic reticulum membrane ($n = 6$, $p = 0.00289$), integral to endoplasmic reticulum membrane ($n = 3, p = 0.00328$), nuclear envelope–endoplasmic reticulum network ($n = 6$, $p = 0.00488$)
C_4	66	Pantothenate biosynthesis ($n = 2$, $p = 0.00246$), pantothenate metabolism ($n = 2$, $p = 0.00245$), transport ($n = 16, p = 0.00332$), localization ($n = 16$, $p = 0.00453$)	Ubiquitin conjugating enzyme activity ($n = 2$, $p = 0.00833$), lipid transporter activity ($n = 2$, $p = 0.00833$)	Golgi vesicle ($n = 2$, $p = 0.00729$)

15.5 CONCLUSIONS

In this chapter we introduced a novel deterministic triclustering algorithm called triCluster, which can mine arbitrarily positioned and overlapping clusters. Depending on different parameter values, triCluster can mine different types of clusters, including those with constant or similar values along each dimension, as well as scaling and shifting expression patterns. triCluster first constructs a range multigraph, which is a compact representation of all similar value ranges in the data set between any two sample columns. It then searches for constrained maximal cliques in this multigraph to yield the set of biclusters for this time slice. Then triCluster constructs another bicluster graph using the biclusters (as vertices) from each time slice. The clique mining of the bicluster graph will give the final set of triclusters. Optionally, triCluster merges/deletes some clusters having large overlaps. We present a useful set of metrics to evaluate the clustering quality, and we evaluate the sensitivity of triCluster to different parameters. We also show that it can find meaningful clusters in real data. Since cluster enumeration is still the most expensive step, in the future we plan to develop new techniques for pruning the search space.

Acknowledgments

This work was supported in part by National Science Foundation career award IIS-0092978, Department of Energy career award DE-FG02-02ER25538, and NSF grants EIA-0103708 and EMT-0432098.

REFERENCES

1. C. C. Aggarwal, C. Procopiuc, J. L. Wolf, P. S. Yu, and J. S. Park. Fast algorithms for projected clustering. In *ACM SIGMOD Conference on Management of Data*, 1999.

2. C. C. Aggarwal and P. S. Yu. Finding generalized projected clusters in high dimensional spaces. In *ACM SIGMOD Conference on Management of Data*, 2000.

3. R. Agrawal, J. Gehrke, D. Gunopulos, and P. Raghavan. Automatic subspace clustering of high dimensional data for data mining applications. In *ACM SIGMOD Conference on Management of Data*, June 1998.

4. Z. Bar-Joseph. Analyzing time series gene expression data. *Bioinformatics*, 20(16): 2493–2503, 2004.

5. A. Ben-Dor, B. Chor, R. Karp, and Z. Yakhini. Discovering local structure in gene expression data: the order-preserving submatrix problem. In *6th Annual International Conference on Computational Biology*, 2002.

6. A. Ben-Dor, R. Shamir, and Z. Yakhini. Clustering gene expression patterns. In *3rd Annual International Conference on Computational Biology* (RECOMB), 1999.

7. Y. Cheng and G. M. Church. Biclustering of expression data. In *8th International Conference on Intelligent Systems for Molecular Biology*, pp. 93–103 , 2000.

8. M. B. Eisen, P. T. Spellman, P. O. Brown, and D. Botstein. Cluster analysis and display of genome-wide expression patterns. *Proc. Natl. Acad. Sci. USA*, 95(25): 14863–14868 , 1998.

9. S. Erdal, O. Ozturk, D. Armbruster, H. Ferhatosmanoglu, and W. C. Ray. A time series analysis of microarray data. In *4th IEEE International Symposium on Bioinformatics and Bioengineering*, May 2004.

10. J. Feng, P. E. Barbano, and B. Mishra. Time-frequency feature detection for timecourse microarray data. In *2004 ACM Symposium on Applied Computing*, 2004.

11. V. Filkov, S. Skiena, and J. Zhi. Analysis techniques for microarray time-series data. In *5th Annual International Conference on Computational Biology*, 2001.

12. J. H. Friedman and J. J. Meulman. Clustering objects on subsets of attributes. *J. R. Stat. Soc. Ser. B*, 66(4):815, 2004.

13. E. Hartuv, A. Schmitt, J. Lange, S. Meier-Ewert, H. Lehrach, and R. Shamir. An algorithm for clustering cdnas for gene expression analysis. In *3rd Annual International Conference on Computational Biology*, 1999.

14. D. Jiang, J. Pei, M. Ramanathany, C. Tang, and A. Zhang. Mining coherent gene clusters from gene-sample-time microarray data. In *10th ACM SIGKDD International Conference on Knowledge Discovery and Data Mining*, 2004.

15. J. Liu and W. Wang. Op-cluster: clustering by tendency in high dimensional spaces. In *3rd IEEE International Conference on Data Mining*, pp. 187–194 , 2003.

16. S. C. Madeira and A. L. Oliveira. Biclustering algorithms for biological data analysis: a survey. *IEEE/ACM Trans. Comput. Biol. Bioinf.*, 1(1):24–45 , 2004.

17. C. S. Möller-Levet, F. Klawonn, K. H. Cho, H. Yin, and O. Wolkenhauer. Clustering of unevenly sampled gene expression time-series data. *Fuzzy Sets Syst.*, 152(1):49–66, 2005.

18. T. M. Murali and S. Kasif. Extracting conserved gene expressionmotifs from gene expression data. In *Pacific Symposium on Biocomputing*, 2003.

19. C. M. Procopiuc, M. Jones, P. K. Agarwal, and T. M. Murali. A Monte Carlo algorithm for fast projective clustering. In *ACMSIGMOD International Conference on Management of Data*, 2002.

20. M. F. Ramoni, P. Sebastiani, and I. S. Kohane. Cluster analysis of gene expression dynamics. *Proc. Natl. Acad. Sci. USA*, 99(14):9121–9126 , July 2002.

21. R. Sharan and R. Shamir. CLICK: a clustering algorithm with applications to gene expression analysis. In *International Conference on Intelligent Systems for Molecular Biology*, 2000.

22. P. T. Spellman, G. Sherlock, M. Q. Zhang, V. R. Iyer, K. Anders, M. B. Eisen, P. O. Brown, D. Botstein, and B. Futcher. Comprehensive identification of cell cycle-regulated genes of the yeast *Saccharomyces cerevisiae* by microarray hybridization. *Mol. Biol. Cell*, 9(12):3273–3297 , Dec. 1998.

23. P. Tamayo, D. Slonim, J. Mesirov, Q. Zhu, S. Kitareewan, E. Dmitrovsky, E. S. Lander, and T. R. Golub. Interpreting patterns of gene expression with self-organizing maps: methods and application to hematopoietic differentiation. *Proc. Natl. Acad. Sci. USA*, 96(6):2907–2912 , 1999.

24. A. Tanay, R. Sharan, and R. Shamir. Discovering statistically significant biclusters in gene expression data. *Bioinfbrmatics*, 18(Suppl.1):S136–S144, 2002.

25. C. Tang, L. Zhang, A. Zhang, and M. Ramanathan. Interrelated two-way clustering: an unsupervised approach for gene expression data analysis. In *2nd IEEE International Symposium on Bioinfbrmatics and Bioengineering* BIBE, 2001.

26. H. Wang, W. Wang, J. Yang, and P. S. Yu. Clustering by pattern similarity in large data sets. In *ACM SIGMOD International Conference on Management of Data*, 2002.

27. E. P. Xing and R. M.Karp. Cliff: clustering high-dim microarray data via iterative feature filtering using normalized cuts. *Bioinfbrmatics*, 17(Suppl 1):S306–S315, 2001.

28. Y. Xu, V. Olman, and D. Xu. Clustering gene expression data using a graph-theoretic approach: an application of minimum spanning trees. *Bioinfbrmatics*, 18(4):536–545 , 2002.

29. J. Yang, W. Wang, H. Wang, and P. Yu. δ-clusters: capturing subspace correlation in a large data set. In *18th International Conference on Data Engineering* ICDE, 2002.

30. K. Y Yeung and W. L. Ruzzo. Principal component analysis for clustering gene expression data. *Bioinformatics*, 17(9):763–774 , 2001.

16

CLUSTERING METHODS IN A PROTEIN–PROTEIN INTERACTION NETWORK

CHUAN LIN, YOUNG-RAE CHO, WOO-CHANG HWANG, PENGJUN PEI, AND AIDONG ZHANG

Department of Computer Science and Engineering, State University of New York at Buffalo, Buffalo, New York

With completion of a draft sequence of the human genome, the field of genetics stands on the threshold of significant theoretical and practical advances. Crucial to furthering these investigations is a comprehensive understanding of the expression, function, and regulation of the proteins encoded by an organism. It has been observed that proteins seldom act as single isolated species in the performance of their functions; rather, proteins involved in the same cellular processes often interact with each other. Therefore, the functions of uncharacterized proteins can be predicted through comparison with the interactions of similar known proteins. A detailed examination of the protein–protein interaction (PPI) network can thus yield significant new understanding of protein function. Clustering is the process of grouping data objects into sets (clusters) which demonstrate greater similarity among objects in the same cluster than in different clusters. Clustering in the PPI network context groups together proteins that share a larger number of interactions. The results of this process can illuminate the structure of the PPI network and suggest possible functions for members of the cluster which were previously uncharacterized.

We begin the chapter with a brief introduction to the properties of protein–protein interaction networks, including a review of the data that have been generated by both experimental and computational approaches. A variety of methods employed to

Knowledge Discovery in Bioinformatics: Techniques, Methods, and Applications, Edited by Xiaohua Hu and Yi Pan
Copyright © 2007 John Wiley & Sons, Inc.

cluster these networks are then presented. These approaches are broadly characterized as either distance- or graph-based clustering methods. Techniques for validating
the results of these approaches are also discussed.

16.1 PROTEIN–PROTEIN INTERACTION

16.1.1 Proteome in Bioinformatics

With the completion of a draft sequence of the human genome, the field of genetics
stands on the threshold of significant theoretical and practical advances. Crucial to
furthering these investigations is a comprehensive understanding of the expression,
function, and regulation of the proteins encoded by an organism [96]. This understanding is the subject of the discipline of proteomics. Proteomics encompasses a
wide range of approaches and applications intended to explicate how complex
biological processes occur at a molecular level, how they differ in various cell types,
and how they are altered in disease states.

Defined succinctly, proteomics is the systematic study of the many and diverse
properties of proteins with the aim of providing detailed descriptions of the structure,
function, and control of biological systems in health and disease [68]. The field has
burst onto the scientific scene with stunning rapidity over the past several years.
Figure 16.1 shows the trend of the number of occurrences of the term *'proteome'*
found in PubMed bioinformatics citations over the past decade. This figure illustrates
strikingly the rapidly increasing role played by proteomics in bioinformatics research
in recent years.

A particular focus of the field of proteomics is the nature and role of interactions
between proteins. Protein–protein interactions play diverse roles in biology and differ
based on the composition, affinity, and lifetime of the association. Noncovalent
contacts between residue side chains are the basis for protein folding, protein
assembly, and protein–protein interaction [65]. These contacts facilitate a variety
of interactions and associations within and between proteins. Based on their diverse
structural and functional characteristics, protein–protein interactions can be

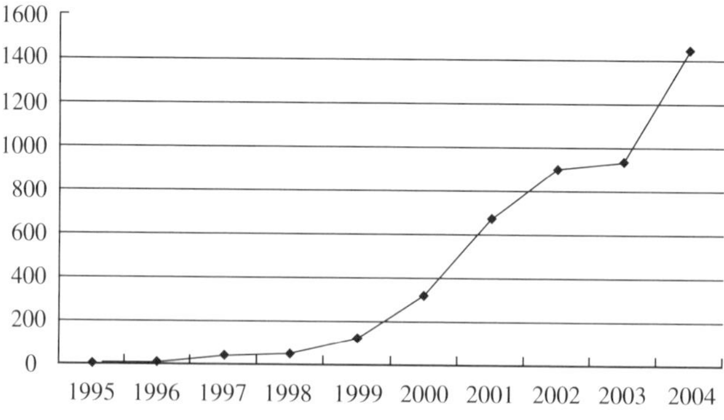

FIGURE 16.1 Results found in PubMed for proteome.

categorized in several ways [64]. On the basis of their interaction surface, they may be homo- or hetero-oligomeric; as judged by their stability, they may be obligate or nonobligate; and as measured by their persistence, they may be transient or permanent. A given protein–protein interaction can fall into any combination of these three categorical pairs. An interaction may also require reclassification under certain conditions; for example, it may be mainly transient in vivo but become permanent under certain cellular conditions.

16.1.2 Significance of Protein–Protein Interaction

It has been observed that proteins seldom act as single isolated species while performing their functions in vivo [91]. The analysis of annotated proteins reveals that proteins involved in the same cellular processes often interact with each other [86]. The function of unknown proteins may be postulated on the basis of their interaction with a known protein target of known function. Mapping protein–protein interactions has not only provided insight into protein function but has facilitated the modeling of functional pathways to elucidate the molecular mechanisms of cellular processes. The study of protein interactions is fundamental to understanding how proteins function within the cell. Characterizing the interactions of proteins in a given cellular proteome will be the next milestone along the road to understanding the biochemistry of the cell.

The result of two or more proteins interacting with a specific functional objective can be demonstrated in several different ways. The measurable effects of protein interactions have been outlined by Phizicky and Fields [74]. Protein interactions can:

- Alter the kinetic properties of enzymes; this may be the result of subtle changes at the level of substrate binding or at the level of an allosteric effect
- Act as a common mechanism to allow for substrate channeling
- Create a new binding site, typically for small effector molecules
- Inactivate or destroy a protein
- Change the specificity of a protein for its substrate through interaction with different binding partners (e.g., demonstrate a new function that neither protein can exhibit alone)

Protein–protein interactions are much more widespread than once suspected, and the degree of regulation that they confer is large. To understand their significance in the cell, one needs to identify the different interactions, understand the extent to which they take place in the cell, and determine the consequences of the interaction.

16.1.3 Experimental Approaches for PPI Detection

In early reviews, physicochemical approaches for detecting protein–protein interactions included site-directed mutagenesis or chemical modification of amino acid groups participating in such interactions [52,66,79,84]. In the following subsections we discuss these bioinformatic and functional proteomic methods. These include predictions of protein–protein interaction via the yeast two-hybrid system, mass spectrometry, and protein microarrays.

Yeast Two-Hybrid System   One of the most common approaches to the detection of pairs of interacting proteins in vivo is the yeast two-hybrid (Y2H) system [7,36]. The Y2H system, which was developed by Fields and Song [23], is a molecular–genetic tool which facilitates the study of protein–protein interactions [1]. The interaction of two proteins transcriptionally activates a reporter gene, and a color reaction is seen on specific media. This indication can track the interaction between two proteins, revealing "prey" proteins which interact with a known "bait" protein.

The yeast two-hybrid system enables both highly sensitive detection of protein–protein interactions and screening of genome libraries to ascertain the interaction partners of certain proteins. The system can also be used to pinpoint protein regions mediating the interactions [37]. However, the classic Y2H system has several limitations. First, it cannot, by definition, detect interactions involving three or more proteins and those depending on posttranslational modifications except those applied to the budding yeast itself [37]. Second, since some proteins (e.g., membrane proteins) cannot be reconstructed in the nucleus, the yeast two-hybrid system is not suitable for the detection of interactions involving these proteins. Finally, the method does not guarantee that an interaction indicated by Y2H actually takes place physiologically.

Recently, numerous modifications of the Y2H approach have been proposed which characterize protein–protein interaction networks by screening each protein expressed in a eukaryotic cell [24]. Drees [19] has proposed a variant that includes the genetic information of a third protein. Zhang et al. [92] have suggested the use of RNA for the investigation of RNA–protein interactions. Vidal [85] used the URA3 gene instead of GAL4 as the reporter gene; this two-hybrid system can be used to screen for ligand inhibition or to dissociate such complexes. Johnson and Varshavsky [43] have proposed a cytoplasmic two-hybrid system that can be used for screening of membrane protein interactions.

Despite the various limitations of the Y2H system, this approach has revealed a wealth of novel interactions and has helped illuminate the magnitude of the protein interactome. In principle, it can be used in a more comprehensive fashion to examine all possible binary combinations between the proteins encoded by any single genome.

Mass Spectrometry Approaches   Another traditional approach to PPI detection is to use quantitative mass spectrometry to analyze the composition of a partially purified protein complex together with a control purification in which the complex of interest is not enriched. Mass spectrometry–based protein interaction experiments have three basic components: bait presentation, affinity purification of the complex, and analysis of the bound proteins [2]. Two large-scale studies [25,35] have been published on the protein–protein interaction network in yeast. Each study attempted to identify all the components that were present in "naturally" generated protein complexes, which requires essentially pure preparations of each complex [49]. In both approaches, bait proteins were generated that carried a particular affinity tag. In the case studied by

Gavin et al. [25], 1739 TAP-tagged genes were introduced into the yeast genome by homologous recombination. Ho et al. [35] expressed 725 proteins modified to carry the FLAG epitope. In both cases the proteins were expressed in yeast cells, and complexes were purified using a single immunoaffinity purification step. Both groups resolved the components of each purified complex with a one-dimensional denaturing polyacrylamide gel electrophoresis (PAGE) step. From the 1167 yeast strains generated by Gavin et al. [25], 589 protein complexes were purified, 232 of which were unique. Ho et al. [35] used 725 protein baits and detected 3617 interactions that involved 1578 different proteins.

Mass spectrometry (MS)–based proteomics can be used not only in protein identification and quantification [16,50,72,89] but also for protein analysis, which includes protein profiling [51], posttranslational modifications (PTMs) [55,56], and in particular, identification of protein–protein interactions.

Compared with two-hybrid approaches, mass spectrometry–based methods are more effective in characterizing highly abundant stable complexes. MS-based approaches permit the isolation of large protein complexes and the detection of networks of protein interactions. The two-hybrid system is more suited to the characterization of binary interactions, particularly to the detection of weak or transient interactions.

Protein Microarray Microarray-based analysis is a relatively high-throughput technology that allows the simultaneous analysis of thousands of parameters within a single experiment. The key advantage of the microarray format is the use of a nonporous solid surface, such as glass, which permits precise deposition of capturing molecules (probes) in a highly dense and ordered fashion. The early applications of microarrays and detection technologies were largely centered on DNA-based applications. Today, DNA microarray technology is a robust and reliable method for the analysis of gene function [12]. However, gene expression arrays provide no information on protein posttranslational modifications (such as phosphorylation or glycosylation) that affect cell function. To examine expression at the protein level and acquire quantitative and qualitative information about proteins of interest, the protein microarray was developed.

A protein microarray is a piece of glass on which various molecules of protein have been affixed at separate locations in an ordered manner, forming a microscopic array [54]. These are used to identify protein–protein interactions, to identify the substrates of protein kinases, or to identify the targets of biologically active small molecules. The experimental procedure for protein microarray involves choosing solid supports, arraying proteins on the solid supports, and screening for protein–protein interactions.

Experiments with the yeast proteome microarray have revealed a number of protein–protein interactions which had not previously been identified through Y2H- or MS-based approaches. Global protein interaction studies were performed with a yeast proteome chip. Ge [26] has described a universal protein array that permits quantitative detection of protein interactions with a range of proteins, nucleic acids, and small molecules. Zhu et al. [95] generated a yeast proteome chip from recombinant protein probes of 5800 open-reading frames.

16.1.4 Computational Methods to Predict Protein–Protein Interaction

The yeast two-hybrid system and other experimental approaches provide a useful tool for the detection of protein–protein interactions occurring in many possible combinations between specified proteins. The widespread application of these methods has generated a substantial bank of information about such interactions. However, the data generated can be erroneous, and these approaches are often not completely inclusive of all possible protein–protein interactions. To form an understanding of the total universe of potential interactions, including those not detected by these methods, it is useful to develop an approach to predict possible interactions between proteins. The accurate prediction of protein–protein interactions is therefore an important goal in the field of molecular recognition.

A number of approaches to PPI prediction are based on the use of genome data. Pellegrini et al. [71] introduced the first such method, which predicts an interaction between two proteins in a given organism if these two proteins have homologs in another organism. A subsequent extension proposed by Marcotte et al. [57,58] detects colocalization of two genes in different genomes. Two proteins in different organisms are predicted to interact if they have consecutive homologs in a single organism. Dandekar et al. [17] used the adjacency of genes in various bacterial genomes to predict functional relationships between the corresponding proteins. Proteins whose genes are physically close in the genomes of various organisms are predicted to interact.

Jasen et al. [40] investigated the relationship between protein–protein interaction and mRNA expression levels by analyzing existing yeast data from a variety of sources and identifying general trends. Two different approaches were used to analyze the two types of available expression data; normalized differences were computed for absolute expression levels, while a more standard analysis of profile correlations was applied to relative expression levels. This investigation indicated that a strong relationship exists between expression data and most permanent protein complexes.

Some researchers have used data-mining techniques to extract useful information from large data sources. Oyama et al. [67] used a method termed *association rules discovery* to identify patterns and other features from accumulated protein–protein interaction data. This research mined data from four different sources. This aggregated data included 4307 unique protein interaction pairs. General rules were derived from 5241 features extracted from the functional, primary-structural, and other aspects of proteins. After transforming the traditional protein-based transaction data into interaction-based transaction data, Oyama was able to detect and articulate 6367 rules. Of these, 5271 rules had at least one feature pertaining to sequences. As this potential had been suggested by other researchers, these results confirmed the efficacy of this method.

As mentioned above, experimental and computational approaches have generated significant quantities of PPI data, but these data sets are typically incomplete, contradictory, and include many false positives. It is therefore necessary for improved accuracy to integrate evidence from many different sources for evaluating protein–protein interactions. Jansen et al. [39] proposed a Bayesian approach for integrating interaction information that allows for the probabilistic combination of multiple data sets and demonstrates its application to yeast data. This approach assesses each source for interactions by comparison with samples of known positives and negatives, yielding a statistical measure of reliability. The likelihood of possible interactions for every protein pair is then predicted by combing each independent data source, weighted according to its reliability. The predictions were validated by TAP (tandem affinity purification) tagging experiments. It was observed that at given levels of sensitivity, the predictions were more accurate than the existing high-throughput experimental data sets.

16.2 PROPERTIES OF PPI NETWORKS

Although reductionism has long been the prevailing paradigm guiding the interpretation of experimental results, it has become increasingly evident that a discrete biological function can only rarely be attributed to an individual molecule. Rather, many biological characteristics arise from complex interactions between numerous cellular constituents, such as proteins, DNA, RNA, and small molecules [4,34,44,46]. Therefore, understanding the structure and dynamics of the complex intercellular web of interactions has become a central focus of biological investigation.

16.2.1 PPI Network Representation

An investigation of protein–protein interaction mechanisms begins with the representation and characterization of the PPI network structure. The simplest representation takes the form of a mathematical graph consisting of nodes and edges (or links) [88]. Proteins are represented as nodes in such a graph; two proteins that interact physically are represented as adjacent nodes connected by an edge.

Degree The *degree* (or connectivity) of a node is the number of other nodes with which it is connected [9]. It is the most elementary characteristic of a node. For example, in the undirected network, Figure 16.2, node A has degree $k = 5$.

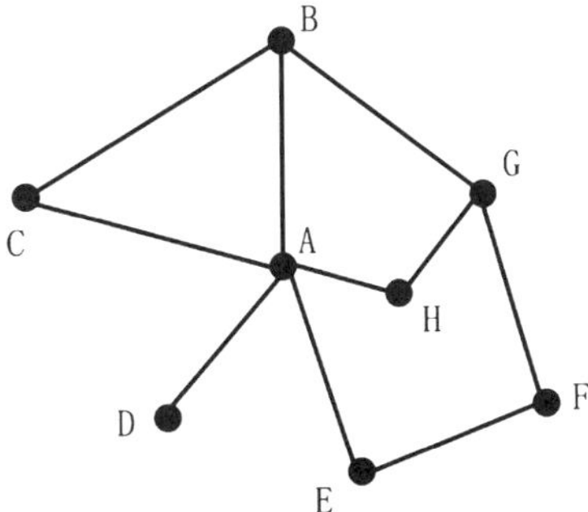

FIGURE 16.2 Graph in which a node has a degree of 5. (Adapted from ref. 9.)

Path, Shortest Path, and Mean Path   The *path* between two nodes is a sequence of adjacent nodes. The number of edges in this path is termed the *path length*, and distances within network are measured in terms of path length. As there are many alternative paths between two nodes, the *shortest path* between the specified nodes refers to the path with the smallest number of links. The *mean path length* of the network represents the average over the shortest paths between all pairs of nodes.

Degree Distribution   Graph structures can be described according to numerous characteristics, including the distribution of path lengths, the number of cyclic paths, and various measures to compute clusters of highly connected nodes [88]. Barabasi and Oltvai [9] introduced the concept of *degree distribution*, $P(k)$, to quantify the probability that a selected node will have exactly k links. $P(k)$ is obtained by tallying the total number of nodes $N(k)$ with k links and dividing this figure by the total number of nodes N. Different network classes can be distinguished by the degree distribution. For example, a random network follows a Poisson distribution. By contrast, a scale-free network has a power-law degree distribution, indicating that a few hubs bind numerous small nodes. Most biological networks are scale-free, with degree distributions approximating a power law, $P(k) \sim k^{-\gamma}$. When $2 \leq \gamma \leq 3$, the hubs play a significant role in the network [9].

Clustering Coefficient   In many networks, if node A is connected to B, and B is connected to C, then A has a high probability of direct linkage to C. Watts [90] quantified this phenomenon using the clustering coefficient, $C_I = 2n_I/k_I(k_I - 1)$, where n_I is the number of links connecting the k_I neighbors of node I to each other. In this coefficient, n_I indicates the number of triangles that pass through node I, and $k_I(k_I - 1)/2$ is the total number of triangles that could pass through node I. For example, in Figure 16.2, $n_A = 1$ and $C_A = 1/10$, while $n_F = 0$, $C_F = 0$.

The average degree, average path length, and average clustering coefficient depend on the number of nodes and links in the network. However, the degree distribution $P(k)$ and clustering coefficient $C(k)$ functions are independent of the size of the network and represent its generic features. These functions can therefore be used to classify various network types [9].

16.2.2 Characteristics of Protein–Protein Networks

Scale-Free Network   Recent publications have indicated that protein–protein interactions have the features of a scale-free network [29,41,53,87], meaning that their degree distribution approximates a power law, $P(k) \sim k^{-\gamma}$. In scale-free networks, most proteins participate in only a few interactions, while a few (termed '*hubs*') participate in dozens of interactions.

Small-World Effect   Protein–protein interaction networks have an characteristic property known as the '*small-world effect*', which states that any two nodes can be connected via a short path of a few links. The small-world phenomenon was first investigated as a concept in sociology [61] and is a feature of a range of networks arising in nature and technology, including the Internet [3], scientific collaboration networks [63], the English lexicon [77], metabolic networks [22], and protein–protein interaction networks [78,87]. Although the small-world effect is a property of random networks, the path length in scale-free networks is much shorter than that predicted by the small-world effect [14,15]. Therefore, scale-free networks are 'ultrasmall'. This short path length indicates that local perturbations in metabolite concentrations could permeate an entire network very quickly.

Disassortativity In protein–protein interaction networks, highly connected nodes (hubs) seldom link directly to each other [59]. This differs from the assortative nature of social networks, in which well-connected people tend to have direct connections to each other. By contrast, all biological and technological networks have the property of disassortativity, in which highly connected nodes are infrequently linked each other.

16.3 CLUSTERING APPROACHES

16.3.1 Significance of Clustering in a PPI Network

A cluster is a set of objects that share some common characteristics. *Clustering* is the process of grouping data objects into sets (clusters) that demonstrate greater similarity among objects in the same cluster than in different clusters. Clustering differs from classification; in the latter, objects are assigned to predefined classes, while clustering defines the classes themselves. Thus, clustering is an unsupervised classification method, which means that it does not rely on training the data objects in predefined classes.

In protein–protein interaction networks, clusters correspond to two types of modules: protein complexes and functional modules. *Protein complexes* are groups of proteins that interact with each other at the same time and place, forming a single multimolecular machine. *Functional modules* consist of proteins that participate in a particular cellular process while binding to each other at a different time and place.

Clustering in protein-protein interaction networks therefore involves identifying protein complexes and functional modules. This process has the following analytical benefits:

1. Clarification of PPI network structures and their component relationships
2. Inference of the principal function of each cluster from the functions of its members
3. Elucidation of possible functions of members in a cluster through comparison with the functions of other members

16.3.2 Challenges of Clustering in PPI Networks

The classic clustering approaches follow a protocol termed '*pattern proximity after feature selection*' [38]. Pattern proximity is usually measured by a distance function defined for pairs of patterns. A simple distance measure can often be used to reflect dissimilarity between two patterns, while other similarity measures can be used to characterize the conceptual similarity between patterns. However, in protein–protein interaction networks, proteins are represented as nodes and interactions are represented as edges. The relationship between two proteins is therefore a simple binary value: 1 if they interact, 0 if they do not. This lack of nuance makes it difficult to define the distance between the two proteins. Additionally, a high rate of false positives and the sheer volume of data render problematic to the reliable clustering of PPI networks.

Clustering approaches for PPI networks can be broadly characterized as distance based or graph based. *Distance-based clustering* uses classic clustering techniques and focuses on the definition of the distance between proteins. *Graph-based clustering* includes approaches that consider the topology of the PPI network. Based on the structure of the network, the density of each subgraph is maximized or the cost of cutoff minimized while separating the graph. In this following section we discuss each of these clustering approaches in greater detail.

16.3.3 Distance-Based Clustering

Distance Measure Based on a Coefficient   As discussed in [30], the distance between two nodes (proteins) in a PPI network can be defined as follows. Let X be a set of n elements and let $d_{ij} = d(i,j)$ be a nonnegative real function $d : X \times X \to R^+$, which satisfies:

1. $d_{ij} > 0$ for $i \neq j$.
2. $d_{ij} = 0$ for $i = j$.
3. $d_{ij} = d_{ji}$ for all i,j, where d is a distance measure and $D = \{d_{ij}\}$ is a distance matrix.
4. If d_{ij} satisfies triangle inequality $d_{ij} \leq d_{ik} + d_{kj}$, then d is a metric.

In PPI network, the binary vectors $X_i = (x_{i1}, x_{i2}, \ldots, x_{iN})$ represent the set of protein purifications for N proteins, where x_{ik} is 1 if the ith protein interacts with kth protein (the kth protein is presented in the ith purification) and 0 otherwise. If a distance can be determined that accounts fully for known protein complexes, unsupervised hierarchical clustering methods can be used to accurately assemble protein complexes from the data. Frequently, a distance can be obtained easily from a simple matching coefficient that calculates the similarity between two elements. The similarity value S_{ij} can be normalized between 0 and 1, and the distance can be derived from $d_{ij} = 1 - S_{ij}$. If the similarity value of two elements is high, the spatial distance between them should be short.

Several suitable measures have been proposed for this purpose. These include: the Jaccard coefficient [32]:

$$S_{mn} = \frac{X_{mn}}{X_{mm} + X_{nn} - X_{mn}} \tag{16.1}$$

the Dice coefficient [32]:

$$S_{mn} = \frac{2X_{mn}}{X_{mm} + X_{nn}} \tag{16.2}$$

the Simpson coefficient [32]:

$$S_{mn} = \frac{X_{mn}}{\min(X_{mm}, X_{nn})} \tag{16.3}$$

the Bader coefficient [8]:

$$S_{mn} = \frac{X_{mn}^2}{X_{mm} \times X_{nn}} \tag{16.4}$$

the Maryland Bridge coefficient [62]:

$$S_{mn} = \frac{1}{2}\left(\frac{X_{mn}}{X_{mm}} + \frac{X_{mn}}{X_{nn}}\right) \tag{16.5}$$

the Korbel coefficient [47]:

$$S_{mn} = \frac{\sqrt{X_{mm}^2 + X_{nn}^2}}{\sqrt{2}\,X_{mm}X_{nn}} X_{mn} \tag{16.6}$$

the correlation coefficient [20]:

$$S_{mn} = \frac{X_{mn} - n\overline{X}_m\overline{X}_n}{\sqrt{(X_{mm} - n\overline{X}_m^{\,2})(X_{nn} - n\overline{X}_n^{\,2})}} \tag{16.7}$$

where $X_{ij} = X_i \cdot X_j$ (dot product of two vectors). The value of S_{mn} ranges from 0 to 1. X_{ij} is equal to the number of bits "on" in both vectors, and X_{ii} is equal to the number of bits "on" in one vector. For example, for the case illustrated in Figure 16.2, the matrix X is

$$X = \begin{bmatrix} 0 & 1 & 1 & 1 & 0 & 0 & 1 & 1 \\ 1 & 0 & 1 & 0 & 0 & 1 & 0 & 0 \\ 1 & 1 & 0 & 0 & 0 & 0 & 0 & 0 \\ 1 & 0 & 0 & 0 & 0 & 0 & 0 & 0 \\ 0 & 0 & 0 & 0 & 0 & 1 & 0 & 1 \\ 0 & 1 & 0 & 0 & 1 & 0 & 1 & 0 \\ 1 & 0 & 0 & 0 & 0 & 1 & 0 & 0 \\ 1 & 0 & 0 & 0 & 1 & 0 & 0 & 0 \end{bmatrix} \tag{16.8}$$

To calculate the distance between A and B, d_{12}, $X_{11} = X_1 \cdot X_1 = 5$, $X_{22} = X_2 \cdot X_2 = 3$, $X_{12} = X_1 \cdot X_2 = 1$. The Jaccard coefficient is calculated as: $S_{12} = 1/(5 + 3 - 1) = 0.1429$; the distance is then $d_{12} = 1 - 0.1429 = 0.8571$.

This group of distance-based approaches uses classic distance measurements, which are not quite suitable for high-dimensional spaces. In a high-dimensional space, the distances between each pair of nodes are almost the same for a large data distribution [10]. Therefore, it is hard to attain ideal clustering results by the simplest distance measurements only.

Distance Measure by Network Distance   There are other definitions based on network distance that give more fine-grained distance measurements for these pairs. In the definition given above, the distance value will be zero for any two proteins not sharing an interaction partner. In [75], each edge in the network is assigned a length of 1. The length of the shortest path (distance) between every pair of vertices in the network is calculated to create an all-pairs-shortest-path distance matrix. Each distance in this matrix is then transformed into an '*association*', defined as $1/d^2$, where d is the shortest-path distance. This transformation emphasizes local associations (short paths) in the subsequent clustering process. The resulting associations range from zero to 1. The association of a vertex with itself is defined as 1, while the association of vertices that have no connecting path is defined as zero. Two vertices that are more widely separated in the network will have a longer shortest-path distance and thus a smaller association. The association value can therefore be served as the similarity measure for two proteins.

In [69], authors consider the paths of various lengths between two vertices in a weighted protein interaction network. The weight of an edge reflects its reliability and lies in the range between zero and 1. The *PathStrength* of a path is defined as the product of the weights of all the edges on the path. Then the *k-length PathStrength between two vertices* is defined as the sum of the PathStrength of all k-length paths between the two vertices. The PathStrength of a path captures the probability that a walk on the path can reach its ending vertex. By summing upon all these paths, the k-length PathStrength between two vertices captures the strength of connections between these two vertices by a k-step walk. Since paths of different lengths should have a different impact on the connection between two vertices, the k-length Path-Strength is normalized by the k-length maximum possible path stength to get the *k-length PathRatio*. Finally, the *PathRatio* measure between two vertices is defined as the sum of the k-length PathRatios between the two vertices for all $k > 1$. Although this measure is applied primarily to assess the reliability of detected interactions and predicting potential interactions that are missed by current experiments, it can also be used as a similarity measure for clustering.

Another network distance measure was developed by Zhou [93,94]. He defined the distance d_{ij} from node i to node j as the average number of steps a Brownian particle takes to reach j from i.

Consider a connected network of N nodes and M edges. Its node set is denoted by $V = \{1, \ldots, N\}$ and its connection pattern is specified by the generalized adjacency matrix A. If there is no edge between node i and node j, $A_{ij} = 0$; if there is an edge between those nodes, $A_{ij} = A_{ji} > 0$, and its value signifies the interaction strength. The set of nearest neighbors of node I is denoted by E_i. As a Brownian particle moves throughout the network, at each time step it jumps from its present position i to a nearest-neighboring position j. When no additional information about the network is known, the jumping probability $P_{ij} = A_{ij} / \sum_{l=1}^{N} A_{il}$ can be assumed. Matrix P is called the *transfer matrix*.

The node–node distance d_{ij} from i to j is defined as the average number of steps needed for the Brownian particle to move from i through the network to j. Using simple linear algebraic calculations, it is obvious that

$$d_{ij} = \sum_{l=1}^{n} \frac{1}{I - B(j)} il \tag{16.9}$$

where I is the $N \times N$ identity matrix and matrix $B(j)$ equals the transfer matrix P, with the exception that $B_{lj}(j) \equiv 0$ for any $l \in V$. The distances from all the nodes in V to node j can thus be obtained by solving the linear algebraic equation

$$[I - B(j)]\{d_{1j}, \ldots, d_{nj}\}^{\mathrm{T}} = \{1, \ldots, 1\}^{\mathrm{T}} \tag{16.10}$$

For example, in the network shown in Figure 16.3, with the set nodes $V = 1, 2, 3, 4$, the adjacency matrix A and transfer matrix P are

$$A = \begin{bmatrix} 0 & 1 & 1 & 1 \\ 1 & 0 & 1 & 0 \\ 1 & 1 & 0 & 0 \\ 1 & 0 & 0 & 0 \end{bmatrix}, \qquad P = \begin{bmatrix} 0 & \frac{1}{3} & \frac{1}{3} & \frac{1}{3} \\ \frac{1}{2} & 0 & \frac{1}{2} & 0 \\ \frac{1}{2} & \frac{1}{2} & 0 & 0 \\ 1 & 0 & 0 & 0 \end{bmatrix}$$

$B(j)$ can be derived from P:

$$B(1) = \begin{bmatrix} 0 & \frac{1}{3} & \frac{1}{3} & \frac{1}{3} \\ 0 & 0 & \frac{1}{2} & 0 \\ 0 & \frac{1}{2} & 0 & 0 \\ 0 & 0 & 0 & 0 \end{bmatrix}, \qquad B(2) = \begin{bmatrix} 0 & 0 & \frac{1}{3} & \frac{1}{3} \\ \frac{1}{2} & 0 & \frac{1}{2} & 0 \\ \frac{1}{2} & 0 & 0 & 0 \\ 1 & 0 & 0 & 0 \end{bmatrix}$$

$$B(3) = \begin{bmatrix} 0 & \frac{1}{3} & 0 & \frac{1}{3} \\ \frac{1}{2} & 0 & 0 & 0 \\ \frac{1}{2} & \frac{1}{2} & 0 & 0 \\ 1 & 0 & 0 & 0 \end{bmatrix}, \qquad B(4) = \begin{bmatrix} 0 & \frac{1}{3} & \frac{1}{3} & 0 \\ \frac{1}{2} & 0 & \frac{1}{2} & 0 \\ \frac{1}{2} & \frac{1}{2} & 0 & 0 \\ 1 & 0 & 0 & 0 \end{bmatrix}$$

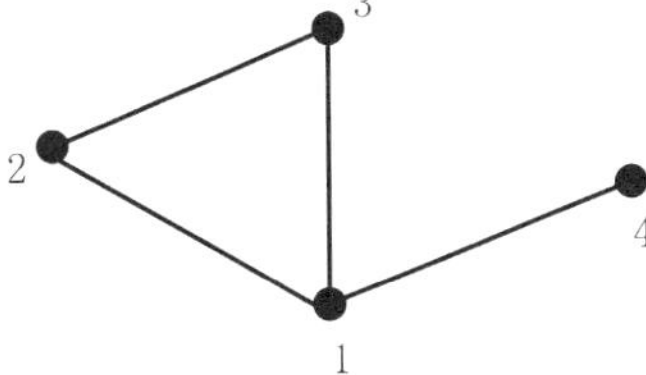

FIGURE 16.3 Example of distance measure by Brownian particle.

The distance between any two nodes can be calculated using equation (16.9):

$$
D = \{d_{ij}\} = \begin{bmatrix} \frac{8}{3} & 2 & 2 & 1 \\ \frac{10}{3} & 4 & \frac{8}{3} & \frac{13}{3} \\ \frac{10}{3} & \frac{8}{3} & 4 & \frac{13}{3} \\ \frac{23}{11} & \frac{27}{11} & \frac{9}{11} & \frac{34}{11} \end{bmatrix}
$$

Based on the distance measure, Zhou [93] defined a dissimilarity index to quantify the relationship between any two nearest-neighboring nodes. Nearest-neighboring vertices of the same community tend to have a small dissimilarity index, whereas those belonging to different communities tend to have a high dissimilarity index.

Given two vertices i and j that are nearest neighbors ($A_{ij} > 0$), the difference in their perspectives regarding the network can be quantitatively measured. The dissimilarity index $\Lambda(i,j)$ is defined by the expression

$$
\Lambda(i,j) = \frac{\sqrt{\sum_{k \neq i,j}^{n} [d_{ik} - d_{jk}]^2}}{n - 2} \tag{16.11}
$$

If two nearest-neighboring vertices i and j belong to the same community, the average distance d_{ik} from i to any another vertex $k(k \neq i,j)$ will be quite similar to the average distance d_{jk} from j to k. This indicates that the perspectives of the network as viewed from i and j will be quite similar. Consequently, $\Lambda(i,j)$ will be small if i and j belong to the same community and large if they belong to different communities.

When this approach is applied to a protein interaction network, clusters of proteins that may be of biological significance can be constructed. Zhou provided three examples of such an application. Most of the proteins in these examples were involved in known functions. It was possible to predict similar biological functions for the few proteins in each cluster that were previously unanalyzed.

UVCLUSTER   The UVCLUSTER [6] approach is informed by the observation that the shortest path distance between protein pairs is typically not very fine grained and that many pairs have the same distance value. This method proposes an iterative approach to distance exploration; unlike other distance-based approaches, it converts the set of primary distances into secondary distances. The secondary distance measures the strength of the connection between each pair of proteins when the interactions for all the proteins in the group are considered. Secondary distance is derived by first applying a hierarchical clustering step based on the affinity coefficient to generate N different clustering results. The number of solutions generated in which any two proteins selected are not in the same cluster is defined as the secondary distance between the two proteins. Defined succinctly, the secondary distance represents the likelihood that the two proteins will not be in the same cluster.

This approach has four steps:

1. A *primary distance d* between any two proteins in a PPI network is measured by the minimum number of steps required to connect them. Each valid step is a known, physical protein–protein interaction. Users are allowed to select groups of proteins to be analyzed either by choosing a single protein and establishing a cutoff distance value or by providing the program with a list of proteins.

2. Next, agglomerative hierarchical clustering is applied to the subtable of primary distances generated in the first step to produce N alternative and equally valid clustering solutions. The user specifies a value for N before starting the analysis. UVCLUSTER first randomly samples the elements of the dataset and then clusters them according to the group average linkage. The agglomerative process ends when the affinity coefficient (AC) is reached, defined as

$$AC = 100 \, \frac{P_m - C_m}{P_m - 1} \qquad (16.12)$$

where C_m (the cluster mean) is the average of the distances for all elements included in the clusters and P_m (the partition mean) is the average value of distances for the entire set of selected proteins. The AC value is selectly by the user at the start of the process.

3. Once the data set of N alternative solutions has been obtained, the number of pairs of elements that appear together in the same cluster is counted. A *secondary distance d'* between two elements is defined as the number of solutions in which those two elements do not appear together in the same cluster, divided by the total number of solutions (N). In effect, the secondary distance iteratively resamples the original primary distance data, thus indicating the strength of the connection between two elements. The secondary distance represents the likelihood that each pair of elements will appear in the same cluster when many alternative clustering solutions are generated.

4. After the generation of secondary distance data, the proteins can be clustered using conventional methods such as UPGMA (unweighted pair group method with arithmetic mean) or neighbor joining. The results of an agglomerative hierarchical clustering process in which UPGMA is applied to the secondary distance data are placed in a second UVCLUSTER output file. A third output file contains a graphical representation of the data in PGM (portable graymap) format. To generate the PGM file, proteins are ordered according to the results described in the second output file.

The use of UVCLUSTER offers four significant benefits. First, the involvement of the secondary distance value facilitates identification of sets of closely linked proteins. Furthermore, it allows the incorporation of previously known information in the discovery of proteins involved in a particular process of interest. Third, guided by the AC value, it can establish groups of connected proteins even when some information is currently unavailable. Finally, UVCLUSTER can compare the relative positions of orthologous proteins in two species to determine whether they retain related functions in both of their interactomes.

Similarity Learning Method   By incorporating very limited annotation data, a similarity learning method is introduced in [70]. The method defines the similarity between two proteins in a probabilistic framework. Edges in the network are regarded as a means of message passing. Each protein propagates its function to neighboring proteins. Meanwhile, each protein receives these function messages from its neighboring proteins to decide its own function. The final probability of a protein having a specific function is therefore a conditional probability defined on its neighbors status of having this function annotation. For a certain functional label in consideration, the probability of a protein A having this function is $P(A)$. Another protein B s probability of having this function by propagation using A as the information source can then be represented as a conditional probability $P(B|A)$. This conditional probability gives the capability of A s function being transferred to B via the network. The similarity between proteins A and B is defined as the product of two conditional probabilities:

$$\text{Similarity}_{AB} = P(A|B)P(B|A)$$

Now the problem of estimating the similarity between two proteins is changed into estimating the two conditional probabilities. For this purpose, a statistic model is defined to predict the conditional probabilities using topological features. Since in most organisms, there are certain amount of annotation data for proteins, some training samples are available. The method uses a two-step approach:

1. *Model training step*. Known annotation data are used to estimate the parameters in the model.
2. *Conditional probability estimation step*. The numerical values of the conditional probabilities are calculated using the model and the parameters estimated in step 1.

The unsupervised clustering method can be applied to the resulting similarity matrix.

Summary   We have reviewed a series of approaches to distance-based clustering. The first category of approaches uses classic distance measurement methods, which offered a variety of coefficient formulas to compute the distance between proteins in PPI networks. The second class of approaches defines a distance measure based on network distance, including the shortest path length, combined strength of paths of various lengths, and the average number of steps that a Brownian particle takes to move from one vertex to another. The third approach type, exemplified by UVCLUS-TER, defines a primary and a secondary distance to establish the strength of the connection between two elements in relation to all the elements in the analyzed data set. The fourth is a similarity learning approach incorporating some annotation data. Although each of these four approaches involves different methods for distance measurement, they all apply classic clustering approaches to the distance computed between proteins.

16.3.4 Graph-Based Clustering

A protein–protein interaction network is an unweighted graph in which the weight of each edge between any two proteins is either 1 or 0. In this section we explore graph-based clustering, another class of approaches to the process of clustering. Graph-based clustering techniques are presented explicitly in terms of a graph, thus converting the process of clustering a data set into such graph-theoretic problems as finding a minimum cut or maximal subgraphs in the graph G.

Finding Dense Subgraphs   The goal of this class of approaches is to identify the densest subgraphs within a graph; specific methods vary in the means used to assess the density of the subgraphs. Five variations on this theme are discussed.

Enumeration of Complete Subgraphs   This approach is to identify all fully connected subgraphs (cliques) by complete enumeration [80]. In general, finding all cliques within a graph is an NP-complete problem. Exceptionally, however, this problem is antimonotonic, meaning that if a subset of set A is not a clique, set A is also not a clique. Because of this property, regions of density can be identified quickly in sparse graphs. In fact, to find cliques of size n, one need only enumerate those cliques that are of size $n - 1$. Assume a process that starts from the smallest statistically significant number, which is 4 in the case depicted in Figure 16.4. All possible pairs of edges in the nodes will be considered. For example, as shown in Figure 16.4, to examine the edge *AB* and *CD*, we must check for edges between *AC, AD, BC,* and *BD*. If these edges all connect, they are considered fully connected, and a clique *ABCD* has thus been identified. To test every identified clique *ABCD*, all known proteins will be selected successively. If for protein *E*, there exist *EA, EB, EC,* and *ED*, the clique will be expanded to *ABCDE*. The end result of this process is the generation of cliques that are fully internally connected.

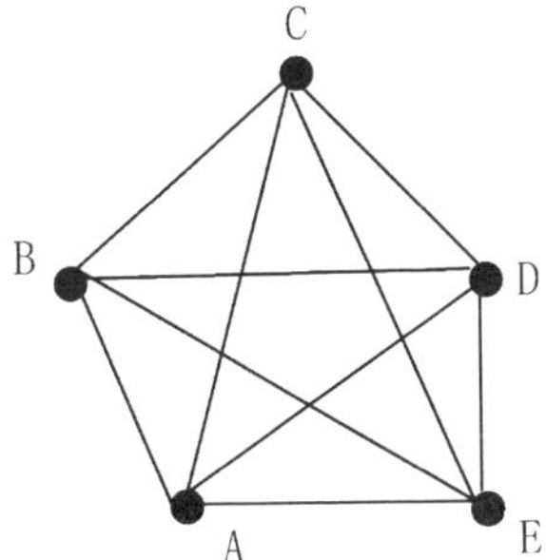

FIGURE 16.4   Example of enumeration of complete subgraphs.

Although this approach is simple, it has several drawbacks. The basic assumption underlying the method—that cliques must be fully internally connected—does not accurately reflect the real structure of protein complexes and modules. Dense subgraphs are not necessarily fully connected. In addition, many interactions in the protein network may fail to be detected experimentally, thus leaving no trace in the form of edges.

Monte Carlo Optimization   Seeking to address these issues, Spirin and Mirny [80] introduced a new approach that searches for highly connected rather than fully connected sets of nodes. This was conceptualized as an optimization problem involving the identification of a set of n nodes that maximizes the object function Q, defined as

$$Q(P) = \frac{2m}{n(n-1)} \tag{16.13}$$

The term m enumerates the edges (interactions) among the n nodes in the subgraph P. In this formula, the function Q characterizes the density of a cluster. If the subset is fully connected, Q equals 1; if the subset has no internal edge , Q equals 0. The goal is to find a subset with n nodes which maximizes the objective function Q.

A Monte Carlo approach is used to optimize the procedure. The process starts with a connected subset S of n nodes. These nodes are randomly picked from the graph and then updated by adding or deleting selected nodes from set S, then remain the nodes that increase function Q of S. These steps are repeated until the maximum Q is identified; this yields an n-node subset with high density.

Another quality measure used in this approach is the sum of the shortest distances between selected nodes. Correspondingly, a similar Monte Carlo approach is applied to minimize this value. This process proceeds as follows. At time $t = 0$, a random set of M nodes is selected. For each pair of nodes i and j from this set, the shortest path L_{ij} between i and j on the graph is calculated. The sum of all shortest paths L_{ij} from this set is denoted as L_0. At each time step, one of M nodes is selected randomly and replaced at random by one from among its neighbors. To assess whether the original node is to be replaced by this neighbor, the new sum of all shortest paths, L_1, is then calculated. If $L_1 < L_0$, the replacement is accepted with probability 1. If $L_1 > L_0$, the replacement is accepted with probability $e^{-(L_1-L_0/T)}$, where T is the effective temperature. At every tenth time step, an attempt is made to replace one of the nodes from the current set with a node that has no edges with the current set. This procedure ensures that the process is not caught in an isolated disconnected subgraph. This process is repeated either until the original set converges to a complete subgraph or for a predetermined number of steps. The tightest subgraph, defined as the subgraph corresponding to the smallest L_0, is then recorded. The clusters recorded are merged and redundant clusters are removed. The use of a Monte Carlo approach allows smaller pieces of the cluster to be identified separately rather than focusing exclusively on the entire cluster. Monte Carlo simulations are therefore well suited to recognizing highly dispersed cliques.

The experiments conducted by Spirin started with the enumeration of all cliques of size 3 and larger in a graph with $N = 3992$ nodes and $M = 6500$ edges. Additionally, 1000 random graphs of the same size and degree distribution were constructed for comparison. Using the approach described above, more than 50 protein clusters of sizes from 4 to 35 were identified. In contrast, the random networks contained very few such clusters. This work indicated that real complexes have many more interactions than the tightest complexes found in randomly rewired graphs. In particular, clusters in a protein network have many more interactions than their counterparts in random graphs.

Redundancies in a PPI Network   Samanta and Liang [76] took a statistical approach to the clustering of proteins. This approach assumes that two proteins that share a significantly larger number of common neighbors than would arise randomly will have close functional associations. This method first ranks the statistical significance of forming shared partnerships for all protein pairs in the PPI network and then combines the pair of proteins with least significance. The p-value is used to rank the statistical significance of the relationship between two proteins. In the next step, the two proteins with smallest p-value are combined and are considered to be in the same cluster. This process is repeated until a threshold is reached. The steps of the algorithm are described in more detail below.

First, the p-values [81] for all possible protein pairs are computed and stored in a matrix. The formula of computing p-value between two proteins is

$$
\begin{aligned}
P(N, n_1, n_2, m) &= \frac{\binom{N}{m}\binom{N-m}{n_1-m}\binom{N-n_1}{n_2-m}}{\binom{N}{n_1}\binom{N}{n_2}} \\
&= \frac{\binom{n_1}{m}\binom{N-n_1}{n_2-m}}{\binom{N}{n_2}} \\
&= \frac{(N-n_1)!(N-n_2)!n_1!n_2!}{N!m!(n_1-m)!(n_2-m)!(N-n_1-n_2+m)!}
\end{aligned}
\tag{16.14}
$$

where N is the number of the proteins in the network, each protein in the pair has n_1 and n_2 neighbors, respectively, and m is the number of neighbors shared by both proteins. This formula is symmetric with respect to interchange of n_1 and n_2. It is a ratio in which the denominator is the total number of ways that two proteins can have n_1 and n_2 neighbors. In the numerator, the first term represents the number of ways by which m common neighbors can be chosen from all N proteins. The second term represents the number of ways by which $n_1 - m$ remaining neighbors can be selected from the remaining $N - m$ proteins. The last term represents the number of ways by which $n_2 - m$ remaining neighbors can be selected, none of which can match any of the n_1 neighbors of the first protein.

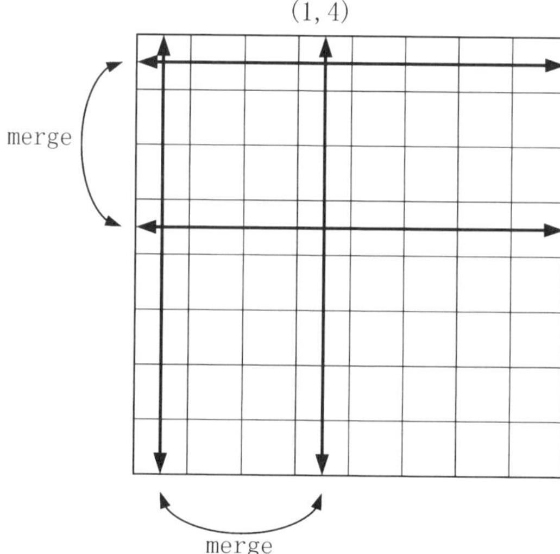

FIGURE 16.5   If the element (m, n) has the lowest p-value, a cluster is formed with proteins m and n. Therefore, rows and columns m and n are merged with a new p-value of the merged row and column as a geometric mean of the separate p-values of the corresponding elements. (Adapted from ref. 76.)

Second, the protein pair with the lowest p-value is designated as the first group in the cluster. As illustrated in Figure 16.5, the rows and columns for these two proteins are merged into one row and one column. The probability values for this new group are the geometric means of the two original probabilities (or the arithmetic means of the log P values). This process is repeated until a threshold is reached, adding elements to increase the size of the original cluster. The protein pair with the second-lowest p-value is selected to generate the next cluster.

As mentioned in Section 3.2, a high rate of false positives typically creates significant noise, which disrupts the clustering of protein complexes and functional modules. This method overcomes this difficulty by using a statistical technique that forms reliable functional associations between proteins from noisy interaction data. The statistical significance of forming shared partnerships for all protein pairs in the interaction network is ranked. This approach is grounded on the hypothesis that two proteins that have a significantly larger number of common interaction pairs in the measured data set than would arise randomly will also have close functional links [76].

To validate this hypothesis, all possible protein pairs were ranked in the order of their probabilities. For comparison, the corresponding probabilities were examined for a random network with the same number of nodes and edges but with different connections. The connections in the random network were generated from a uniform distribution. The comparison suggests that the associations in the real data set contain biologically meaningful information. It also indicates that such low-probability associations did not arise simply from the scale-free nature of the network.

Molecular Complex Detection (MCODE)   MCODE, proposed by Bader and Hogue [8], is an effective approach to detecting densely connected regions in large protein–protein interaction networks. This method weights a vertex by local neighborhood density, chooses a few seeds with high weight, and isolates the dense regions according to given parameters. The MCODE algorithm operates in three steps: vertex weighting, complex prediction, and optional postprocessing to filter or add proteins to the resulting complexes according to certain connectivity criteria.

In the first step, all vertices are weighted based on their local network density using the highest k-core of the vertex neighborhood. The core-clustering coefficient of a vertex v is defined to be the density of the highest k-core of the vertices connected directly to v and also v itself (which is called the immediate neighborhood of v). Compared with the traditional clustering coefficient, the core-clustering coefficient amplifies the weighting of heavily interconnected graph regions while removing the many less-connected vertices that are usually part of a biomolecular interaction network. For each vertex v, the weight of v is

$$w = k \times d \tag{16.15}$$

where d is the density of the highest k-core graph from the set of vertices, including all the vertices connected directly with v and vertex v itself. For example, using the example provided in Figure 16.2, the two-core weight of node A is $(2)[(2)(5)/(5)(5-1)] = 1$. It should be noted that node D is not included in the two-core node set because the degree of node D is 1.

The second step of the algorithm is the prediction of molecular complexes. With a vertex-weighted graph as input, a complex with the highest-weighted vertex is selected as the seed. Once a vertex is included, its neighbors are inspected recursively to determine if they are part of the complex. Then the seed is expanded to a complex until a threshold is encountered. The algorithm assumes that complexes cannot overlap (this condition is more fully addressed in the third step), so a vertex is not checked more than once. This process stops when, as governed by the specified threshold, no additional vertices can be added to the complex. The vertices included in the complex are marked as having been examined. This process is repeated for the next-highest unexamined weighted vertex in the network. In this manner, the densest regions of the network are identified. The vertex weight threshold parameter defines the density of the resulting complex.

Postprocessing occurs optionally in the third step of the algorithm. Complexes are filtered out if they do not contain at least a two-core node set. The algorithm may be run with the "fluff" option, which increases the size of the complex according to a given fluff parameter between 0.0 and 1.0. For every vertex v in the complex, its neighbors are added to the complex if they have not yet been examined and if the neighborhood density (including v) is higher than the given fluff parameter. Vertices that are added by the fluff parameter are not marked as examined, so there can be overlap among predicted complexes with the fluff parameter set.

Evaluated using the Gavin et al. [25] and MIPS [60] data set, MCODE effectively finds densely connected regions of a molecular interaction network based solely on connectivity data. Many of these regions correspond to known molecular complexes.

Summary   In this subsection we introduced a series of graph-based clustering approaches which are structured to maximize the density of subgraphs. The first approach examined seeks to identify fully connected subgraphs within the network. The second approach improves on this method by optimizing a density function for finding highly connected rather than fully connected subgraphs. The third approach merges pairs of proteins with the lowest *p*-values, indicating that those proteins have a strong relationship, to identify the dense subgraphs within the network. The final approach discussed assigns each vertex a weight to represent its density in the entire graph and uses the vertex with the highest weight as the seed to generate to a dense subgraph. These approaches all use the topology of the graph to find a dense subgraph within the network and to maximize the density of each subgraph.

Finding a Minimum Cut   A second category of graph-based clustering approaches generates clusters by trimming or cutting a series of edges to divide the graph into several unconnected subgraphs. Any edge which is removed should be the least important (minimum) in the graph, thus minimizing the informational cost of removing the edges. Here, the least important is based on the structure of the graph. It doesn't mean the interaction between these two proteins is not important. This subsection will present several techniques which are based upon this method.

Highly Connected Subgraph (HCS) Algorithm   The HCS method [33] is a graph-theoretic algorithm that separates a graph into several subgraphs using minimum cuts. The resulting subgraphs satisfy a specified density threshold. Despite its interest in density, this method differs from approaches discussed earlier which seek to identify the densest subgraphs. Rather, it exploits the inherent connectivity of the graph and cuts the most unimportant edges to find highly connected subgraphs.

Some graph-theoretic concepts should be defined at this point. The *edge-connectivity* $k(G)$ of a graph G is the minimum number k of edges whose removal results in a disconnected graph. If $k(G) = l$, G is termed an *l-connected* or *l-connectivity graph*. For example, in Figure 16.6, the graph G is a two-connectivity graph because we need at least cut two edges (dashed lines in graph) to produce a disconnected graph. A *highly connected subgraph* (HCS) is defined as a subgraph whose *edge-connectivity* exceeds half the number of vertices. For example, in Figure 16.6, graph G_1 is a *highly connected subgraph* because its *edge-connectivity* $k(G) = 3$ is more than half of the vertices number. A *cut* in a graph is a set of edges whose removal disconnects the graph. A *minimumcut* (abbreviated *mincut*) is a cut with a minimum number of edges. Thus, a cut S is a minimum cut of a nontrivial graph G iff $|S| = k(G)$. The length of a path between two vertices consists of the number of edges in the path. The distance $d(u, v)$ between vertices u and v in graph G is the minimum length of their connecting path, if such a path exists; otherwise $d(u, v) = \infty$. The diameter of a connected graph G, denoted $\mathrm{diam}(G)$, is

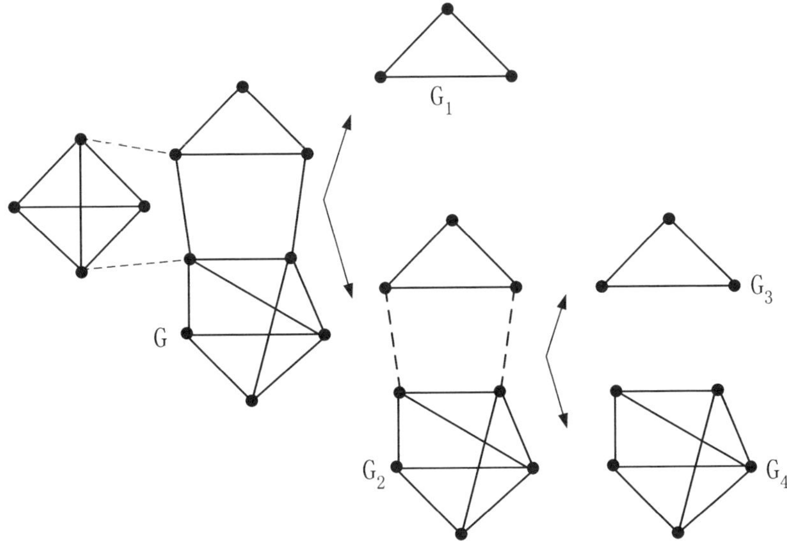

FIGURE 16.6 Example of applying the HCS algorithm to a graph. Minimum cut edges are denoted by dashed lines. (Adapted from ref. 33.)

the longest distance between any two vertices in G. The degree of vertex v in a graph, denoted $\deg(v)$, is the number of edges incident to the vertex.

The algorithm identifies highly connected subgraphs as clusters. The HCS algorithm is detailed below, and Figure 16.6 contains an example of its application. Graph G is first separated into two subgraphs G_1 and G_2, with G_1 being a *highly connected subgraph* and G_2, not. Subgraph G_2 is separated into subgraphs G_3 and G_4. This process produces three *highly connected subgraphs* G_1, G_3, and G_4, which are considered clusters.

HCS$(G(V, E))$ algorithm

```
begin
  (H, H̄, C) ← MINCUT (G)
  if G is highly connected
    then return(G)
  else
    HCS (H)
    HCS (H̄)
end
```

The HCS algorithm generates solutions with desirable properties for clustering. The algorithm has low polynomial complexity and is efficient in practice. Heuristic improvements made to the initial formulation have allowed this method to generate useful solutions for problems with thousands of elements in a reasonable computing time.

Restricted Neighborhood Search Clustering Algorithm (RNSC)   In [45], King et al. proposed a cost-based local search algorithm based on the tabu search metaheuristic [31]. In the algorithm, a clustering of a graph $G = (V, E)$ is defined as a partitioning of the node set V. The process begins with an initial random or user-input clustering and defines a cost function. Nodes are then randomly added to or removed from clusters to find a partition with minimum cost. The cost function is based on the number of invalid connections. An invalid connection incident with v is a connection that exists between v and a node in a different cluster, or alternatively, a connection that does not exist between v and a node u in the same cluster as v. The process begins with an initial random or user-input clustering and defines a cost function. Nodes are then randomly added to or removed from clusters to find a partition with minimum cost. The cost function is based on the number of invalid connections.

Consider a node v in a graph G, and a clustering C of the graph. Let α_v be the number of invalid connections incident with v. The naive cost function of C is then defined as

$$C_n(G, C) = \frac{1}{2}\sum_{v \in V} \alpha_v \tag{16.16}$$

where V is the set of nodes in G. For a vertex v in G with a clustering C, let β_v be the size of the following set: v itself, any node connected to v, and any node in the same cluster as v. This measure reflects the size of the area that v influences in the clustering. The scaled cost function of C is defined as

$$C_n(G, C) = \frac{|V| - 1}{3}\sum_{v \in V} \frac{\alpha_v}{\beta_v} \tag{16.17}$$

For example, in Figure 16.7, if the eight vertices are grouped into two clusters as shown, the naive cost function $C_n(G, C) = 2$, and the scaled cost function $C_n(G, C) = \frac{20}{9}$.

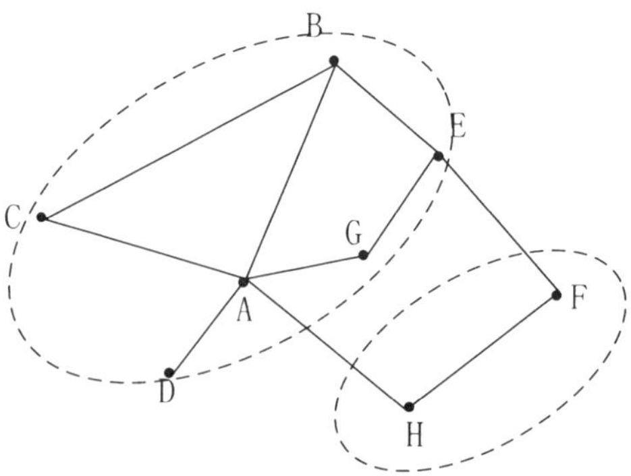

FIGURE 16.7   Example of the RNSC approach.

Both cost functions seek to define a clustering scenario in which the nodes in a cluster are all connected to one another and there are no other connections between two clusters. The RNSC approach searches for a low-cost clustering solution by optimizing an initial state. Starting with an initial clustering defined randomly or by user input, the method iteratively moves a node from one cluster to another in a random manner. Since the RNSC is randomized, different runs on the same input data will result in different clustering results. To achieve high accuracy in predicting true protein complexes, the RNSC output is filtered according to a maximum p-value selected for functional homogeneity, a minimum density value, and a minimum size. Only clusters that satisfy these three criteria are presented as predicted protein complexes.

Superparamagnetic Clustering (SPC) The SPC method uses an analogy to the physical properties of an inhomogenous ferromagnetic model to find tightly connected clusters in a large graph [11,27,28]. Every node on the graph is assigned a Potts spin variable $S_i = 1, 2, \ldots, q$. The value of this spin variable S_i engages in thermal fluctuations that are determined by the temperature T and the spin values of the neighboring nodes. Two nodes connected by an edge are likely to have the same spin value. Therefore, the spin value of each node tends to align itself with that of the majority of its neighbors.

The SPC procedure proceeds via the following steps:

1. Assign to each point $\vec{x_i}$ a q-state Potts spin variable S_i.
2. Find the nearest neighbors of each point according to a selected criterion; measure the average nearest-neighbor distance a.
3. Calculate the strength of the nearest-neighbor interactions using

$$J_{ij} = J_{ji} = \frac{1}{\hat{K}} \exp\left(-\frac{||\vec{x_i} - \vec{x_j}||^2}{2a^2}\right) \tag{16.18}$$

   where $\hat{K}$ is the average number of neighbors per site.
4. To calculate the susceptibility χ, use an efficient Monte Carlo procedure with

$$\chi = \frac{N}{T}(\langle m^2 \rangle - \langle m \rangle^2), \qquad m = \frac{(N_{\max}/N)q - 1}{q - 1} \tag{16.19}$$

   where $N_{\max} = \max\{N_1, N_2, \ldots, N_q\}$ and N_μ is the number of spins with the value μ.
5. Identify the range of temperatures corresponding to the superparamagnetic phase, between T_{fs}, the temperature of maximal χ, and the (higher) temperature T_{ps}, where χ diminishes abruptly. Cluster assignment is performed at $T_{\mathrm{clus}} = (T_{fs} + T_{ps})/2$.

6. Once the J_{ij} have been determined, the spin–spin correlation function can be obtained by a Monte Carlo procedure. Measure at $T = T_{\text{clus}}$ the spin–spin correlation function, $\langle \delta_{S_i,S_j} \rangle$, for all pairs of neighboring points $\vec{x_i}$ and $\vec{x_j}$.

7. Clusters are identified according to a thresholding procedure. If $\langle \delta_{S_i,S_j} \rangle > \theta$, points $\vec{x_i}$ and $\vec{x_j}$, are defined as "friends." Then all mutual friends (including friends of friends, etc.) are assigned to the same cluster.

The SPC algorithm is robust in conditions with noise and initialization errors and has been shown to identify natural and stable clusters with no requirement for prespecifying the number of clusters. Additionally, clusters of any shape can be identified.

Markov Clustering (MCL) The MCL algorithm was designed specifically for application to simple and weighted graphs [82] and was initially used in the field of computational graph clustering [83]. The MCL algorithm finds cluster structures in graphs by a mathematical bootstrapping procedure. The MCL algorithm simulates random walks within a graph by the alternation of expansion and inflation operations.

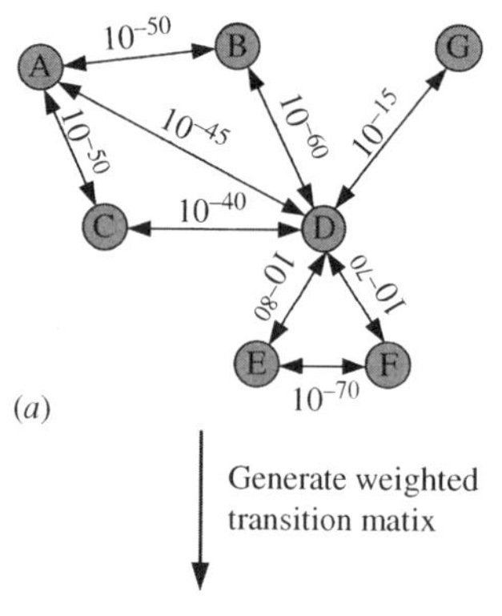

Generate weighted transition matix

	A	B	C	D	E	F	G
A	100	50	50	45	0	0	0
B	50	100	0	60	0	0	0
C	50	0	100	40	0	0	0
D	45	60	40	100	80	70	15
E	0	0	0	80	100	70	0
F	0	0	0	70	70	100	0
G	0	0	0	15	0	0	100

(b)

Transform weights into column-wise transition probablities

	A	B	C	D	E	F	G
A	0.42	0.24	0.20	0.11	0.00	0.00	0.00
B	0.20	0.48	0.24	0.15	0.00	0.00	0.00
C	0.20	0.00	0.40	0.10	0.00	0.00	0.00
D	0.18	0.28	0.16	0.24	0.32	0.29	0.13
E	0.00	0.00	0.00	0.19	0.40	0.29	0.00
F	0.00	0.00	0.00	0.17	0.28	0.42	0.00
G	0.00	0.00	0.00	0.04	0.00	0.00	0.87

(c)

FIGURE 16.8 (*a*) Example of a protein–protein similarity graph for seven proteins (A to F); circles represent proteins (nodes), and lines (edges) represent BLASTp similarities detected with *E*-values (also shown). (*b*) Weighted transition matrix for the seven proteins shown in part (*a*). (*c*) associated column stochastic Markov matrix for the seven proteins shown in part (*a*). (Adapted from ref. 21.)

Expansion refers to taking the power of a stochastic matrix using the normal matrix product. *Inflation* corresponds to taking the Hadamard power of a matrix (taking powers entrywise), followed by a scaling step, so that the resulting matrix is again stochastic.

Enright et al. [21] employed the MCL algorithm for the assignment of proteins to families. A protein–protein similarity graph is represented as described in Section 16.2 and as illustrated in Figure 16.8*a*. Nodes in the graph represent proteins that are desirable clustering candidates, while edges within the graph are weighted according to a sequence similarity score obtained from an algorithm such as BLAST [5]. Therefore, the edges represent the degree of similarity between these proteins.

A Markov matrix [shown in Figure 16.8*b*] is then constructed in which each entry in the matrix represents a similarity value between two proteins. Diagonal elements are set arbitrarily to a "neutral" value and each column is normalized to produce a column total of 1. This Markov matrix is then provided as input to the MCL algorithm.

As noted above, the MCL algorithm simulates random walks within a graph by alternating two operators: expansion and inflation. The structure of the MCL algorithm is described by the flowchart in Figure 16.9. After parsing and normalization of the similarity matrix, the algorithm starts by computing the graph of random walks of

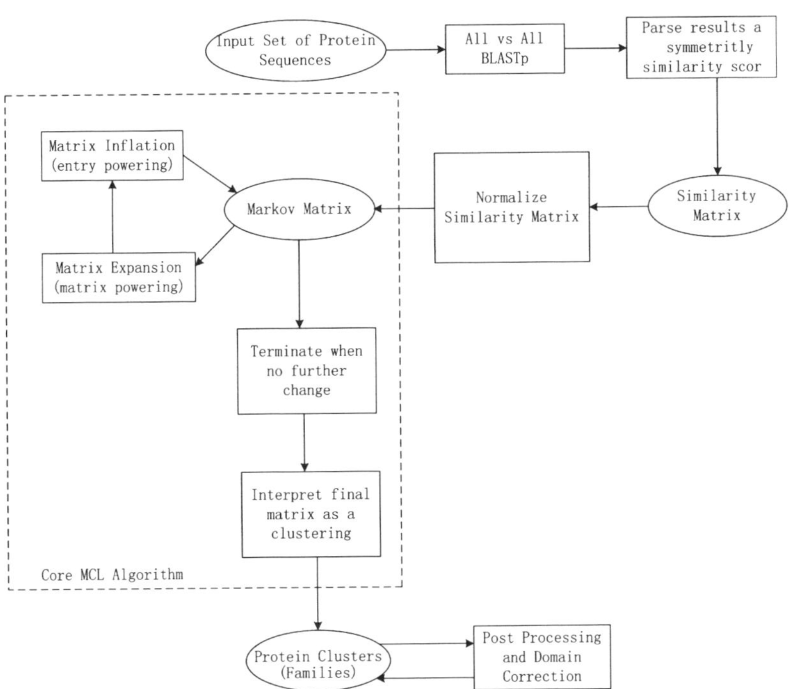

FIGURE 16.9 Flowchart of the TRIBE-MCL algorithm. (From ref. 21 with permission from Oxford University Press.)

an input graph, yielding a stochastic matrix. It then uses iterative rounds of the expansion operator, which takes the squared power of the matrix, and the inflation operator, which raises each matrix entry to a given power and then rescales the matrix to return it to a stochastic state. This process continues until there is no further change in the matrix. The final matrix is interpreted as protein clusters with some postprocessing and domain correction.

Given a matrix $M \in R^{k \times k}, M > 0$, and a real number, $r > 1$, the column stochastic matrix resulting from inflating each of the columns of M with power coefficient r is denoted by $\Gamma_r M$, and Γ_r represents the inflation operator with power coefficient r. Formally, the action of $\Gamma_r : R^{k \times k} \to R^{k \times k}$ is defined by

$$(\Gamma_r M)_{pq} = \frac{(M_{pq})^r}{\sum_{i=1}^{k}(M_{iq})^r} \qquad (16.20)$$

Each column j of a stochastic matrix M corresponds to node j of the stochastic graph associated with the probability of moving from node j to node i. For values of $r > 1$, inflation changes the probabilities associated with the collection of random walks departing from one particular node by favoring more probable over less probable walks.

Here expansion and inflation are used iteratively in the MCL algorithm to strengthen the graph where it is strong and to weaken where it is weak until equilibrium is reached. At this point, clusters can be identified according to a threshold. If the weight between two proteins is less than the threshold, the edge between them can be deleted. An important advantage of the algorithm is its "bootstrapping" nature, retrieving cluster structure via the imprint made by this structure on the flow process. Additionally, the algorithm is fast and very scalable, and its accuracy is not compromised by edges between different clusters. The mathematics underlying the algorithm is indicative of an intrinsic relationship between the process it simulates and cluster structure in the input graph.

Line Graph Generation   Pereira-Leal et al. [73] expressed the network of proteins (e.g., nodes) connected by interactions (e.g., edges) as a network of connected interactions. Figure 16.10*a* exemplifies an original protein interaction network graph in which the nodes represent proteins and the edges represent interactions. Periera-Leal's method generates from this an associated line graph, such as that depicted in Figure 16.10*b*, in which edges now represent proteins and nodes represent interactions. This simple procedure is commonly used in graph theory.

First, the protein interaction network is transformed into a weighted network, where the weights attributed to each interaction reflect the degree of confidence attributed to that interaction. Confidence levels are determined by the number of experiments as well as the number of different experimental methodologies that support the interaction. Next, the network connected by interactions is expressed as a network of interactions, which is known in graph theory as a line graph. Each interaction is condensed into a node that includes the two interacting proteins. These nodes are then linked by shared protein content. The scores for the original constituent

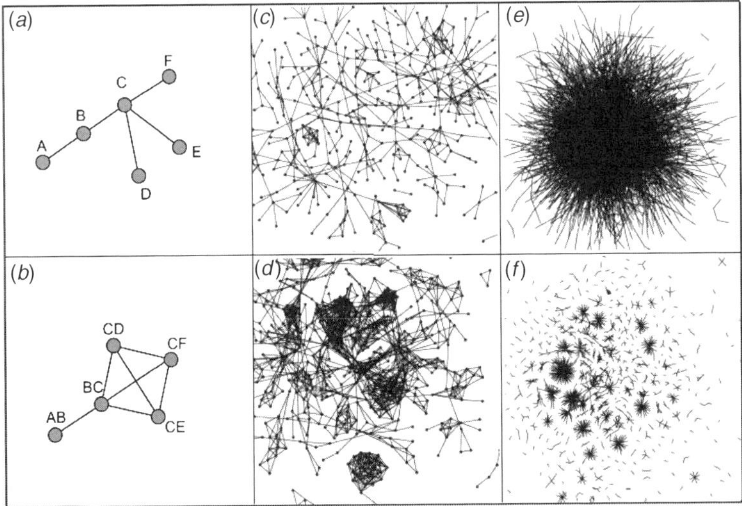

FIGURE 16.10 Transforming a network of proteins to a network of interactions. (*a*) Schematic representation illustrating a graph representation of protein interactions; nodes correspond to proteins and edges to interactions. (*b*) Schematic representation illustrating the transformation of the protein graph connected by interactions to an interaction graph connected by proteins. Each node represents a binary interaction, and edges represent shared proteins. Note that labels that are not shared correspond to terminal nodes in part (*a*): in this particular case, A, D, E, and F in edges AB, CD, CE, and CF. (*c*) Graph illustrating a section of a protein network connected by interactions. (*d*) Graph illustrating the increase in structure as an effect of transforming the protein graph in (*c*) to an interaction graph. (*e*) Graph representation of yeast protein interactions in DIP. (*f*) Graph representing a pruned version of (*e*) with the interactions reconstituted after transformation and clustering. These graphs were produced using BioLayout. (From ref. 73 with permission from Wiley-Liss, Inc., a Wiley Company.)

interactions are then averaged and assigned to each edge. Finally, an algorithm for clustering by graph flow simulation, TribeMCL [21], is used to cluster the interaction network and then to reconvert the identified clusters from an interaction–interaction graph back to a protein–protein graph for subsequent validation and analysis.

This approach focuses on the structure of the graph itself and what it represents. It has been included here among the graph-based minimum cutting approaches because it employs the MCL method of clustering. This approach has a number of attractive features. It does not sacrifice informational content, because the original bidirectional network can be recovered at the end of the process. Furthermore, it takes into account the higher-order local neighborhood of interactions. Additionally, the graph it generates is more highly structured than the original graph. Finally, it produces an overlapping graph partitioning of the interaction network, implying that proteins may be present in multiple functional modules. Many other clustering approaches cannot place elements in multiple clusters. This represents a significant inability on the part of those approaches to represent the complexity of biological systems, where proteins may participate in multiple cellular processes and pathways.

Pereira-Leal's group used the protein interaction network derived from the yeast subset of the database of interacting proteins, which consists of 8046 physical interactions involving 4081 yeast proteins. For each protein in a cluster, the research team obtained manually derived regulatory and metabolic classifications (KEGG), automatic functional classifications (GQFC), and cellular localization information (LOC) from KEGG, GeneQuiz, and MIPS, respectively. On average, the coverage of clusters is 20 regulatory and metabolic roles in KEGG, 45 classes in GeneQuiz, and 48 classes in MIPS.

Summary   In this subsection we profiled a selection of graph-based clustering approaches that minimize the cost of cutting edges. The first approach discussed defines a highly connected subgraph and then performs a minimum cut repeatedly until all subgraphs are highly connected. The second approach searches the space of partitions of all nodes efficiently and assigns each a cost function related to cutting the edges in the graph. Identification of the lowest-cost partitions becomes synonymous with finding those clusters with minimum cutting. The third approach assigns each node a Potts spin value and computes the spin–spin correlation function. If the correlation between two spins exceeds a threshold, the two proteins are assigned to the same cluster. The MCL algorithm, which was the fourth approach presented, uses iterative rounds of expansion and inflation to promote flow through the graph where it is strong and to remove flow where it is weak. Clusters are then generated via minimum cutting. The final approach discussed transforms the network of proteins connected by interactions into a network of connected interactions and then uses the MCL algorithm to cluster the interaction network. The first two approaches use the topology of the network to remove the edges in the network; in these methods, the edges have no weight. The other approaches assign each edge a weight that represents the similarity of two proteins; edges with low weights are then cut.

16.4   VALIDATION

So far, we have reviewed a series of approaches to clustering within protein–protein interaction networks. These approaches aim to find functional modules to predict unannotated protein functions based on the structure of an annotated PPI network. However, disparate results can be generated using different approaches, and even from the repeated application of a given approach with different parameters. Therefore, these solutions must be compared carefully with predicted results to select the approach and parameters that provide the best outcome. Validation is a process of evaluating the performance of the clustering or prediction results derived from different approaches. In this section we introduce several basic validation approaches for clustering used in proteomics.

A survey performed by Jiang et al. [42] of clustering of gene expression data revealed three main components to cluster validation: evaluation of performance based on ground truth, an intuitive assessment of cluster quality, and an assessment of the reliability of the cluster sets. These components are also relevant to the evaluation of clustering performance in proteomics.

16.4.1 Validation Based on Agreement with Annotated Protein Function Databases

Clustering results can be compared with ground truth derived from various protein domain databases, such as InterPro, the structural classification of protein (SCOP) database, and the Munich Information Center (MIPS) hierarchical functional categories [13,21,48]. These databases are collections of well-characterized proteins that have been expertly classified into families based on their folding patterns and a variety of other information.

In Jiang's et al. [42] work, some simple validation methods are listed that use construction of an $n \times n$ matrix C based on the clustering results, where n is the number of data objects. $C_{ij} = 1$ if object pairs O_i and O_j belong to the same cluster, and $C_{ij} = 0$ otherwise. Similarly, a matrix P is built based on the ground truth. Several indices are defined to measure the degree of similarity between C and P.

However, simply counting matches while comparing each predicted cluster against each complex in a data set does not provide a robust evaluation. In cases where each cluster corresponds to a purification, a maximal number of matches will be found, which leads to maximally redundant results. Krause et al. [48] defined the criteria to assess the fit of the clustering results to the benchmark data set:

1. The number of clusters matching ground truth should be maximal.
2. The number of clusters matching an individual complex should be 1.
3. Each cluster should map to one complex only. Clusters matching more than one complex are possibly predicted too inclusively.
4. Complexes should have an average size and size distribution similar to that of the data set.

Application of these criteria allows a more accurate comparison between clustering results and ground truth, as a one-to-one correspondence is required between predicted clusters and complexes.

16.4.2 Validation Based on the Definition of Clustering

Clustering is defined as the process of grouping data objects into sets by degree of similarity. Clustering results can be validated by computing the homogeneity of predicted clusters or the extent of separation between two predicted clusters. The quality of a cluster C increases with higher homogeneity values within C and lower separation values between C and other clusters.

The homogeneity of clusters may be defined in various ways; all measure the similarity of data objects within cluster C.

$$H_1(C) = \frac{\sum_{O_i, O_j \in C, O_i \neq O_j} \text{Similarity}(O_i, O_j)}{\| C \| \cdot (\| C \| - 1)} \tag{16.21}$$

$$H_2(C) = \frac{1}{\| C \|} \sum_{O_i \in C} \text{Similarity}(O_i, \overline{O}) \tag{16.22}$$

H_1 represents the homogeneity of cluster C by the average pairwise object similarity within C. H_2 evaluates the homogeneity with respect to the 'centroid' of the cluster C, where $\bar{O}$ is the centroid of C.

Cluster separation is defined analogously from various perspectives to measure the dissimilarity between two clusters C_1 and C_2. For example:

$$S_1(C_1, C_2) = \frac{\sum_{O_i \in C_1, O_j \in C_2} \text{Similarity}(O_i, O_j)}{\| C_1 \| \cdot \| C_2 \|} \tag{16.23}$$

$$S_2(C_1, C_2) = \text{Similarity}(\overline{O}_1, \overline{O}_2) \tag{16.24}$$

16.4.3 Validation Based on the Reliability of Clusters

The performance of clustering results can also be validated by the *reliability* of clusters, which refers to the likelihood that the cluster structure has not arisen by chance. The significance of the derived clusters is typically measured by the p-value.

In [13], Bu et al. mapped 76 uncharacterized proteins in 48 quasi-cliques in the MIPS hierarchical functional categories. Each protein was assigned a function according to the main function of its hosting quasi-clique. For each cluster, p-values were calculated to measure the statistical significance of functional category enrichment. The p-value is defined as

$$p = 1 - \sum_{i=0}^{k-1} \frac{\binom{C}{i}\binom{G-C}{n-i}}{\binom{G}{n}} \tag{16.25}$$

where C is the total number of proteins within a functional category and G is the total number of proteins within the graph. The authors regarded as significant those clusters with p-values smaller than $0.01/N_C$ (here N_C is the number of categories).

16.4.4 Validation for Protein Function Prediction:
Leave-One-Out Method

Deng et al. [18] used a leave-one-out method to measure the accuracy of clustering predictions. This method randomly selects a protein with known functions and hypothesizes its functions to be unknown. Prediction methods are then used to predict its functions, and these are compared with the actual functions of the protein. The process is then repeated for K known proteins, $P_i, \ldots, P_K$. Let n_i be the number of functions for protein P_i in YPD, m_i be the number of predicted functions for protein

P_i, and k_i be the overlap between these functions. The specificity (SP) and sensitivity (SN) can be defined as

$$SP = \frac{\sum_i^K k_i}{\sum_i^K m_i} \tag{16.26}$$

$$SN = \frac{\sum_i^K k_i}{\sum_i^K n_i} \tag{16.27}$$

Trials using MIPS and other data sets have produced results that are very consistent with those of the distributions of expression correlation coefficients and reliability estimations.

16.5 CONCLUSIONS

In this chapter we reviewed a set of clustering approaches which have yielded promising results in application to protein–protein interaction networks. Clustering approaches for PPI networks can be broadly differentiated between the classic distance-based methods and the more recently developed graph-based approaches. Given a network comprised of proteins and their interactions, distance-based clustering approaches assign weights to each protein pair based on their interactions and use classic clustering techniques to generate predicted clusters. With graph-based approaches, the PPI network is viewed as an unweighted network. Clustering algorithms are employed to identify subgraphs with maximal density or with a minimum cost of cutoff based on the topology of the network. Clustering a PPI network permits a better understanding of its structure and the interrelationship of constituent components. More significantly, it also becomes possible to predict the potential functions of unannotated proteins by comparison with other members of the same cluster.

REFERENCES

1. http://www.plbio.kvl.dk/ dacoj3/resource/yeast_2H.htm.
2. R. Aebersold and M. Mann. Mass spectrometry-based proteomics. *Nature*, 422:198–207, 2003.
3. R. Albert and A. Barabasi. Statistical mechanics of complex networks. *Rev. Mod. Phys.*, 74:47–97, 2002.
4. U. Alon. Biological networks: the tinkerer as an engineer *Science*, 301:1866–1867, 2003.
5. S. F. Altschul, T. L. Madden, A. A. Schaffer, J. Zhang, Z. Zhang, W. Miller, and D. J. Lipman. Gapped BLAST and PSI-BLAST: a new generation of protein database search programs. *Nucleic Acids Res.*, 25:3389–3402, 1997.
6. V. Arnau, S. Mars, and I. Marin. Iterative cluster analysis of protein interaction data *Bioinformatics*, 21:364–378, 2005.
7. D. Auerbach, S. Thaminy, M. O. Hottiger, and I. Stagljar. Post-yeast-two hybrid era of interactive proteomics: facts and perspectives *Proteomics*, 2:611–623, 2002.

8. G. D. Bader and C. W. Hogue. An automated method for finding molecular complexes in large protein interaction networks. *BMC Bioinf.*, 4:2, 2003.

9. A. L. Barabasi and Z. N. Oltvai. Network biology: understanding the cell's functional organization. *Nat. Rev.*, 5:101–113, 2004.

10. K. Beyer, J. Goldstein, R. Ramakrishnan, and U. Shaft. When is "nearest neighbor" meaningful? In *Proc. 7th International Conference on Database Theory* (ICDT), 1999.

11. M. Blatt, S. Wiseman, and E. Domany. Superparamagnetic clustering of data. *Phys. Rev. Lett.*, 76:3251–3254, 1996.

12. D. H. Blohm and A. Guiseppi-Elie. *Curr. Opin. Microbiol.*, 12:41–47, 2001.

13. D. Bu, Y. Zhao, L. Cai, H. Xue, X. Zhu, et al. Topological structure analysis of the protein–protein interaction network in budding yeast. *Nucleic Acids Res.*, 31:2443–2450, 2003.

14. F. Chung and L. Lu. The average distances in random graphs with given expected degrees. *Proc. Natl. Acad. Sci.*, 99:15879–15882, 2002.

15. R. Cohen and S. Havlin. Scale-free networks are ultra small. *Phys. Rev. Lett.*, 90:058701, 2003.

16. T. P. Conrads, H. J. Issaq, and T. D. Veenstra. New tools for quantitative phosphoproteome analysis. *Biochem. Biophys. Res. Commun.*, 290:885–890, 2002.

17. T. Dandekar. et al. Conservation of gene order: a fingerprint of proteins that physically interact. *Trends Biochem. Sci.*, 23:324–328, 1998.

18. M. Deng, F. Sun, and T. Chen. Assessment of the reliability of protein–protein interactions and protein function prediction. *Pac. Symp. Biocomput.*, pp. 140–151, 2003.

19. B. L. Drees. Progress and variations in two-hybrid and three-hybrid technologies. *Curr. Opin. Chem. Biol.*, 3:64–70, 1999.

20. M. B. Eisen, P. T. Spellman, P. O. Brown, and D. Botstein. Cluster analysis and display of genome-wide expression patterns. *Proc. Natl. Acad. Sci.*, 95:14863–14868, 1998.

21. A. J. Enright, S. van Dongen, and C. A. Ouzounis. An efficient algorithm for large-scale detection of protein families. *Nucleic Acids Res.*, 30:1575–1584, 2002.

22. D. A. Fell and A. Wagner. The small world of metabolism. *Nat. Biotechnol.*, 18: 1121–1122, 2000.

23. S. Fields and O. Song. A novel genetic system to detect protein–protein interactions. *Nature*, 340(6230):245–246, 1989.

24. M. Fransen, C. Brees, K. Ghys, L. Amery, et al. *Mol. Cell Proteom.*, 2:611–623, 2002.

25. A. C. Gavin, et al. Functional organization of the yeast proteome by systematic analysis of protein complexes. *Nature*, 415:141–147, 2002.

26. H. Ge. *Nucleic Acids Res.*, 28:1–7, 2000.

27. G. Getz, E. Levine, and E. Domany. Coupled two-way clustering analysis of gene microarray data. *Proc. Natl. Acad. Sci.*, 97:12079–12084, 2000.

28. G. Getz, M. Vendruscolo, D. Sachs, and E. Domany. Automated assignment of SCOP and CATH protein structure classifications from FSSP scores. *Proteins*, 46:405–415, 2002.

29. L. Giot, et al. A protein interaction map of *Drosophila melanogaster. Science*, 302: 1727–1736, 2003.

30. G. Glazko, A. Gordon, and A. Mushegian. The choice of optimal distance measure in genome-wide data sets. 2005.

31. F. Glover. Tabu search. *ORSA J. Comput.*, 1:190–206, 1989.

32. D. S. Goldberg and F. P. Roth. Assessing experimentally derived interactions in a small world. *Proc. Natl. Acad. Sci.*, 100:4372–4376, 2003.

33. E. Hartuv and R. A. Shamir. Clustering algorithm based graph connectivity. *Inf. Process. Lett.*, 76:175–181, 2000.

34. L. H. Hartwell, J. J. Hopfield, S. Leibler, and A. W. Murray. From molecular to modular cell biology. *Nature*, 402:C47–C52, 1999.

35. Y. Ho et al. Systematic identification of protein complexes in *Saccharomyces cerevisiae* by mass spectrometry. *Nature*, 415:180–183, 2002.

36. T. Ito, T. Chiba, R. Ozawa, M. Yoshida, et al. A comprehensive two-hybrid analysis to explore the yeast protein interactome. *Proc. Natl. Acad. Sci.*, 98:4569–4574, 2001.

37. T. Ito, K. Ota, H. Kubota, Y. Yamaguchi, T. Chiba, K. Sakuraba, and M. Yoshida. Roles for the two-hybrid system in exploration of the yeast protein interactome. *Mol. Cell Proteom.*, 1:561–566, 2002.

38. A. Jain, M. Murty, and P. Flynn. Data clustering: a review. *ACM Comput. Surv.*, 31: 264–323, 1999.

39. R. A. Jansen et al. A Bayesian networks approach for predicting protein–protein interactions from genomic data. *Science*, 302:449–453, 2003.

40. R. Jansen et al. Relating whole-genome expression data with protein–protein interactions. *Genome Res.*, 12:37–46, 2002.

41. H. Jeong, S. P. Mason, A. -L. Barabási, and Z. N. Oltvai. Lethality and centrality in protein networks. *Nature*, 411:41–42, 2001.

42. D. Jiang, C. Tang, and A. Zhang. Cluster analysis for gene expression data: a survey. *IEEE Trans. Knowledge Data Eng.*, 16:1370–1386, 2004.

43. N. Johnsson and A. Varshavsky. Split ubiquitin as a sensor of protein interactions in vivo. *Proc. Natl. Acad. Sci.*, 91:10340–10344, 1994.

44. S. Jones and J. M. Thornton. Principles of protein–protein interactions. *Proc. Natl. Acad. Sci.*, 93:13–20, 1996.

45. A. D. King, N. Przulj, and I. Jurisica. Protein complex prediction via cost-based clustering. *Bioinformatics*, 20:3013–3020, 2004.

46. E. V. Koonin, Y. I. Wolf, and G. P. Karev. The structure of the protein universe and genome evolution. *Nature*, 420:218–223, 2002.

47. J. O. Korbel, B. Snel, M. A. Huynen, and P. Bork. SHOT: a Web server for the construction of genome phylogenies. *Trends Genet.*, 18:159–162, 2002.

48. R. Krause, C. von Mering, and P. Bork. A comprehensive set of protein complexes in yeast: mining large scale protein–protein interaction screens. *Bioinformatics*, 19:1901–1908, 2003.

49. A. Kumar and M. Snyder. Protein complexes take the bait. *Nature*, 415:123–124, 2002.

50. B. Kuster, P. Mortensen, J. S. Andersen, and M. Mann. Mass spectrometry allows direct identification of proteins in large genomes. *Protemics*, 1:641–650, 2001.

51. E. Lasonder et al. Analysis of the *Plasmodium falciparum* proteome by high-accuracy mass spectrometry. *Nature*, 419:537–542, 2002.

52. J. Lebowitz, M. S. Lewis, and P. Schuck. Modern analytical ultracentrifugation in protein science: a tutorial review. *Protein Sci.*, 11:2067–2079, 2002.

53. S. Li et al. A map of the interactome network of the metazoan. *Science*, 303:540–543, 2004.

54. G. MacBeath and S. L. Schreiber. Printing proteins as microarrays for high-throughput function determination. *Science*, 289:1760–1763, 2000.

55. M. Mann et al. Analysis of protein phosphorylation using mass spectrometry: deciphering the phosphoproteome. *Trends Biotechnol.*, 20:261–268, 2002.

56. M. Mann and O. N. Jensen. Proteomic analysis of post-translational modifications. *Nat. Biotechnol.*, 21:255–261, 2003.

57. E. M. Marcotte et al. Detecting protein function and protein–protein interactions from genome sequences. *Science*, 285:751–753, 1999.

58. E. M. Marcotte et al. Detecting protein function and protein–protein interactions from genome sequences. *Nature*, 402:83–86, 1999.

59. S. Maslov and K. Sneppen. Specificity and stability in topology of protein networks. *Science*, 296:910–913, 2002.

60. H. W. Mewes et al. MIPS: analysis and annotation of proteins from whole genomes. *Nucleic Acids Res.*, 32:D41–D44, 2004.

61. S. Milgram. The small world problem. *Psychol. Today*, 2:60, 1967.

62. B. Mirkin and E. V. Koonin. A top-down method for building genome classification trees with linear binary hierarchies. *Bioconsensus*, 61:97–112, 2003.

63. M. E. Newman. Network construction and fundamental results. *Proc. Natl. Acad. Sci.*, 98:404–409, 2001.

64. I. M. A. Nooren and J. M. Thornton. Diversity of protein–protein interactions. *EMBO J.*, 22:3486–3492, 2003.

65. Y. Ofran and B. Rost. Analyzing six types of protein–protein interfaces. *J. Mol. Biol.*, 325:377–387, 2003.

66. D. E. Otzen and A. R. Fersht. Analysis of protein–protein interactions by mutagenesis: direct versus indirect effects. *Protein Eng.*, 12:41–45, 1999.

67. T. Oyama et al. Extraction of knowledge on protein–protein interaction by association rule discovery. *Bioinformatics*, 18:705–714, 2002.

68. S. D. Patterson and R. H. Aebersold. Proteomics: the first decade and beyond. *Nat. Genet.*, 33:311–323, 2003.

69. P. Pei and A. Zhang. A topological measurement for weighted protein interaction network. In *Proc. IEEE Computer Society Bioinformatics Conference* (CSB'05), pp. 268–278, 2005.

70. P. Pei and A. Zhang. A two-step approach for clustering proteins based on protein interaction profile. In *Proc. 5th IEEE International Symposium on Bioinformatics and Bioengineering* (BIBE'05), pp. 201–209, 2005.

71. M. Pellegrini et al. Assigning protein functions by comparative genome analysis: protein phylogenetic profiles. *Proc. Natl. Acad. Sci.*, 96:4285–4288, 1999.

72. J. Peng, J. E. Elias, C. C. Thoreen, L. J. Licklider, and S. P. Gygi. Evaluation of multidimensional chromatography coupled with tandem mass spectrometry (LC/LC-MS/MS) for large-scale protein analysis: the yeast proteome. *J. Proteome Res.*, 10:1021, 2002.

73. J. B. Pereira-Leal, A. J. Enright, and C. A. Ouzounis. Detection of functional modules from protein interaction networks. *Proteins: Struct. Funct. Bioinf.*, 54:49–57, 2004.

74. E. M. Phizicky and S. Fields. Protein–protein interactions: methods for detection and analysis. *Microbiol. Rev.*, 59:94–123, 1995.

75. A. W. Rives and T. Galitski. Modular organization of cellular networks. *Proc. Natl. Acad. Sci.*, 100(3):1128–33, 2003.

76. M. P. Samanta and S. Liang. Redundancies in large-scale protein interaction networks. *Proc. Natl. Acad. Sci.*, 100:12579–12583, 2003.

77. M. Sigman and G. A. Cecchi. Global organization of the WordNet lexicon. *Proc. Natl. Acad. Sci.*, 99:1742–1747, 2002.

78. R. V. Sole, R. Pastor-Satorras, E. Smith, and T. B. Kepler. A model of large-scale proteome evolution. *Adv. Complex Syst.*, 5:43–54, 2002.

79. F. Spinozzi, D. Gazzillo. A. Giacometti, P. Mariani, and F. Carsughi. Interaction of proteins in solution from small angle scattering: a perturbative approach. *J. Biophys.*, 82:2165–2175, 2002.

80. V. Spirin and L. A. Mirny. Protein complexes and functional modules in molecular networks. *Proc. Natl. Acad. Sci.*, 100:12123–12128, 2003.

81. S. Tavazoie, D. Hughes, M. J. Campbell, R. J. Cho, and G. M. Church. Systematic determination of genetic network architecture. *Nat. Genet.*, pp. 281–185, 1999.

82. S. Van Dongen. A new cluster algorithm for graphs. Technical Report INS-R0010. Center for Mathematics and Computer Science (CWI), Amsterdam, Netherlands, 2000.

83. S. Van Dongen. Performance criteria for graph clustering and markov cluster experiments. Technical Report INS-R0012. Center for Mathematics and Computer Science (CWI). Amsterdam, Netherlands, 2000.

84. A. V. Veselovsky, Y. D. Ivanov, A. S. Ivanov, and A. I. J. Archakov. Protein–protein interactions: mechanisms and modification by drugs. *Mol. Recognit.*, 15:405–422, 2002.

85. M. Vidal. *The Two-Hybrid System*, p. 109. Oxford University Press, New York, 1997.

86. C. von Mering, R. Krause, B. Snel, M. Cornell, and S. G. Oliver. Comparative assessment of large-scale data sets of protein–protein interactions. *Nature*, 417:399–403, 2002.

87. A. Wagner. The yeast protein interaction network evolves rapidly and contains few redundant duplicate genes. *Mol. Biol. Evol.*, 18:1283–1292, 2001.

88. A. Wagner. How the global structure of protein interaction networks evolves. *Proc. R. Soc. London*, 270:457–466, 2003.

89. M. P. Washburn, D. Wolters, and J. R. Yates. Large-scale analysis of the yeast proteome by multidimensional protein identification technology. *Nat. Biotechnol.*, 19:242–247, 2001.

90. D. J. Watts. *Small Worlds*, p. Princeton University Press, Princeton, NJ, 1999.

91. M. Yanagida. Functional proteomics: current achievements. *J. Chromatogr. B*, 771: 89–106, 2002.

92. B. Zhang, B. Kraemer, S. SenGupta, S. Fields, and M. Wickens. Yeast three-hybrid system to detect and analyze interactions between RNA and protein. *Methods Enzymol.*, 306: 93–113, 1999.

93. H. Zhou. Distance, dissimilarity index, and network community structure. *Phys. Rev.*, E67:061901, 2003.

94. H. Zhou. Network landscape from a Brownian particle's perspective. *Phys. Rev.*, E67:041908, 2003.

95. H. Zhu, M. Bilgin, R. Bangham, D. Hall. et al. *Science*, 293:2101–2105, 2001.

96. H. Zhu, M. Bilgin, and M. Snyder. Proteomics. *Annu. Rev. Biochem.*, 72:783–812, 2003.

INDEX